ADVANCE DICTIONARY OF PHYSICS

TIGER DICTIONARIES

DICTIONARY OF PHYSICS

D.R. BROWN

K.S. PAPERBACKS
NEW DELHI - 110 002

[This text has been published under special arrangement with IPH, Delhi, India]

Published by

K.S. PAPERBACKS

New Delhi - 110 002

Marketed by

SHUBHAM BOOK DISTRIBUTORS

88, Chander Nagar Market

Alambagh, Lucknow-226 005 (U.P.)

Ph. 2453269

e-mail:shubhambooks@sify.com

ISBN–978-81-89261-85-6

edition 2007

DICTIONARY OF PHYSICS

Printed in India

Preface

This dictionary has been compiled and edited to bring together the various concepts of Physics. The terms included in this dictionary have been presented in brief analytical phrases thereby avoiding the more comprehensive type of treatment appropriate to larger reference book. The extent of each entry has been ascertained by the need for definition of meaning, function and physics relevance. A very systematic/scientific analysis of each entry is given. This dictionary will be useful for students, scholars and professionals from different disciplines such as physics, science, engineering science, biophysics and environmental science.

A/D

Analog to Digital. Used to refer to the conversion of analog data to its digital equivalent.

Ampere (amp)

Standard unit to measure the strength of an electric current. One amp is the amount of current produced by an electromotive force of one volt acting through the resistance of one ohm. The ampere is 10^{-1} of the theoretical electromagnetic unit of current. Named for the French physicist Andre Marie Ampere.

Absolute Humidity

In a system of moist air, the ratio of the mass of water vapor present to the volume occupied by the mixture; that is, the density of the water vapor component. Absolute humidity is normally expressed in grams of water vapor in a cubic meter of air (25 g/m^3).

Absolute Uncertainty

The uncertainty in a measured quantity is due to inherent variations in the measurement process itself. The uncertainty in a result is due to the combined and accumulated effects of these measurement uncertainties which were used in the calculation of that result. When these uncertainties are expressed in the same units as the quantity itself they are called *absolute uncertainties*. Uncertainty values are usually attached to the quoted value of an experimental measurement or result, one common format being: (quantity) ± (absolute uncertainty in that quantity).

Absorber

Any material that stops ionizing radiation. Lead, concrete, and steel attenuate gamma rays. A thin sheet of paper or metal will stop or absorb alpha particles and most beta particles.

Absorption

1. The process in which radiant energy is retained by a substance. A further process always results from absorption, that is, the irreversible conversion of the absorbed radiation into some other form of energy within and according to the nature of the absorbing medium. The absorbing medium itself may emit radiation, but only after an energy conversion has occurred
2. The transfer of energy to a medium, such as body tissues, as a radiation *beam* passes through the medium.
3. What happens when wave passes through a medium and gives up some of its energy.

Acceleration

1. Acceleration is defined as the rate of change of velocity. Acceleration is inherently a vector quantity, and an object will have non-zero acceleration if its speed and/or direction is changing. The average acceleration is given by

$$\vec{a}_{average} = \vec{a} = \frac{\Delta \vec{v}}{\Delta t} = \frac{\vec{v}_2 - \vec{v}_1}{t_2 - t_1}$$

   where the small arrows indicate the vector quantities. The operation of subtracting the initial from the final velocity must be done by vector addition since they are inherently vectors. The units for acceleration can be implied from the definition to be meters/second divided by seconds, usually written m/s^2.
2. Acceleration is the rate at which the velocity vector changes
3. The rate of change of velocity; the slope of the tangent line on a *v-t* graph.

Accelerator

1. A device (i.e., machine) used to produce high-energy high-speed *beams* of *charged* particles, such as *electrons* protons, or heavy ions, for research in high-energy and nuclear physics, synchrotron radiation research, medical therapies, and some industrial applications. The accelerator at SLAC is an *electron accelerator*
2. A machine used to accelerate particles to high speeds (and thus high energy compared to their rest-mass energy).

Accretion Disk

In a binary system containing a star and a compact object (white dwarf, neutron star, or black hole) gas may flow from the star to the compact object. According to the theoretical model, the gas will spiral in and fall to the surface of the compact object creating a flow of matter in the shape of a disk. It is generally believed that this model explains many features of X-ray *pulsars.*

Accurate

Conforming closely to some standard. Having very small error of any kind.

Acid Rain

Acids form when certain atmospheric gases (primarily carbon dioxide, sulfur dioxide, and nitrogen oxides) come in contact with water in the atmosphere or on the ground and are chemically converted to acidic substances. Oxidants play a major role in several of these acid-forming processes. Carbon dioxide dissolved in rain is converted to a weak acid (carbonic acid). Other gases. primarily oxides of sulfur and nitrogen, are converted to strong acids (sulfuric and nitric acids). Although rain is naturally slightly acidic because of carbon dioxide, natural emissions of sulfur and nitrogen oxides, and certain organic acids, human activities can make it much more acidic. Occasional pH readings of well below 2.4 (the acidity of vinegar) have been reported in industrialized areas. The principal natural phenomena that contribute acid-producing gases to the atmosphere are emissions from volcanoes and from biological processes that occur on the land, in wetlands, and in the oceans. The effects of acidic deposits have been detected in glacial ice thousands of years old in remote parts of the globe. Principal human sources are industrial and power-generating plants and transportation vehicles. The gases may be carried hundreds of miles in the atmosphere before they are converted to acids and deposited. Since the industrial revolution, emissions of sulfur and nitrogen oxides to the atmosphere have increased. Industrial and energy-generating facilities that burn fossil fuels, primarily coal, are the principal sources of increased sulfur oxides. These sources, plus the transportation sector, are the major originators of increased nitrogen oxides. The problem of acid rain not only has increased with population and industrial growth, it has become more widespread. The use of tall smokestacks to reduce local pollution has contributed to the spread of acid rain by releasing gases into regional atmospheric circulation. The same remote glaciers that provide evidence of natural variability in acidic deposition show in their more recently formed layers, the increased deposition caused by human activity during the past half century.

Acquisition of Signal

The time you begin receiving a signal from a spacecraft. For polar-orbiting satellites, radio reception of the APT signal can begin only when the polar-orbiting satellite is above the horizon of a particular location. This is determined by both the satellite and its particular path during orbit across the reception range of a ground station.

Action

This technical term is a historic relic of the 17th century, before energy and momentum were understood. In modern terminology, action has the dimensions of energy × time. Planck's constant has those dimensions, and is therefore sometimes called *Planck's quantum of action*. Pairs of measurable quantities whose product has dimensions of energy×time are called *conjugate quantities* in quantum mechanics, and have a special relation to each other, expressed in

Heisenberg's uncertainty principle. Unfortunately the word *action* persists in textbooks in meaningless statements of Newton's third law: 'Action equals reaction.' This statement is useless to the modern student, who hasn't the foggiest idea what action is.

Active System (Active Sensor)

A remote-sensing system that transmits its own radiation to detect an object or area for observation and receives the reflected or transmitted radiation. Radar is an example of an active system.

Acute Dose

An acute dose means a person received a radiation dose over a short period of time.

Advanced Very High Resolution Radiometer (AVHRR)

A five-channel scanning instrument that quantitatively measures electromagnetic radiation, flown on *NOAA* environmental satellites. AVHRR remotely determines cloud cover and surface temperature. Visible and infrared detectors observe vegetation, clouds, lakes, shorelines, snow, and ice. Automatic Picture Transmissions are derived from this instrument.

Aerosol

Particules of liquid or solid dispersed as a suspension in gas.

Afforestation

The act or process of establishing a forest, especially on land not previously forested.

Air Friction

Air friction, or air drag, is an example of *fluid friction*. Unlike the standard model of *surface friction*, such friction forces are velocity dependent. The velocity dependence may be very complicated, and only special cases can be treated analytically. At very low speeds for small particles, air resistance is approximately proportional to velocity and can be expressed in the form

$$f_{drag} = -bv$$

Air Mass

Large body of air often hundreds or thousands of miles across, containing air of a similar temperature and humidity. Sometimes the differences between air masses are hardly noticeable, but if colliding air masses have very different temperatures and humidity values, storms can erupt.

Air Pollution

The existence in the air of substances in concentrations that are determined unacceptable to human health and the environment. Contaminants in the air we breathe come mainly from manufacturing industries, electric power plants, exhaust from automobiles, buses, and trucks.

Air Pressure

The weight of the atmosphere over a particular point, also called barometric pressure. Average air exerts approximately 14.7 pounds (6.8 kg) of force on every square inch (or 101,325 newtons on every square meter) at sea level.

AIR

Airborne Imaging Radar.

Albedo

The term albedo (Latin for white) is commonly used to applied to the overall average *reflection coefficient* of an object. For example, the albedo of the Earth is 0.39 (Kaufmann) and this affects the equilibrium temperature of the Earth. The *greenhouse effect* by trapping infrared radiation, can lower the albedo of the earth and cause *global warming*. The albedo of an object will determine its visual brightness when viewed with reflected light. For example, the planets are viewed by reflected sunlight and their brightness depends upon the amount of light received from the sun and their albedo. *Mercury* receives the maximum amount of sunlight, but its albedo is only 0.1 so it is not as bright as it would be with a higher albedo.

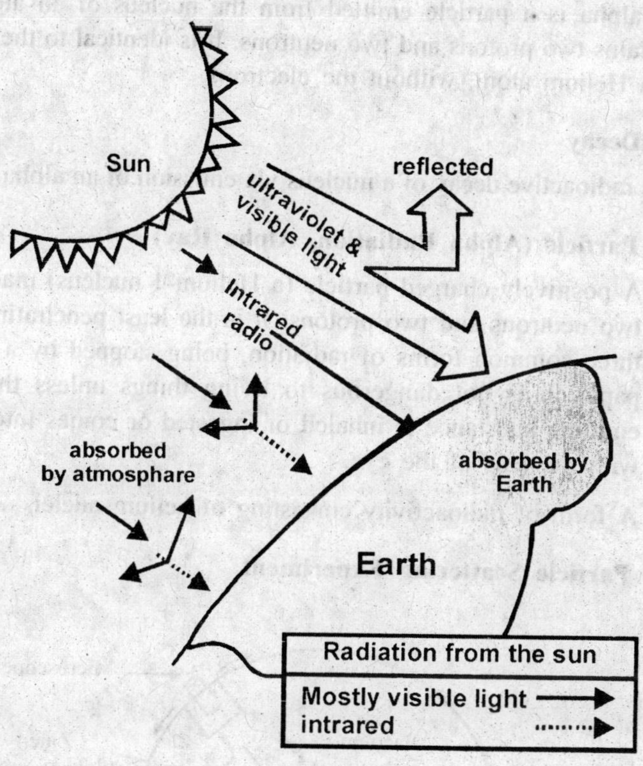

Algorithm

A mathematical relation between an observed quantity and a variable used in a step-by-step mathematical process to calculate a quantity. In the context of remote sensing, algorithms generally specify how to determine higher-level data products from lower-level source data. For example, algorithms prescribe how atmospheric temperature and moisture profiles are determined from a set of radiation observations originally sensed by satellite sounding instruments.

Alkaline

Substance capable of neutralizing acid, with a pH greater than 7.0.

Alphas

An alpha is a particle emitted from the nucleus of an atom, that contains two protons and two neutrons. It is identical to the nucleus of a Helium atom, without the electrons.

Alpha Decay

The radioactive decay of a nucleus via emission of an alpha particle.

Alpha Particle (Alpha Radiation, Alpha Ray)

1. A positively charged particle (a Helium-4 nucleus) made up of two neutrons and two protons. It is the least penetrating of the three common forms of radiation, being stopped by a sheet of paper. It is not dangerous to living things unless the alpha-emitting substance is inhaled or ingested or comes into contact with the lens of the eye.
2. A form of radioactivity consisting of helium nuclei.

Alpha Particle Scattering Experiment

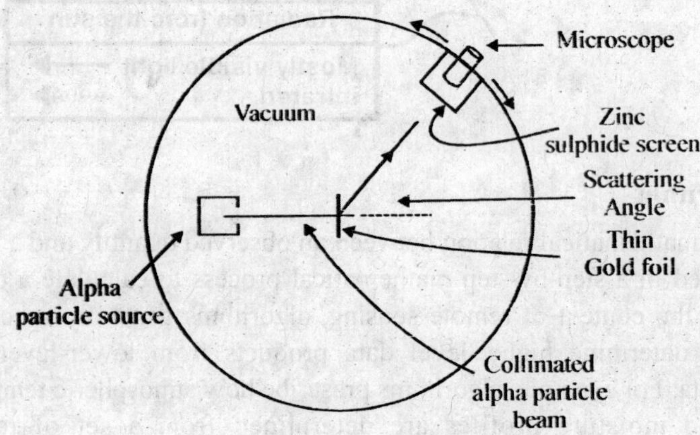

The classic experiment leading to the discovery of the nucleus and carried out by Geiger and Marsden under the direction of Rutherford. The essential features of the apparatus are shown in the diagram above. The alpha particle beam is narrow and parallel. The vacuum is essential; alpha particles travel no more than a few centimetres in air. The gold foil has to be very thin. The detector was a zinc sulphide screen, and scintillation were observed through a microscope.

Altimeter

An active instrument used to measure the altitude of an object above a fixed level. For example, a laser altimeter can measure height from a spacecraft to an icesheet. That measurement, coupled with radial orbit knowledge, will enable determination of the topography.

AM Radio

AM radio uses the electrical image of a sound source to modulate the amplitude of a *carrier* wave. At the receiver end in the *detection process,* that image is stripped back off the carrier and turned back into sound by a *loudspeaker.*

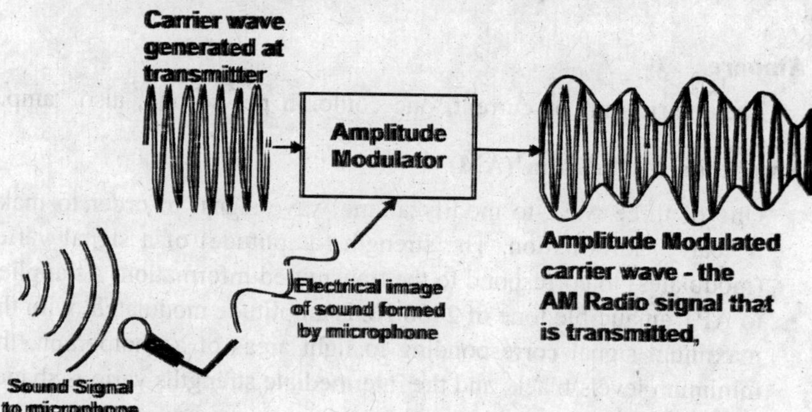

Ammeter

1. A device for measuring electrical current.
2. Ammeters measure electric current. Digital ammeters are accurate and cheap. There is no point in wondering how they work unless you have a special interest. They must be placed in series with the current to be measured, so that all the current flows

through the ammeter and none bypasses it. The diagram below shows how a voltmeter and ammeter should be connected when measuring the current in, and the potential different across, a resistor R.

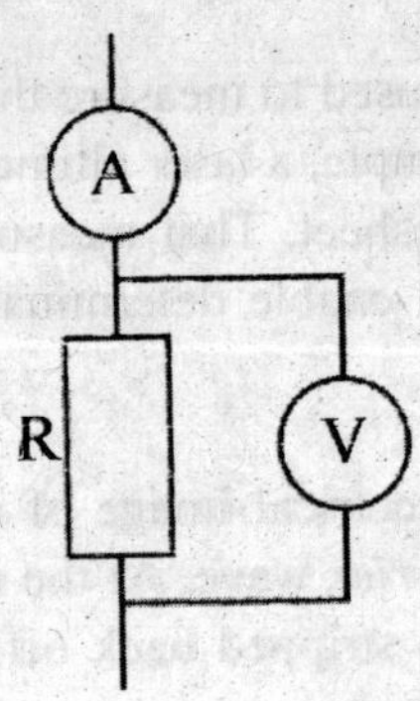

Ampere

The metric unit of current, one coulomb per second; also "amp."

Amplitude Modulation (AM)

One of three ways to modify a sine wave signal in order to make it "carry" information. The strength (amplitude) of a signal varies (modulates) to correspond to the transmitted information. As applied to APT an audible tone of 2400 Hz is amplitude modulated, with the maximum signal corresponding to light areas of a photograph, the minimum levels black, and the intermediate strengths various shades of gray.

Amplitude

1. The amount of vibration, often measured from the center to one side; may have different units depending on the nature of the vibration.

2. The magnitude of the displacement of a wave from a mean value. For a simple harmonic wave, it is the maximum

displacement from the mean. For more complex wave motion, amplitude is usually taken as one-half of the mean distance (or difference) between maxima and minima.

Analog

Transmission of a continuously variable signal as opposed to a discretely variable signal. Compare with digital. A system of transmitting and receiving information in which one value (*i.e.,* voltage, current, resistance, or, in the APT system, the volume level of the video tone) can be compared directly to the information (in the APT system, the white, black, and gray values) in the image.

Ancillary Data

Data other than instrument data required to perform an instrument's data processing. Ancillary data includes such information as orbit and/or attitude data time information spacecraft engineering data and calibration information.

Anemometer

Instrument used to measure wind speed usually measured either from the rotation of wind driven cups or from wind pressure through a tube pointed into the wind. (Fig. on next page)

Angstrom

A unit of length equal to 10^{-10} meter.

anemometer

Angular and Linear Momentum

Angular momentum and linear momentum are examples of the *parallels* between linear and rotational motion. They have the same form and are subject to the fundamental constraints of *conservation laws*, the *conservation of momentum* and the *conservation of angular momentum*

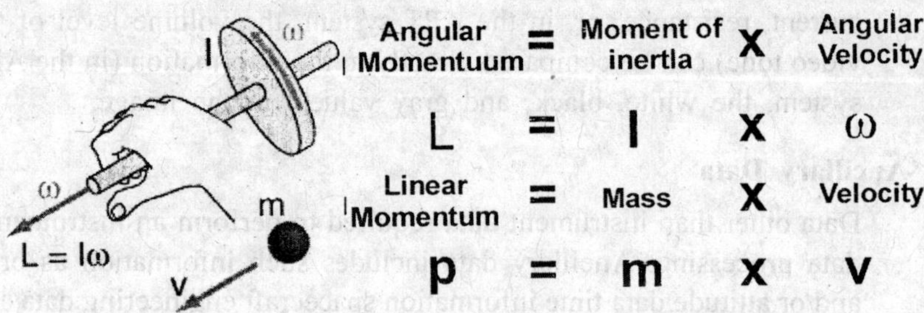

Angle

Angle is an indefinable idea, it is a part of our language. But the magnitude of an angle can be defined, the unit depending upon the particular definition chosen. The commonly used unit, the degree, is defined as one complete rotation divided by 360. The diagram defines the magnitude of an angle in radians :

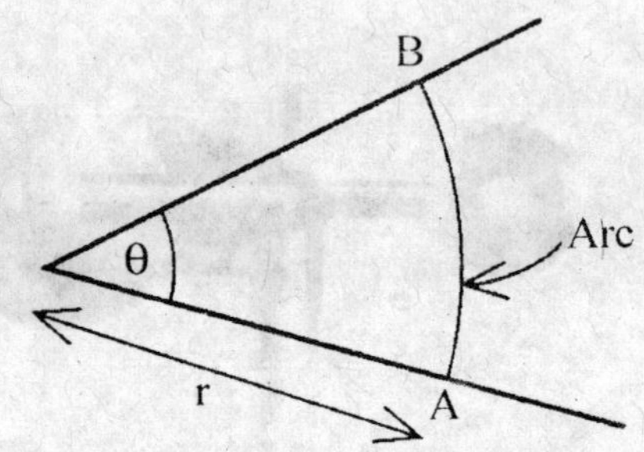

$$\text{angle in radians} = \frac{\text{arc length}}{\text{radius}}$$

$$\theta = \frac{AB}{r}$$

Angstrom

Abbreviated *Å*. A unit of length equal to 10^{-8} cm (one-hundredth of a millionth of a centimeter). An Angstrom is on the order of the size of an atom.

Angular Magnification

1. It is the ratio of the angular size of the object as seen through the instrument to the angular size of the object as seen *with the 'naked eye'*. The 'naked eye' view is *without* use of the optical instrument, but under optimal viewing conditions. Certain 'gotchas' lurk here. What are 'optimal' conditions? Usually this means the conditions in which the object's details can be seen most clearly. For a small object held in the hand, this would be when the object is brought as close as possible and still seen clearly, that it, to the near point of the eye, about 25 cm for normal eyesight. For a distant mountain, one can't bring it close, so when determining the magnification of a telescope, we assume the object is very distant, or at infinity. And what is the 'optimal' position of the image? For the simple magnifier, in which the magnification depends strongly on the image position, the image is best seen at the near point of the eye, 25 cm. For the telescope, the image size doesn't change much as you fiddle with the focus, so you likely will put the image at infinite distance for relaxed viewing. The microscope is an intermediate case. Always striving for greater resolution, the user may pull the image close, to the near point, even though that doesn't increase its size very much. But usually, users will place the image farther away, at the distance of a meter or two, or even at infinity. But, because the object is very near the focal point, the magnification is only weakly dependent on image position. Some texts express angular

magnification as the ratio of the angles, some express it as the ratio of the tangents of the angles. If all of the angles are small, there's negligible difference between these two definitions. However, if you examine the derivation of the formula these books give for the magnification of a telescope fo/fe, you realize that they must have been using the tangents. The tangent form of the definition is the traditionally correct one, the one used in science and industry, for nearly all optical instruments which are designed to produce images which preserve the linear geometry of the object.

2. The factor by which an image's apparent angular size is increased (or decreased).

Angular Momentum

A measure of rotational motion; a conserved quantity for a closed system.

Annihilation

1. A process in which a particle meets its corresponding antiparticle and both disappear. The energy appears in some other form, perhaps as a different particle and its antiparticle (and their energy), perhaps as many mesons, perhaps as a single neutral boson. The produced particles may be any combination allowed by conservation of energy and momentum and of all the charge types.
2. A process in which a particle meets its corresponding antiparticle and both disappear. Their energy and momentum appears in some other form, producing other particles together with their *anti-particles* and providing their motion.

Anomaly

1. The deviation of (usually) temperature or precipitation in a given region over a specified period from the normal value for the same region.
2. The angular distance of an Earth satellite (or planet) from its *perigee* or perihelion) as seen from the center of the Earth (sun).

Antenna

A wire or set of wires used to send and receive electromagnetic waves. Two primary features must be considered when selecting antennas: beamwidth, or the "width" of the antenna pattern (wide beamwidth suggests the ability to receive signals from a number of different directions), and gain, or the increase in signal level. Generally beamwidth or gain can be increased only at the expense of the other. Gain can be increased by multiplying the number of antenna elements, although this adds "directionality" that reduces beamwidth. Important antenna considerations include the following: (1) The physical size of antenna components is determined by the frequency of the trans-missions it will receive—the higher the frequency the shorter the elements. At high frequencies, use of a satellite dish will compensate for the reduced amount of energy intercepted by shortened components. (2) The antenna design should fit the type of radio frequency (RF) signal polarization it will receive. The orientation of radio waves in space is a function of the orientation of the elements of the transmitting antenna. A circularly polarized wave rotates as it propagates through space. Antennas can be designed for either right or left-handed circular polarization. Earth-based communication antennas are either vertical or horizontal in polarization, and not suited for space communication. Police and cellular phone transmissions use vertical polarization because a simple vertical whip antenna is the easiest sort of omnidirectional antenna to mount on a vehicle. (3) The antenna needs to produce sufficient signal gain to produce noise-free reception. (4) The antenna should be clear of conductive objects such as power lines, phone wires, etc., so height above the ground becomes important. *Basic antenna components are:* (1) Driven element—the parts connected to and receiving power from the receiver/transmitter. (2) Parasitic elements—the parts dependent upon resonance rather than connection to a power source. A director or parasitic element that rein forces radiation on a line pointing to it from the driven element. A reflector or parasitic element that rein forces radiation on a line pointing from it to the driven element. A fundamental form of antenna is a single wire whose length approximately equals half the transmitting wavelength. Known as a dipole antenna, it is the unit from which many more complex forms of antennas are constructed. One of the most common forms of VHF antenna is the Yagi/beam, named for the Japanese scientist who first described the principles of combining a basic dipole (driven element) and parasitic elements. A common TV antenna is an example of this type. A Yagi/beam antenna is directional and therefore includes a rotator to aim (direct) the antenna. An omnidirectional antenna has a wide beamwidth and consequently does not require "tracking" (aiming the antenna toward the signal source). An example of an omnidirectional antenna is the turnstile antenna, a variation of the standard dipole antenna well suited for space communications. The quadrifilar helix antenna is omnidirectional and an inherently excellent antenna for ground station use. Quadrifilars are also used on NOAA's polar-orbiting environmental satellites. The parabolic reflector or satellite dish antenna collects RF signals on a passive dish-shaped surface. A feedhorn antenna—a simple dipole antenna mounted in a resonant tube structure (cylinder with one open end)—transfers the RF energy to a transmission line. The bigger the dish, the greater the amount of RF energy intercepted, and therefore the greater the gain from the signal.

quadrifilar helix

parabolic reflector

Antenna Array

An ordered assembly of elementary antennae spaced apart and fed in such a manner that the resulting radiation is concentrated in one or more directions.

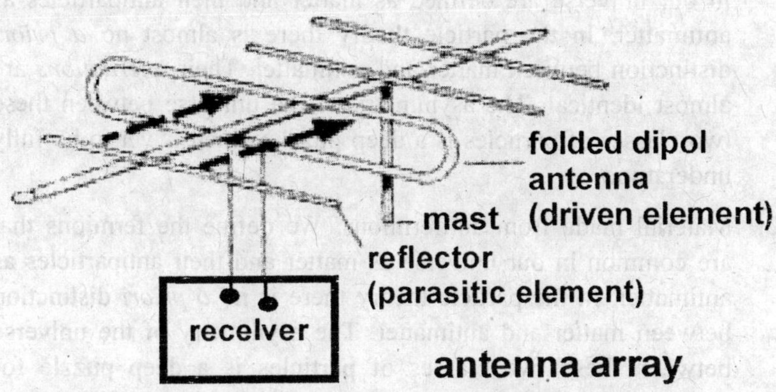

Antenna Beam

The focused pattern of electromagnetic radiation that is either received or transmitted by an antenna.

Anticipated Future of the Sun

Hydrogen burnup at center

COLLAPSE

Increased temperature and kinetic energy

Faster fusion at intermediate radii, outward expansion, surface cools and reddens

RED GIANT
Increased luminosity because of area, loss of much mass

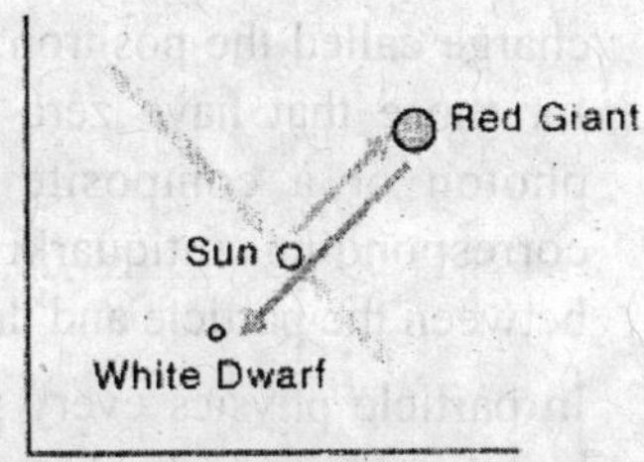

Core heats to 10^8 K, helium fuses to carbon. Helium flash (few years)

Contraction, perhaps variability

WHITE DWARF

Anticyclone

A high pressure area where winds blow clockwise in the Northern Hemisphere and counter-clockwise in the Southern Hemisphere.

Antimatter

1. A material made from *antiparticles* The particles that are common in our universe are defined as matter and their antiparticles as antimatter. In the particle theory there is almost no *a priori* distinction between matter and antimatter. Their *interactions* are almost identical. The asymmetry of the universe between these two classes of particles is a deep puzzle which is yet to be fully understood.
2. Material made from antifermions. We define the fermions that are common in our universe as matter and their antiparticles as antimatter. In the particle theory there is no *a priori* distinction between matter and antimatter. The asymmetry of the universe between these two classes of particles is a deep puzzle for which we are not yet completely sure of an explanation.

Antiparticle

1. For every fermion type there is another fermion type that has exactly the same mass but the opposite value of all other charges (quantum numbers). This is called the antiparticle. For example, the antiparticle of an electron is a particle of positive electric charge called the positron. Bosons also have antiparticles except for those that have zero value for all charges, for example a photon or a composite boson made from a quark and its corresponding antiquark. In this case there is no distinction between the particle and the antiparticle, they are the same object.
2. In particle physics every particle with any type of *charge* label has a corresponding antiparticle type. Any particle and its antiparticle have identical mass and *spin* but opposite charges. For example the antiparticle of an *electron* is a *positron*. It has exactly the same mass as an electron but positive charge. Some particles are their own antiparticles, the antiparticle of a photon is a photon for instance. *Conserved quantities* such as *baryon* number and *lepton* number are further types of "charges" that are reversed for particle and antiparticle. Thus an *electron* and an electron neutrino both have electron number +1 while their

antiparticles the positron and the anti-electron-neutrino have electron number –1.

Antiquark

The antiparticle of a quark.

Apparent Horizon

When matter falls inward to form a black hole it is not always easy to see where the event horizon might be. It might appear at one time that a light ray is capable of escaping but infalling matter might eventually prevent it from doing so. The apparent horizon is a surface on which outgoing light rays are just trapped, and cannot expand outward. It is a stronger condition than the event horizon, and the apparent horizon always lies inside the event horizon, or coincides with it. This situation is analgous to a man running through a corridor filled with doors. He is trying to run outward, but the doors are closing in sequence from the outside in. How many doors will he be able to pass through before he is blocked by a closed door? The door that is closest to him that is currently closed is analgous to the apparent horizon. The door that he will actually reach before he cannot travel further is analogous to the event horizon.

Apogee (aka apoapsis or apifocus)

On an elliptical orbit path, point at which a satellite is farthest from the Earth.

Aquifer

Layer of water-bearing permeable rock, sand, or gravel capable of providing significant amounts of water.

Arc Degree

A unit of angular measure in which there are 360 arc degrees in a full circle.

Arc Second

1. Abbreviated *arcsec*. A unit of angular measure in which there are

60 arc seconds in 1 arc minute and therefore 3600 arc seconds in 1 *arc degree.* One arc second is equal to about 725 *km* on the Sun.

2. The size of a celestial object expressed in terms of the **angle** that it covers (or "subtends") when viewed from Earth. For example, the moon subtends an angle of 1/2 a degree. One degree of arc is defined as equivalent to 60 minutes of arc (or "arc minutes"). Arc minutes are further divided into arc seconds, such that there 60 × 60 or 3600 arc seconds per degree. So the moon's apparent size can also be expressed as 1/2 degree x 3600 = 1800 arc seconds. If the the distance to an object is also known, then its angular size can be used to calculate its diameter in miles or kilometers.

Archimedes' Principle

1. Archimedes' principle states that a body that is wholly or partially immersed in a fluid is acted on by an upward force equal to the weight of fluid displaced. The upward forced is called an upthust. The resultant force on the body is the difference between the upthrust and the body's weight. This difference is zero for a floating body. The first part of the diagram shows a stone suspended in air from a newton balance. The weight of the stone in air, W_{air}, is recorded. The second part of the diagram shows the same stone suspended from the same newton balance but immersed in water. The weight of the stone wholly immersed in water, W_{water}, is recorded. The water pushes up on the stone with a force equal to the weight of water whose place has been taken up by the stone. This is shown in the free-body force diagram. Upthrust on stone = $W_{air} - W_{water}$ = Weight of water displaced.

   It is easy to calculate the masses of the stone and of the displaced water in kilograms from their weights in newtons. The volume of the displaced water can be calculated from its mass and the known density of water. This equals the volume of the stone.

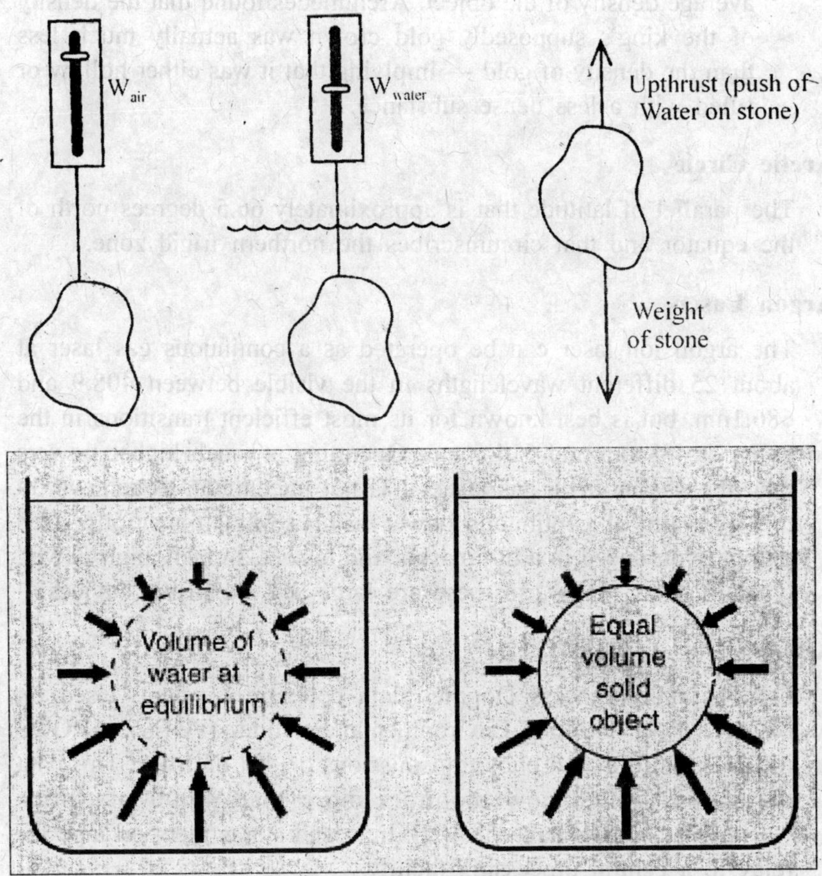

2. The *buoyant force* on a submerged object is equal to the weight of the fluid displaced. This principle is useful for determining the volume and therefore the *density* of an irregularly shaped object by measuring its *mass* in air and its effective mass when submerged in water (density = 1 gram per cubic centimeter). This effective mass under water will be its actual mass minus the mass of the fluid displaced. The difference between the real and effective mass therefore gives the mass of water displaced and allows the calculation of the volume of the irregularly shaped object (like the *king's crown* in the Archimedes story). The mass divided by the volume thus determined gives a measure of the

average density of the object. Archimedes found that the density of the king's supposedly gold crown was actually much less than the density of gold — implying that it was either hollow or filled with a less dense substance.

Arctic Circle

The parallel of latitude that is approximately 66.5 degrees north of the equator and that circumscribes the northern frigid zone.

Argon Laser

The argon ion laser can be operated as a continuous gas laser at about 25 different wavelengths in the visible between 408.9 and 686.1nm, but is best known for its most efficient transitions in the green at 488 nm and 514.5 nm. Operating at much higher powers than the *helium-neon* gas laser, it is not uncommon to achieve 30 to 100 watts of continuous power using several transitions. This output is produced in a hot plasma and takes extremely high power, typically 9 to 12 kW, so these are large and expensive devices.

ARGOS

French random-access Doppler data collection system. Used on NOAA's Polar-Orbiting Environmental Satellites (POES). ARGOS receives platform and buoy transmissions on 401.65 MHz. This data collection system now monitors more than 4,000 platforms worldwide, outputs data via VHF link, and stores them on tape for relay to a central processing facility.

Argument of Perigee (aka ARGP or w)

One of the six Keplerian elements, it gives the rotation of the satellite on the orbit. The argument (argument meaning angle) of perigee—perigee is the point on an orbital path when the satellite is closest to the earth—is the angle (measured from the center of the Earth) from the ascending node to perigee. Example: When ARGP = 0 degrees, the perigee occurs at the same place as the ascending node. That means that the satellite would be closest to Earth just as it rises up over the equator. When ARGP = 180 degrees, apogee would occur at the same place as the descending node. This means that the satellite would be farthest from earth just as it rises over the equator.

Artificial Intelligence

Neural networks. The branch of computer science that attempts to program computers to respond as if they were thinking—capable of reasoning, adapting to new situations, and learning new skills. Examples of artificial intelligence programs include those that can locate minerals underground and understand human speech.

Ascending Node

The point in an orbit (longitude) at which a satellite crosses the equatorial plane from south to north.

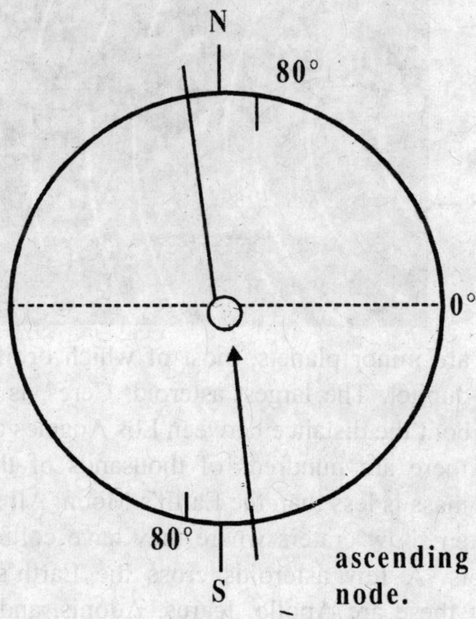

Aspect Ratio

The ratio of image width to image height. Weather Facsimile (WEFAX) images have a 1:1 aspect ratio (square); a conventional TV aspect ratio is 4:3 (rectangle).

Aspirator Emonstration

A common device which makes use of the *Bernoulli principle* is the aspirator. When water is forced through a smooth constriction, the fluid velocity increases, lowering the pressure below atmospheric pressure. This can entrain air into the tube at the constricton and be used to partially evacuate the air from an attached volume.

Asteroids

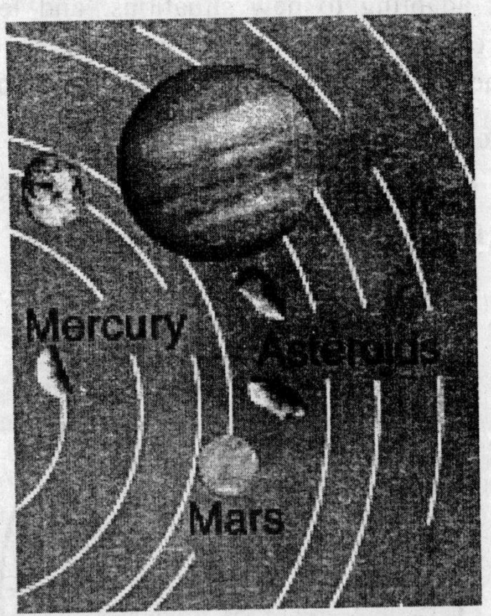

Asteroids are minor planets, most of which orbit the Sun between Mars and Jupiter. The largest asteroid, Ceres, is 800 kilometers in diameter,about the distance between Los Angeles and San Francisco. Although there are hundreds of thousands of these objects, their combined mass is less than the Earth's Moon. All solid bodies in our Solar Systen show craters where they have collided with asteroids and comets. A few asteroids cross the Earth's orbit. The most notable of these are Apollo, Icarus, Adonis, and Eros. Hector has an orbit similar to that of Jupiter, and Hidalgo has a large elliptical orbit which extends from about the orbit of Mars out well beyond the orbit of Jupiter. Asteroid Gaspra was used in slingshot maneuver with Galileo in 1991 on the way toward ultimate rendevouz with

Jupiter. Galileo also provided a close view of asteroid Ida. The vast majority of the meteorites which hit the Earth are thought to come from the asteroid belt.

Astronomical Unit (AU)

The distance from the Earth to the sun. On average, the sun is 149,599,000 kilometers from Earth.

Astrophysics

1. The physics of astronomical objects such as stars and galaxies.

2. *Astrophysics* is the branch of astronomy that deals with the physics of the universe, including the physical properties (luminosity, density, temperature and chemical composition) of astronomical objects such as stars, galaxies, and the interstellar medium, as well as their interactions. The study of cosmology is theoretical astrophysics at its largest scale; conversely, since the energies involved in cosmology, especially the Big Bang, are the largest known, the observations of the cosmos also serve as the laboratory for physics at its smallest scales as well. In practice,

virtually all modern astronomical research involves a substantial amount of physics. The name of a school's doctoral program ("Astrophysics" or "Astronomy") in many places like the United States often has to do more with the department's history than with the contents of the programs.

ATLAS (Atmospheric Laboratory for Applications and Science) Mission

The focus of ATLAS is to study the chemistry of the Earth's upper atmosphere (mainly the stratosphere/mesosphere) and the solar radiation incident on the Earth system (both total solar irradiance and spectrally resolved radiance, especially ultraviolet). Science operations onboard ATLAS 1 (March 1992) and ATLAS 2 (March-April, 1993) began a comprehensive and systematic collection of data that will help establish benchmarks for atmospheric conditions and the sun's stability.

Atmosphere

The air surrounding the Earth, described as a series of shells or layers of different characteristics. The atmosphere, composed mainly of nitrogen and oxygen with traces of carbon dioxide, water vapor, and other gases, acts as a buffer between Earth and the sun. The layers, troposphere, stratosphere, mesosphere, thermosphere, and the exosphere, vary around the globe and in response to seasonal changes. Troposphere stems from the Greek word tropos, which means turning or mixing. The troposphere is the lowest layer of the Earth's atmosphere, extending to a height of 8-15 km, depending on latitude. This region, constantly in motion, is the most dense layer of the atmosphere and the region that essentially contains all of Earth's weather. Molecules of nitrogen and oxygen compose the bulk of the troposphere. The tropopause marks the limit of the troposphere and the beginning of the stratosphere. The temperature above the tropopause increases slowly with height up to about 50 km. The stratosphere and stratopause stretch above the troposphere to a height of 50 km. It is a region of intense interactions among radiative, dynamical, and chemical processes, in which horizontal mixing of gaseous components proceeds much more rapidly than vertical mixing. The stratosphere is warmer than the upper troposphere, primarily because of a stratospheric ozone layer that absorbs solar ultraviolet energy. The mesosphere, 50 to 80 km above the Earth, has diminished ozone concentration and radiative cooling becomes relatively more important. The temperature begins to decline again (as it does in the troposphere) with altitude. Temperatures in the upper mesosphere fall to -70° to -140° Celsius, depending upon latitude and season. Millions of meteors burn up daily in the mesosphere as a result of collisions with some of the billions of gas particles contained in that layer The collisions create enough heat to burn the falling objects long before they reach the ground. The stratosphere and mesosphere are referred to as the middle atmosphere. The mesopause, at an altitude of about 80 km, separates the mesosphere from the thermosphere—the outermost layer of the Earth's atmosphere. The thermosphere, from the Greek thermo for heat, begins about 80 km above the Earth. At these high altitudes, the residual atmospheric gases sort into strata according to molecular mass. Thermospheric temperatures increase with altitude due to absorption of highly energetic solar radiation by the small amount of residual oxygen still present. Temperatures can rise to 2,000° C. Radiation causes the scattered air particples in this layer to become charged electrically, enabling radio waves to bounce off and be received beyond the horizon. At the exosphere, beginning at 500 to 1,000 km above the Earth's surface, the atmosphere blends into space. The few particles of gas here can reach 4,500°F (2,500° C) during the day.

Atmospheric Infrared Sounder

Advanced sounding instrument selected to fly on the EOS-PM 1 mission (intermediate-sized, sun-synchronous, morning satellite) in the year 2000. It will retrieve vertical temperature and moisture profiles in the troposphere and stratosphere. Designed to achieve temperature retrieval accuracy of 1°C with a 1 km vertical resolution, it will fly with two operational microwave sounders. The three instruments will constitute an advanced operational sounding system, relative to the TIROS Operational Vertical Sounder (TOVS) currently flying on NOAA polar-orbiting satellites.

Atmospheric Pressure

The amount of force exerted over a surface area, caused by the weight of air molecules above it. As elevation increases, fewer air molecules are present. Therefore, atmospheric pressure always decreases with increasing height. A column of air; 1 square inch in cross section, measured from sea level to the top of the atmosphere would weigh approximately 14.7 lb/in^2. The standard value for atmospheric pressure at sea level is: 29.92 inches or 760 mm of mercury 1013.25 millibars(mb) or 101,325 pascals (Pa).

Atmospheric Radiation Measurements Program (ARM)

U.S. Department of Energy program for the continual, ground-based measurements of atmospheric and meteorological parameters over approximately a ten-year period. The program will study radiative forcing and feedbacks, particularly the role of clouds. The general program goal is to improve the performance of climate models, particularly general circulation models of the atmosphere.

Atmospheric Response Variables

Variables that reflect the response of the atmosphere to external forcing (*e.g.*, temperature, pressure, circulation, and precipitation).

Atmospheric Windows

The range of wavelengths at which water vapor; carbon dioxide, or other atmospheric gases only slightly absorb radiation. Atmospheric windows allow the Earth's radiation to escape into space unless clouds absorb the radiation.

Atoll

A coral island consisting of a ring of coral surrounding a central lagoon. Atolls are common in the Indian and Pacific Oceans.

Atom

1. A particle of matter indivisible by chemical means. It is the fundamental building block of elements.

2. The basic unit of one of the chemical elements.

Atomic Clocks

Very accurate clocks can be constructed by locking an electronic oscillator to the frequency of an atomic transition. The frequencies associated with such transitions are so reproducible that the definition of the second is now tied to the frequency associated with a transition in cesium-133: 1 second = 9,192, 631,770 cycles of the standard Cs-133 transition

Atomic Mass

1. The mass of an atom.
2. The mass of a neutral atom of a nuclide. The atomic weight of an atom is the weight of the atom based o n a scale where 12C = 12. The atomic weight of an element is the weighted average of each isotope.

Atomic, Molecular, and Optical Physics

Atomic, molecular, and optical physics is the study of matter-matter and light-matter interactions on the scale of single atoms or structures containing a few atoms. The three areas are grouped together because of their interrelationships, the similarity of methods used, and the commonality of the *energy* scales that are relevant. Physicists sometimes abbreviate the field as *AMO physics*. *Atomic physics* is distinct from *nuclear physics*, despite their association in the public consciousness. Atomic physics is not concerned with the nuclear processes studied in nuclear physics, although properties of the nucleus can be important in atomic physics (e.g., *hyperfine structure*). Molecular physics focuses on multi-atomic structures and their internal and external interactions with matter and light. Optical physics is distinct from optics in that it tends to focus not on the control of classical light fields by macroscopic objects, but on the fundamental properties of optical fields and their interactions with matter in the microscopic realm. All three areas include both *classical* and quantum treatments.

Atomic Number

1. The number assigned to each element on the basis of the number of protons found in the element's nucleus.
2. The number of protons in an atom's nucleus; determines what element it is.
3. The number of *protons* in the *nucleus.*

Atomic Weight (Atomic Mass)

Approximately the sum of the number of protons and neutrons found in the nucleus of an atom.

Attenuation

1. The decrease in the magnitude of current, voltage, or power of a signal in transmission between points. Attenuation may be expressed in decibels, and can be caused by interferences such as rain, clouds, or radio frequency signals.

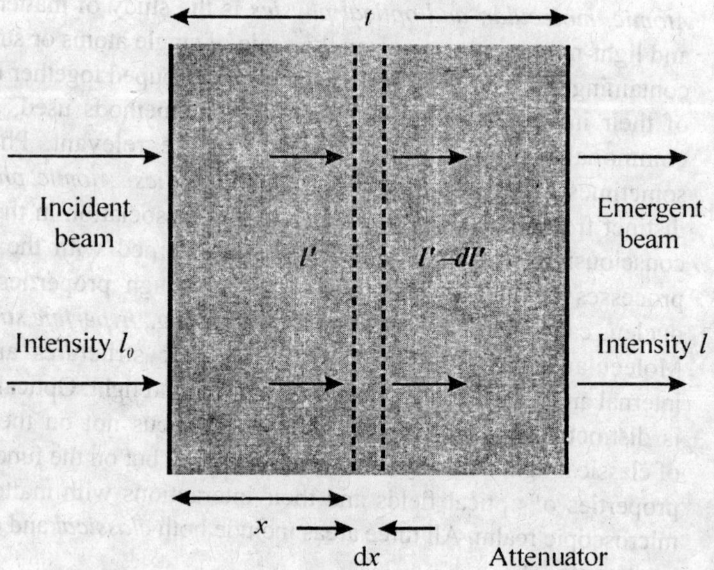

2. The process by which a compound is reduced in concentration over time, through adsorption, degradation, dilution, and/or trans-

formation. Radiologically, it is the reduction of the *intensity* of radiation upon passage through a medium. The attenuation is caused by *absorption* and scattering.

3. Attenuation is the weakening of a beam of particles (or of electromagnetic radiation) by any material placed in its path. It may be due to scattering or to absorption processes. Either way, the degree of the attenuation is given by an attenuation coefficient.

Attractive

Describes a force that tends to pull the two participating objects together.

Audio Frequencies

Frequencies that the human ear can hear (usually 30 to 20,000 cycles per second).

Automatic Picture Transmission (APT)

System developed to make real-time reception of satellite images possible whenever an APT-equipped satellite passes within range of an environmental satellite ground station. Transmission (analog video format) consists of an amplitude-modulated audible tone that can be displayed as an image on a computer monitor when received by an appropriate ground station. APT images are transmitted by polar-orbiting satellites such as the TIROS-N/NOAA satellites and Russia's METEOR, which orbit 500-900 miles above the Earth, and offer both visible and infrared images. An APT image has thousands of squares called picture elements or pixels. Each pixel represents a four-km square.

Avogadro's Constant

Avogadro's constant has the unit *mole*$^{-1}$. It is *not* merely a number, and should *not* be called *Avogadro's number.* It *is* ok to say that the number of particles *in* a gram-mole is 6.02 x 10^{23}. Some older books call this value *Avogadro's number*, and when that is done, no units are attached to it. This can be confusing and misleading to students who are conscientiously trying to learn how to balance units in equations. One *must* specify whether the value of Avogadro's constant is expressed for a gram-mole or a kilogram-mole. A few books prefer a kilogram-mole. The unit name for a gram-mole is simply *mol*. The unit name for a kilogram-mole is *kmol*. When the kilogram-mole is used, Avogadro's constant should be written: 6.02252×10^{26} $kmol^{-1}$. The fact that Avogadro's constant has units further convinces us that it is not 'merely a number.' Though it seems inconsistent, the SI base unit is the gram-mole. As Mario Iona reminds me, SI is *not* an MKS system. Some textbooks still prefer to use use the kilogram-mole, or worse, use it *and* the gram-mole. This affects their quoted values for the universal gas constant and the Faraday Constant. Is Avogadro's constant just a number? What about those textbooks which say 'You could have a mole of stars, grains of sand, or people.' In science we *do* use entities which are *just* numbers, such as, e, 3, 100, etc. Though these are *used* in science, their definitions are *independent of science*. No experiment of science can ever determine their value, except approximately. Avogadro's constant, however, *must* be determined experimentally, for example by *counting* the number of atoms in a crystal. The value of Avogadro's number found in handbooks is an *experimentally* determined number. You won't discover its value experimentally by counting stars, grains of sand, or people. You find it only by counting atoms or molecules in something of known relative molecular mass. And you won't find it playing any role in any equation or theory about stars, sand, or people. The reciprocal of Avogadro's constant is numerically equal to the unified atomic mass unit, u, that is, 1/12 the mass of the carbon 12 atom. $1\ u = 1.66043 \times 10^{-27}\ kg = 1/6.02252 \times 10^{23}\ mole^{-1}$.

Axis

An arbitrarily chosen point used in the definition of angular momentum. Any object whose direction changes relative to the axis is considered to have angular momentum. No matter what axis is chosen, the angular momentum of a closed system is conserved.

Axisymmetry

An axisymmetric system looks the same if we change our point of view by rotating our position about an axis. Since a symmetry in physics is an operation that leaves our system unchanged, an object that does not look different after a rotation about an axis has "axis-symmetry." If we are looking at a soup can within our imaginary sphere and put our imginary axis through the center of its two flat faces we will find that it is axisymmetric if we take the label off of it, but is not axisymmetric if we leave the label on.

Azimuth

The direction, in degrees referenced to true north, that an antenna must be pointed to receive a satellite signal (compass direction). The angular distance is measured in a clockwise direction.

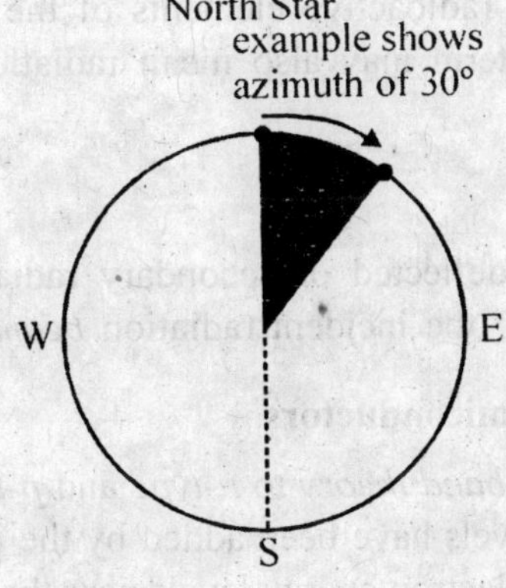

B

Background Radiation

The radiation of man's natural environment originating primarily from the naturally radioactive elements of the earth and from the cosmic rays. The term may also mean radiation extraneous to an experiment.

Backscattering

Primary radiation deflected or secondary radiation emitted in the general direction of the incident radiation *beam*

Bands for Doped Semiconductors

The application of *band theory* to *n-type* and *p-type* semiconductors shows that extra levels have been added by the impurities. In *n*-type material there are electron energy levels near the top of the band gap so that they can be easily excited into the conduction band. In *p*-type material, extra holes in the band gap allow excitation of valence band electrons, leaving mobile holes in the valence band.

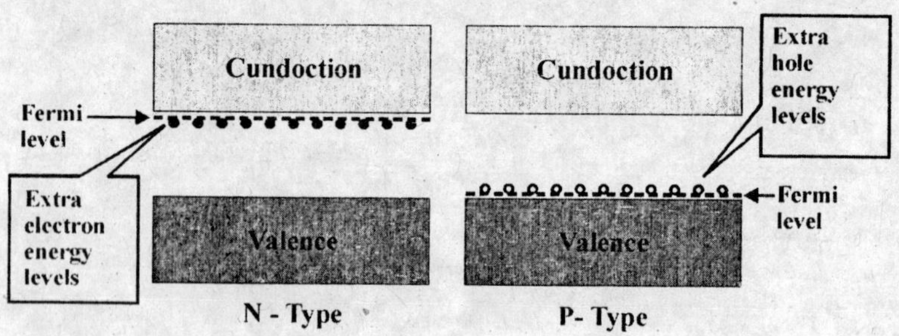

Band Theory of Solids

A useful way to visualize the difference between *conductors insulators* and *semiconductors* is to plot the available energies for electrons in the materials. Instead of having *discrete energies* as in the case of free atoms, the available energy states form *bands* Crucial to the conduction process is whether or not there are electrons in the conduction band. In insulators the electrons in the valence band are separated by a large gap from the conduction band, in conductors like metals the valence band overlaps the conduction band, and in semiconductors there is a small enough gap between the valence and conduction bands that thermal or other excitations can bridge the gap. With such a small gap, the presence of a small percentage of a *doping* material can increase conductivity dramatically.

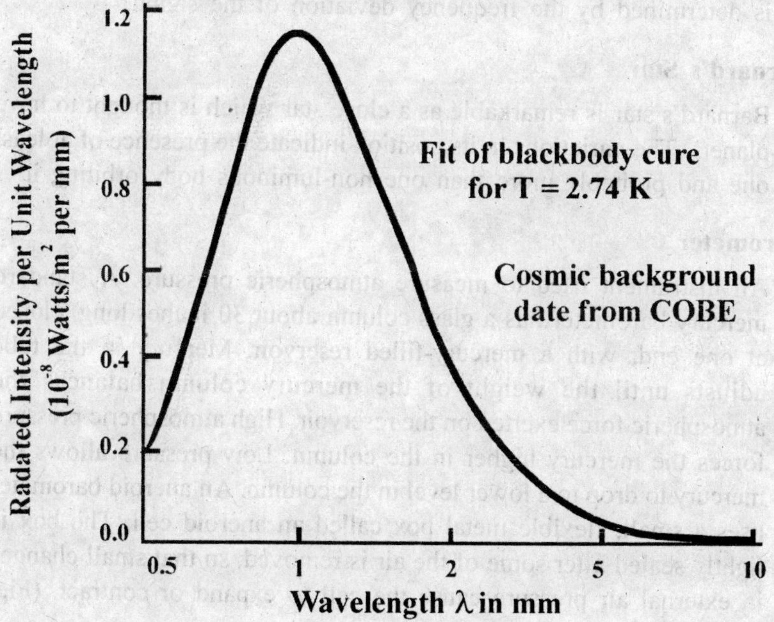

Band

1. In radio a continuous sequence of broadcasting frequencies within given limits. 2. In radiometry, a relatively narrow region of the electromagnetic spectrum to which a remote sensor responds; a

multispectral sensor makes measurements in a number of spectral bands. 3. In spectroscopy spectral regions where atmospheric gases absorb (and emit) radiation, e.g., the 15 μm carbon dioxide absorption band, the 6.3 μm water vapor absorption band, and the 9.6 μm ozone absorption band.

Bandwidth

The total range of frequency required to pass a specific modulated signal without distortion or loss of data. The ideal bandwidth allows the signal to pass under conditions of maximum AM or FM adjustment. (Too narrow a bandwidth will result in loss of data during modulation peaks. Too wide a bandwidth will pass excessive noise along with the signal.) In FM, radio frequency signal bandwidth is determined by the frequency deviation of the signal.

Barnard's Star

Barnard's star is remarkable as a close star which is thought to have planets. The variations in its position indicate the presence of at least one and probably more than one non-luminous body orbiting it.

Barometer

An instrument used to measure atmospheric pressure. A standard mercury barometer has a glass column about 30 inches long, closed at one end, with a mercury-filled reservoir. Mercury in the tube adjusts until the weight of the mercury column balances the atmospheric force exerted on the reservoir. High atmospheric pressure forces the mercury higher in the column. Low pressure allows the mercury to drop to a lower level in the column. An aneroid barometer uses a small, flexible metal box called an aneroid cell. The box is tightly sealed after some of the air is removed, so that small changes in external air pressure cause the cell to expand or contract. (Fig. on next page)

Barrier

Radiation-absorbing material, such as lead or concrete, used to reduce radiation exposure. A primary barrier attenuates useful *beam* to the required degree. A secondary barrier attenuates stray radiation to the required degree.

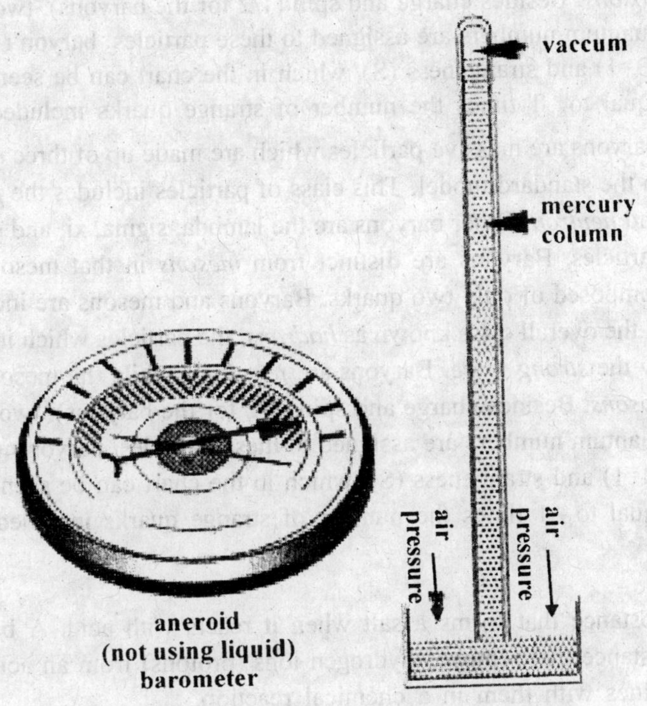

Baryon

1. A *hadron* made from a basic structure of three quarks. The *proton* and the *neutron* are both *baryon.* The antiproton and the antineutron are antibaryons.
2. A hadron made from three quarks. The proton (uud) and the neutron (udd) are both baryons. They may also contain additional quark-antiquark pairs.
3. Baryons are massive particles which are made up of three *quarks* in the standard model. This class of particles includes the *proton* and *neutron.* Other baryons are the lambda, sigma, xi, and omega particles. Baryons are distinct from *mesons* in that mesons are

composed of only two quarks. Baryons and mesons are included in the overall class known as *hadrons,* the particles which interact by the *strong force.* Baryons are *fermions*, while the mesons are *bosons*. Besides charge and spin (1/2 for the baryons), two other quantum numbers are assigned to these particles: baryon number (B=1) and strangeness (S), which in the chart can be seen to be equal to –1 times the number of strange quarks included.

4. Baryons are massive particles which are made up of three *quarks* in the standard model. This class of particles includes the *proton* and *neutron*. Other baryons are the lambda, sigma, xi, and omega particles. Baryons are distinct from *mesons* in that mesons are composed of only two quarks. Baryons and mesons are included in the overall class known as *hadrons*, the particles which interact by the *strong force*. Baryons are *fermions*, while the mesons are *bosons*. Besides charge and spin (1/2 for the baryons), two other quantum numbers are assigned to these particles: baryon number (B=1) and strangeness (S), which in the chart can be seen to be equal to –1 times the number of strange quarks included.

Base

A substance that forms a salt when it reacts with acid. A base is a substance that removes hydrogen ions (protons) from an acid and combines with them in a chemical reaction.

Baud

Unit of signaling speed. The speed in bauds is the number of discrete conditions or signal events per second. If each signal event represents only one bit condition, baud is the same as bits per second.

Bay

A wide area of water extending into land from a sea or lake. The measure of the "width" of an antenna pattern, measured in degrees of arc. Generally an antenna with low gain has a wide pattern, receiving signals well from a number of different directions.

Beam

1. A unidirectional or approximately unidirectional flow of electromagnetic radiation or particles.

2. The particle stream produced by an accelerator usually clustered in bunches.

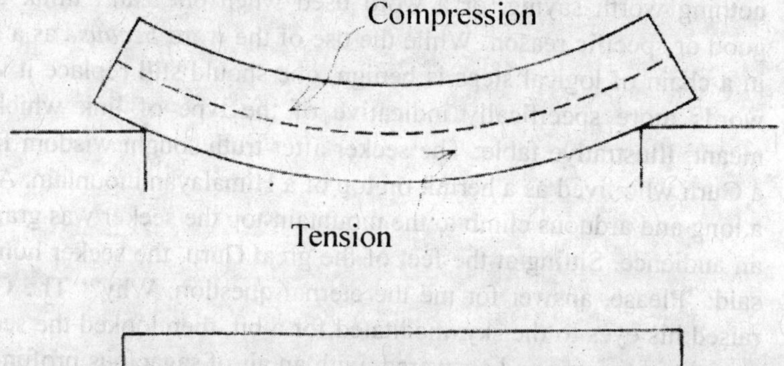

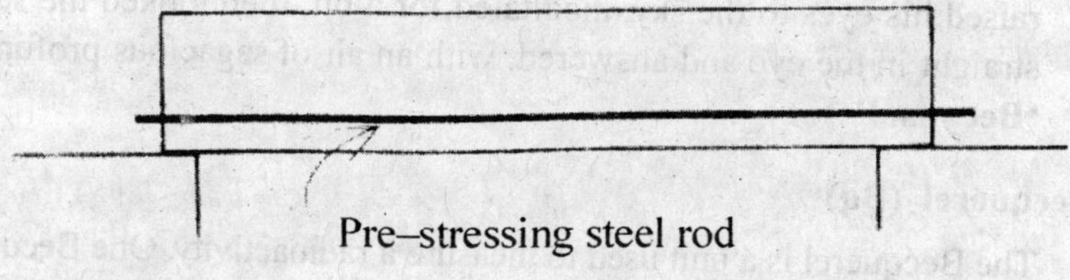

3. Beams are long bars of *rigid* material used for bridging gaps. A beam is possible the simplest example of a structure. The exaggerated sagging of the beam in the diagram highlights that, than in place, the lower half of a beam is in tension whilst the upper half is in compression. The tension along the lower half of the beam can only be withstood by a material which is strong and the deflection will be small only for a material with a high Young's modulus. This is why steel is favoured for beams. Where concrete is used, the beam is reinforced with steel rods along its lower half. Concrete beams are even stronger if tensioned steel rods are used to pre-stress the concrete. A low value for the bending implies high stiffness.

Bearing

The combination of antenna azimuth and elevation required to point (aim) an antenna at a spacecraft. The bearing for geostationary (*i.e.*, GOES) satellites is constant. The bearing for polar-orbiting satellites varies continuously.

Because

Here's a word best avoided in physics. Whenever it appears one can be almost certain that it's a *filler* word in a sentence which says nothing worth saying, or a word used when one can't think of a good or specific reason. While the use of the word *because* as a link in a chain of logical steps is benign, one should still replace it with words more specifically indicative of the type of link which is meant. Illustrative fable: The seeker after truth sought wisdom from a Guru who lived as a hermit on top of a Himalayan mountain. After a long and arduous climb to the mountain-top the seeker was granted an audience. Sitting at the feet of the great Guru, the seeker humbly said: 'Please, answer for me the eternal question: Why?' The Guru raised his eyes to the sky, meditated for a bit, then looked the seeker straight in the eye and answered, with an air of sagacious profundity, **'Because!'**

Becquerel (Bq)

The Becquerel is a unit used to measure a radioactivity. One Becquerel is that quantity of a radioactive material that will have 1 transformations in one second. Often radioactivity is expressed in larger units like: thousands (kBq), one millions (MBq) or even billions (GBq) of a becquerels. As a result of having one Becquerel being equal to one transformation per second, there are 3.7×10^{10} Bq in one curie.

Beginners All-purpose Symbolic Instruction Code (BASIC)

A most popular and widespread "high level" language for microcomputers. BASIC uses a sequence of English-like commands and statements.

Betelgeuse

Betelgeuse is a prominent example of a red supergiant star. It is located at the shoulder of Orion. It forms part of the winter triangle seen from North America. Betelgeuse has a luminosity about 10,000 times that of the Sun and its radius is calculated to be about 370 times that of the sun. If it were positioned at the center of our sun, its radius would extend out past the radius of Mars, about 2 astronomical units. Most stars on the main sequence have about the same size as the sun. It has a surface temperature of about 3000 K as determined from its blackbody radiation curve. Betelgeuse is about 310 light years away from the Earth. Its celestial coordinates are RA=5 h 52 m , dec=7° 24' .

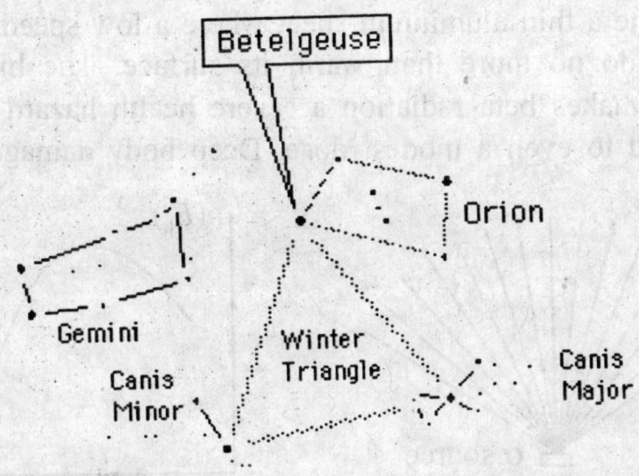

Betas

A beta is a high speed particle, identical to an electron, that is emitted from the nucleus of an atom

Beta Decay

The radioactive decay of a nucleus via the reaction $n \rightarrow p + e^{-} +$ or $p \rightarrow n + e^{+} + n$; so called because an electron or antielectron is also known as a beta particle.

Beta Particle (Beta Radiation, Beta Ray)

An electron of either positive charge (ß+) or negative charge (ß-), which has been emitted by an atomic nucleus or neutron in the process of a transformation. Beta particles are more penetrating than alpha particles but less than gamma rays or x-rays.

Beta Radiation

Beta Radiation consist primarily of electrons emitted from radioactive nuclei but the term also cover streams of positrons (positively charged electrons). The properties of a beam of *beta particles* seem not to be quite the same as those of a beam of electrons in a cathode ray tube, for instance. This is because electrons in a cathode ray or fine beam tube are accelerated through a few thousand volts, whereas electrons expelled from a nucleus during *beta decay* have about a thousand times as much energy. Beta particles are deflected by magnetic fields, but a much stronger field is needed to give even a modest deflection with the lower speed electron beam. Beta particles penetrate a thin aluminium sheet where a low speed electron beam would do no more than warm its surface. The high penetrating power makes beta radiation a severe health hazard to any animal exposed to even a modest dose. Deep body damage is possible.

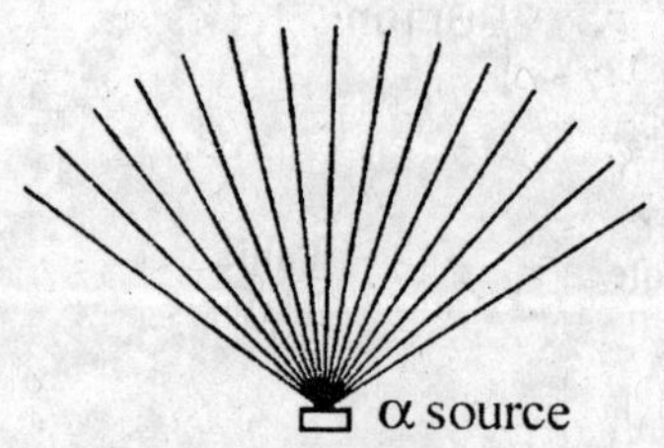

Range in air of alpha particles from a pure alpha emitter

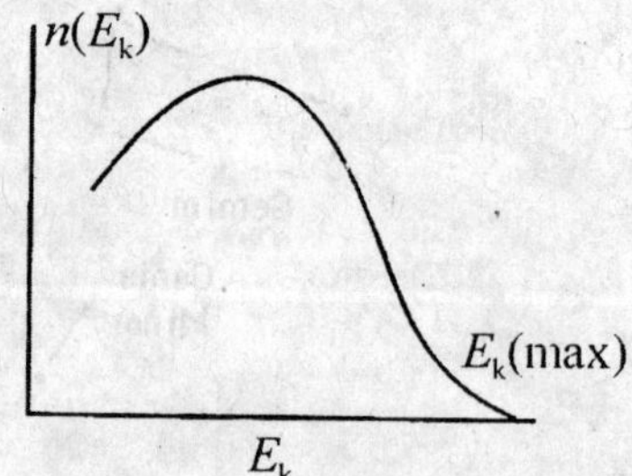

Number of beta particles with energy E_k, $n(E_k)$, against energy when emitted E_k

Beta Particle

A form of radioactivity consisting of electrons.

B-factory

An accelerator designed to maximize the production of B mesons. The properties of the B mesons are then studied with special detectors.

Bhabha Scattering

Scattering of positrons by electrons.

Big Bang

1. The "fireball" of cosmic creation. Modern cosmology is founded on the "Big Bang" model in which all the known universe is thought have have emerged some 13-20 billion years ago from an unimaginably hot, dense state born of a singularity.

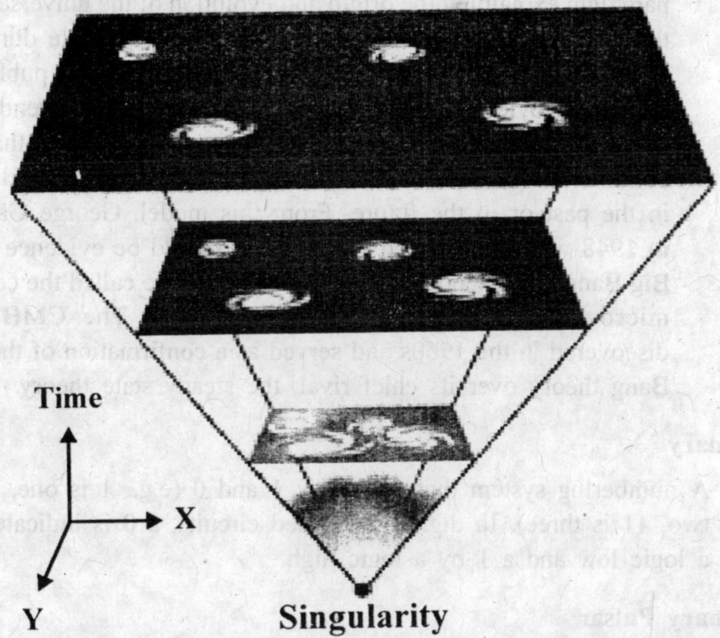

2. The theory of an expanding universe that begins as an infinitely dense and hot medium. The initial instant is called the Big Bang.
3. According to the *Big Bang* theory, the universe originated in an infinitely dense and physically paradoxical singularity. Space has expanded with the passage of time, objects being moved farther away from each other. In cosmology, the *Big Bang* is the scientific theory that describes the early development and shape of the universe. The central idea is that the theory of general relativity can be combined with the observations on the largest scales of

galaxies receding from each other to extrapolate the conditions of the universe back or forward in time. A natural consequence of the Big Bang is that in the past the universe had a higher temperature and a higher density. The term "Big Bang" is used both in a narrow sense to refer to a point in time when the observed expansion of the universe (Hubble's law) began, and in a more general sense to refer to the prevailing cosmological paradigm explaining the origin and evolution of the universe. The term "Big Bang" was coined in 1949 by Fred Hoyle during a BBC radio program, *The Nature of Things*; the text was published in 1950. Hoyle did not subscribe to the theory and intended to mock the concept. One consequence of the Big Bang is that the conditions of today's universe are different from the conditions in the past or in the future. From this model, George Gamow in 1948 was able to predict that there should be evidence for a Big Bang in a phenomenon that would later be called the cosmic microwave background radiation (CMB). The CMB was discovered in the 1960s and served as a confirmation of the Big Bang theory over its chief rival, the steady state theory.

Binary

A numbering system that uses only 1 and 0 (e.g., 1 is one, 10 is two, 11 is three). In digital integrated circuits, a 0 is indicated by a logic low and a 1 by a logic high.

Binary Pulsar

1. A source that pulsates in the radio or x-ray spectrum is called a "pulsar" and it is generally believed that a pulsar is a *neutron star* (although some of the pulsars with longer periods might be *white dwarfs*). A binary pulsar is a binary star system (a system where two stars orbit each other), where one of the two is a pulsar.
2. Hulse and Taylor won the Nobel Prize in 1993 for the discovery of the first binary pulsar in 1974. It has a period of 59 milli-seconds but shows an orbital period of 7 hours and 45 minutes. Discovered at Arecibo, it was an important test of general

relativity. There have been about 40 binary pulsars discovered to date. An exciting close binary was reported in Nature in December 2003 and in Science in early 2004. With the cumbersome designation PSR J0737-3039A, it is composed of pulsars with an eccentric orbit of period just 2.4 hours! The most active of the pulsars spins 44 times per second and its companion just once in 2.8 seconds. Irion in Science described the pair as "two pulsars in a tight orbital embrace, blasting each other with radiation as they spiral toward a mutual doom." General relativity calculations reportedly suggest a convergence of the two pulsars by about 7 millimeters/day with a projected crash in about 85 million years. At just 2000 light years distance, this binary pulsar is relatively close. It's orbit is almost edge-on from the Earth, optimum for viewing. Part of the promise of this dramatic pair is information about relativistic theories of the gravitational interaction. The discovery of this binary pulsar is credited to the 64-meter Parkes radio telescope in New South Wales, Australia. The measurement of the slower period of the companion is credited to Jodrell Bank Observatory in Macclesfield, U.K.

Bioassay

A measurement of the effects of a substance on living organisms.

Biodegradation

Decomposition of material by microorganisms.

Biogeochemical Cycles

Movements through the Earth system of key chemical constituents essential to life, such as carbon, nitrogen, oxygen, and phosphorus.

Biomass

The amount of living material in unit area or volume, usually expressed as mass or weight.

Biome

Well-defined terrestrial environment (e.g., desert, tundra, or tropical forest). The complex of living organisms found in an ecological region.

Biosphere

Part of the Earth system in which life can exist, between the outer portion of the geosphere and the inner portion of the atmosphere.

Biota

The plant and animal life of a region or area.

Birefringent Materials

Crystalline materials may have different *indices of refraction* associated with different *crystallographic directions*. A common situation with mineral crystals is that there are two distinct indices of refraction, and they are called birefringent materials. If the y- and z- directions are equivalent in terms of the crystalline forces, then the x-axis is unique and is called the optic axis of the material. The propagation of light along the optic axis would be independent of its polarization; it's electric field is everywhere perpendicular to the optic axis and it is called the ordinary- or o-wave. The light wave with E-field parallel to the optic axis is called the extraordinary- or e-wave. Birefringent materials are used widely in optics to produce *polarizing prisms* and retarder plates such as the *quarter-wave plate*. Putting a birefringent material between *crossed polarizers* can give rise to *interference colors*

Bit Rate

The speed at which bits are transmitted, usually expressed in bits per second.

Bit

A contraction of "binary digit." The basic element of a two-element (binary) computer language.

Black Hole

1. A black hole is a region of spacetime enclosed by an *event*

horizon. A black hole is formed by the collapse of massive objects. If the heat and pressure supplied by the fusion of the material within the star is less than the gravitational pull inward, the object may collapse to form a *white dwarf*, a *neutron star*, or (if it is massive enough) a black hole. A black hole is termed "black" because nothing can escape from within it, not even light. Everything that passes through the event horizon is gone from the observable universe.

2. A region of space that has so much mass concentrated in it that there is no way for a nearby object to escape its gravitational pull.

Blizzard

A severe weather condition characterized by low temperatures and strong winds (greater than 35 mph) bearing a great amount of snow, either falling or blowing. When these conditions persist after snow has stopped falling, it is called a ground blizzard.

Boltzmann Constant

The *Boltzmann constant* (k or k_B) is the physical constant relating temperature to energy.It is named after the Austrian physicist Ludwig Boltzmann, who made important contributions to the theory of statistical mechanics, in which this constant plays a crucial role. Its experimentally determined value is (in SI units, 2002 CODATA value) $k = 1.380\ 6505(24) \times 10^{-23}$ J/K. The digits in parentheses are the uncertainty (standard deviation) in the last two digits of the measured value. In principle, the Boltzmann constant could be a derived physical constant, as its value is determined by other physical constants and in the definition of unit of absolute temperature. However, calculating the Boltzmann constant from first principles is far too complex to be done with current knowledge. In Planck's system of natural units, the natural unit temperature is such that the Boltzmann constant is one. The universal gas constant R is simply the Boltzmann constant multiplied by Avogadro's number. The gas constant is more useful when calculating numbers of particles in moles.

Boson

1. A particle that has integer intrinsic angular momentum (spin) measured in units of h-bar (spin = 0, 1, 2, ...). All particles are eitherfermions or bosons. The particles associated with all the fundamental interactions (forces) are bosons. Composite particles with even numbers of fermion constituents (quarks) are also bosons.
2. The general name for any particle with a spin of an integer number (0,1 or 2...) of quantum units of angular momentum. The *carrier particles* of all *interactions* are bosons. *Mesons* are also bosons.

Bottom Quark (b)

The fifth flavour of quark (in order of increasing mass), with electric charge of -1/3.

Bound State

This is a state in which a particle is confined within a composite system, for example an atom or a nucleus, because it does not have enough energy to escape. An *electron* in a atom is bound because of its electrical attraction to the nucleus, which makes the mass of the atom slightly less than the sum of the masses of the electron plus the rest of the atom without that electron.

Boundaries

Lines indicating the limits of countries, states, or other political jurisdictions, or different air masses.

Brachytherapy

This involves placing the source of radiation directly within the tumor and employs radioactive plaques, needles, tubes, wires, or small "seeds" made of radionuclides. These radioactive materials are placed over the surface of the tumor or implanted within the tumor, or placed within a body cavity surrounded by the tumor.

Bremsstrahlung

1. X-rays emitted when a *charged* particle (such as an electron) is decelerated by passing through matter. The word bremsstrahlung is German for "braking radiation".

2. *Radiation* that is emitted when a *free electron* is deflected by an *ion*, but the free electron is not captured by the ion. Generally, it is a type of radiation emitted when high energy electrons are accelerated. (German for braking radiation)

British Thermal Unit

The amount of heat needed to raise the temperature of one pound of water by one degree Fahrenheit.

Bubble Chamber

A chamber filled with liquid at low pressure chosen so that small bubbles form along the path of any *charged* particle. After each beam pulse a photographic record is made of the chamber and then it is depressurized to clear the bubbles.

Bulk Elastic Properties

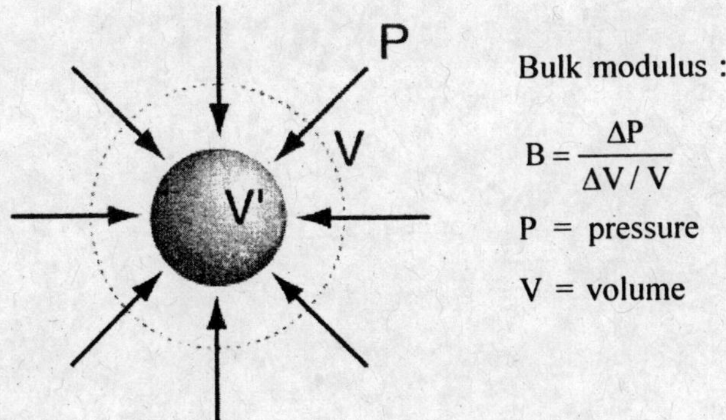

The bulk *elastic* properties of a material determine how much it will compress under a given amount of external *pressure.* The ratio of the change in pressure to the fractional volume compression is called the bulk modulus of the material. A representative value for the bulk modulus for steel is and that for water is The reciprocal of the bulk modulus is called the compressibility of the substance. The amount of compression of solids and liquids is seen to be very small.

The bulk modulus of a solid influences the *speed of sound* and other mechanical waves in the material. It also is a factor in the amount of energy stored in solid material in the Earth's crust. This buildup of elastic energy can be released violently in an earthquake, so knowing bulk moduli for the Earth's crust materials is an important part of the study of earthquakes.

Bus

The basic frame of a satellite system that includes the propulsion and stabilization systems, but not the instruments or data systems.

Byte

A unit of eight bits of data or memory in microcomputer systems.

C

Calcite

The mineral calcite, also known as Iceland spar, is a widely used material in optics because of its *birefringence*. Its birefringence is so large that a calcite crystal placed over a dot on a page will reveal two distinct images of the dot. One image will remain fixed as the crystal is rotated, and that ray through the crystal is called the "ordinary ray" since it behaves just as a ray through glass. However, the other image will rotate with the crystal, tracing out a small circle around the ordinary image. This ray is called the "extraordinary ray".

Calibration

Act of comparing an instrument's measuring accuracy to a known standard.

Calorie

The amount of heat needed to raise the temperature of one gram of water at 15° centigrade one degree centigrade.

Calorimeter

In particle physics, any device that can measure the energy deposited in it by particles (originally a device that measured heat energy deposited, thus a calorie-meter).

Canal

A man-made watercourse designed to carry goods or water.

Canopy

The layer formed naturally by the leaves and branches of trees and plants.

Canyon

A large but narrow gorge with deep sides.

Capacitance

1. The *capacitance* of a capacitor is measured by this procedure: Put equal and opposite charges on its plates and then measure the potential between the plates. Then C = |Q/V|, where Q is the charge on *one* of the plates. Capacitors for use in circuits consist of two conductors (plates). We speak of a capacitor as 'charged' when it has charge Q on one plate, and -Q on the other. Of course the net charge of the entire object is zero; that is, the charged capacitor hasn't had net charge added to it, but has undergone an internal separation of charge. Unfortunately this process is usually called *charging* the capacitor, which is misleading because it suggests adding charge to the capacitor. In fact, this process usually consists of moving charge from one plate to the other. The capacity of a single object, say an isolated sphere, is determined by considering the *other plate* to be an infinite sphere surrounding it. The object is given charge, by moving charge from the infinite sphere, which acts as an infinite charge reservoir ('ground'). The potential *of the object* is the potential between the object and the infinite sphere. Capacitance depends only on the geometry of the capacitor's physical structure and the dielectric constant of the material medium in which the capacitor's electric field exists. The size of the capacitor's capacitance is the same whatever the charge and potential (assuming the dielectric constant doesn't change). This is true even if the charge on both plates is reduced to zero, and therefore the capacitor's potential is zero. If a capacitor with charge on its plates has a capacitance of, say, 2 microfarad, then its capacitance is also 2 microfarad when the plates have no charge. This should remind us that C = |Q/V| is not by itself the *definition* of capacitance, but merely a formula which allows us to relate the capacitance to the charge and potential *when* the capacitor plates have equal and opposite charge on them. A common misunderstanding about electrical capacitance is to

assume that capacitance represents the maximum amount of charge a capacitor can store. That is misleading because capacitors don't store charge (their total charge being zero) but their plates have equal and opposite charge. It is wrong because the maximum charge one may put on a capacitor plate is determined by the potential at which dielectric breakdown occurs. We probably should avoid the phrase 'charged capacitor' or 'charging a capacitor'. Some have suggested the alternative expression 'energizing a capacitor' because the process is one of giving the capacitor electrical potential energy by rearranging charges in it.

2. Capacitance is the ability of a capacitor to store potential difference or *voltage* for a given amount of stored charge. The SI unit of capacitance is the farad.

$$C = \frac{Q}{V}$$

here C is the capacitance Q is the charge V is the potential difference. A capacitor has a capacitance of one farad when one coulomb of charge causes a potential difference of one volt across the capacitor. Since the farad is a very large unit, values of capacitors are usually expressed in microfarads (ìF), nanofarads (nF) or picofarads (pF). The above equation is only accurate for values of Q which are much larger than the electron charge e = $1.602 \cdot 10^{-19}$ C. For example, if a capacitance of 1 pF is charged to a voltage of 1 µV, the equation would predict a charge Q = 10^{-19} C, but this is impossible as it is smaller than the charge on a single electron.

Capacitor/inductor Duality

In mathematical terms, the ideal capacitor can be considered as an inverse of the ideal inductor, because the voltage-current equations of the two devices can be transformed into one another by exchanging the voltage and current terms. Just as two or more inductors can be magnetically coupled to make a transformer, two or more charged conductors can be electrostatically coupled to make a capacitor. The *mutual capacitance* of two conductors is defined as the current that flows in one when the voltage across the other changes by unit voltage in unit time.

Capacity

This word is used in names of quantities which express the *relative* amount of some quantity with respect to a another quantity upon which it depends. For example, heat capacity is dU/dT, where U is the internal energy and T is the temperature. Electrical capacity, or *capacitance* is another example: $C = |dQ/dV|$, where Q is the magnitude of charge on each capacitor plate and V is the potential diference between the plates.

Cape (or point)

A piece of land extending into water.

Capillary Action

Capillary action is the result of *adhesion* and *surface tension*. Adhesion of water to the walls of a vessel will cause an upward force on the liquid at the edges and result in a meniscus which turns upward. The surface tension acts to hold the surface intact, so instead of just the edges moving upward, the whole liquid surface is dragged upward.

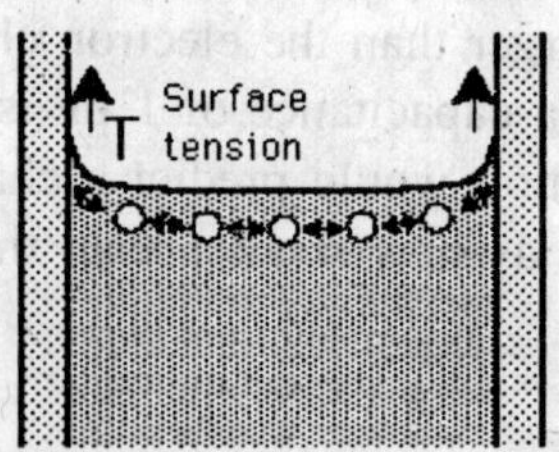

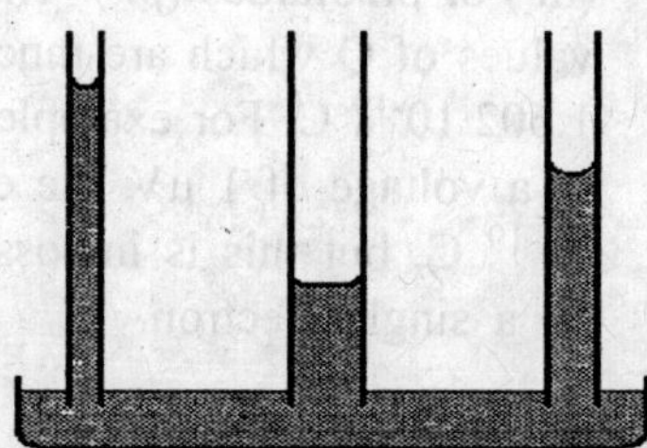

Carbon Cycle

All parts (reservoirs) and fluxes of carbon. The cycle is usually thought of as four main reservoirs of carbon interconnected by pathways of exchange. The reservoirs are the atmosphere, terrestrial biosphere (usually includes freshwater systems), oceans, and sediments (includes fossil fuels). The annual movements of carbon,

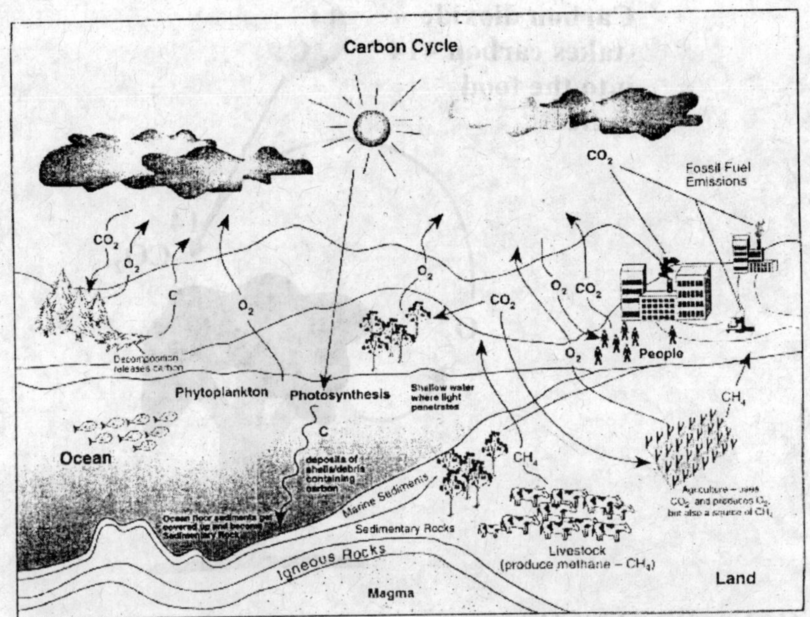

the carbon exchanges between reservoirs, occur because of various chemical, physical, geological, and biological processes. The ocean contains the largest pool of carbon near the surface of the Earth, but most of that pool is not involved with rapid exchange with the atmosphere.

Carbon Dating

Presuming the *rate of production* of carbon-14 to be constant, the activity of a sample can be directly compared to the equilibrium activity of living matter and the age calculated. Various tests of *reliability* have confirmed the value of carbon data, and many *examples* provide an interesting range of application. Carbon-14 decays with a halflife of about 5730 years by the emission of an electron of energy 0.016 MeV. This changes the atomic number of the nucleus to 7, producing a nucleus of nitrogen-14. At equilibrium with the atmosphere, a gram of carbon shows an activity of about 15 decays per minute.

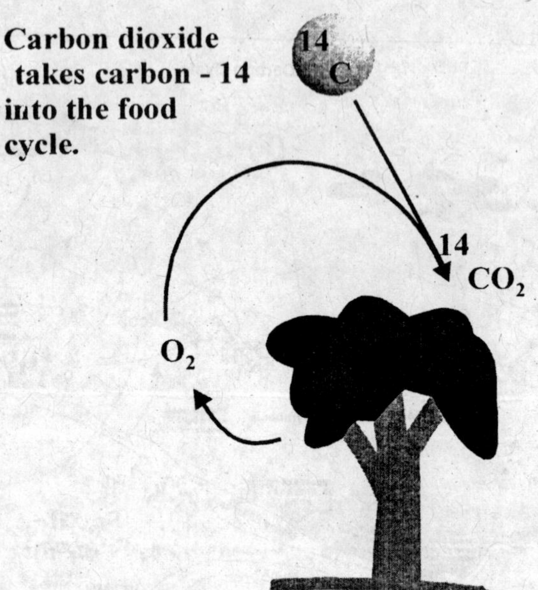

Carbon Dioxide (CO_2)

A minor but very important component of the atmosphere, carbon dioxide traps infrared radiation. Atmospheric CO_2 has increased about 25 percent since the early 1800s, with an estimated increase of 10 percent since 1958 (burning fossil fuels is the leading cause of increased CO_2, deforestation the second major cause). The increased amounts of CO_2 in the atmosphere enhance the greenhouse effect, blocking heat from escaping into space and contributing to the warming of Earth's lower atmosphere.

Carbon-Nitrogen-Oxygen Cycle

In stars more massive than the sun (>1.1 Solar masses), this cycle is the primary process which converts hydrogen into helium. ^{12}C serves as a catalyst, an ingredient which is necessary for the reaction but is not consumed.

Carrier Particle

A fundamental *boson* associated with quantum excitations of the force field corresponding to some *interaction. Gluons* are carrier particles for strong interactions (color force fields), photons are carrier particles of *electromagnetic interactions,* and the W and Z bosons are carrier particles for weak interactions.

Carrier

Radio frequency capable of being modulated with some type of information. Carrying Capacity. The steady-state density of a given species that a particular habitat can support.

Catalog Number

A five-digit number assigned to a cataloged orbiting object. This number may be found in the NASA Satellite Situation Report and on the NASA Prediction Bulletins.

Cathode Ray Tube (CRT)

1. A television picture tube for image display.
2. An *evacuated* tube containing an anode and a cathode that generates cathode rays (electrons) when operated at a high voltage. The cathode rays produce an image on a screen when they strike phosphors on the screen, causing them to glow.

Cathode Ray

The mysterious ray that emanated from the cathode in a vacuum tube; shown by Thomson to be a stream of particles smaller than atoms.

Cavity Resonator

A *cavity resonator* uses resonance to amplify a wave. The cavity has interior surfaces which reflect one type of wave. When a wave that is resonant with the cavity enters, it bounces back and forth within the cavity, with low loss. As more wave energy enters the cavity, it combines with and reinforces the standing wave, increasing its intensity. Some common examples of cavity resonators include the klystron tube in a microwave oven the tube of a flute, and the body of a violin (this latter also being an example of a Helmholtz resonator). In a laser, light is amplified in a cavity resonator which is usually composed of two or more mirrors. Thus an *optical cavity*, also known as a resonator, is a cavity with walls which reflect electromagnetic waves (light). This will allow standing wave modes to exist with little loss outside the cavity.

Centre of Gravity

Centre of Gravity is he point through which the weight of a rigid body acts. No matter how much the body tumbles or turns, it is a fixed point relative to the body. If a rigid body is imagined to consist of innumerable tiny masses, the centre of gravity is the point about which the sum of the moments of the weights of all these tiny masses is zero. A right body resting on a surface is in stable equilibrium if any small rotation raises the centre of gravity. It is in unstable equilibrium if such a rotation lowers the centre of gravity. It is in neutral equilibrium if such a rotation neither raises nor lower the centre of gravity. The three cases are illustrated in the diagram above. The centre of gravity of a rigid body coincides with its *centre of mass* in a uniform gravitational field.

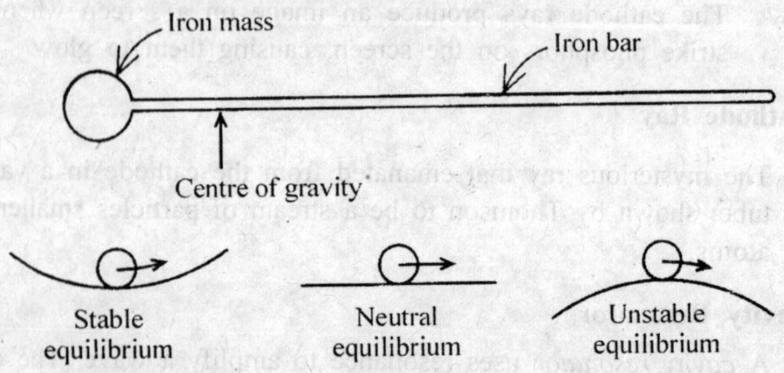

Center of Mass

The balance point of an object.

Centigrade

Temperature scale proposed by Swedish astronomer Anders Celsius in 1742. A mixture of ice and water is zero on the scale; boiling water is designated as 100 degrees. A degree is defined as one hundredth of the difference between the two reference points, resulting in the term centigrade (100th part). To convert centigrade to Fahrenheit; multiply the centigrade temperature by 1.8 and add 32°. F = 9/5C + 32. To convert Fahrenheit to centigrade: subtract 32° from the Fahrenheit temperature and divide the quantity by 1.8. C = (F -32) / 1.8.

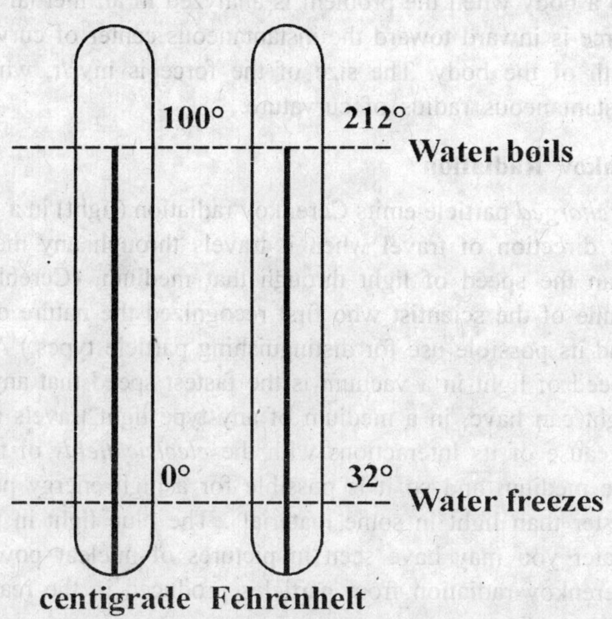

Central Processing Unit (CPU)

Main part of a computer consisting of an arithmetic logic unit and a control unit.

Centrifugal Force

When a non-inertial rotating coordinate system is used to analyze motion, Newton's law F = m*a* is *not* correct unless one adds to the real forces a *fictitious force* called the *centrifugal force*. The centrifugal force required in the non-inertial system is equal and opposite to the *centripetal force* calculated in the inertial system. Since the centrifugal and centripetal forces are concepts used in *two different* formulations of the problem, they can not in any sense be considered a pair of reaction forces. Also, they act on the same body, not different bodies.

Centripetal Force

The *centripetal force* is the radial component of the net force acting on a body when the problem is analyzed in an inertial system. The force is inward toward the instantaneous center of curvature of the path of the body. The size of the force is mv^2/r, where r is the instantaneous radius of curvature.

Cerenkov Radiation

A *charged* particle emits Cerenkov radiation (light) in a cone around its direction of travel when it travels through any medium faster than the speed of light through that medium. (Cerenkov - is the name of the scientist who first recognized the nature of this effect and its possible use for distinguishing particle types.) Although the speed of light in a vacuum is the fastest speed that any particle or light can have, in a medium of any type light travels more slowly because of its interactions with the *electric fields* of the atoms in the medium and so it is possible for a high energy particle to be faster than light in some material . The blue light in the pools of water you may have seen in pictures of nuclear power plants is Cerenkov radiation from particles produced in the reactor.

Cetacean Sound

Orcas produce a wide variety of clicks, whistles and pulsed calls. They vary in frequency from 1 to 25 kHz. Individual pods of whales have their own distinctive dialect of calls, similar to songbirds. Some such calls are known to be stable over a period of 10 years. Humpback whales produce a variety of moans, snores, and groans that are repeated to form what we might call songs. The frequency of these songs range from about 40 Hz to 5 kHz. Singing whales are usually solitary males who exhibit it in a shallow smooth-bottomed area where sound propagates well. They are interprוted as territorial and mating calls. Whales are also known to produce some very intense low frequency sounds which they may use to stun or disorient small fish for prey. Bottlenose dolphins produce sounds in the range 7 to 15 kHz which are continuously variable in pitch. In addition, they produce short burst from 20 to 170 kHz, presumably for better echolocation. A dolphin's clicks come from small knobs near its blowhole. There are no vocal cords.

CGS

1. The system of units based upon the fundamental metric units: centimeter, gram and second.
2. Centimeter-Gram-Second (abbreviated *cm-gm-sec* or *cm-g-s*). The system of measurement that uses these units for distance, mass, and time.

Chaos Theory

Chaos theory, in mathematics and physics, deals with the behavior of certain nonlinear dynamical systems that (under certain conditions) exhibit the phenomenon known as chaos, most famously characterised by sensitivity to initial conditions. Examples of such systems include the atmosphere, the solar system, plate tectonics, turbulent fluids, economies, and population growth. Systems that exhibit mathematical chaos are deterministic and thus orderly in some sense; this technical use of the word *chaos* is at odds with common parlance, which suggests complete disorder. When we say that chaos theory studies deterministic systems, it is necessary to mention a related field of physics called quantum chaos theory that studies non-deterministic systems following the laws of quantum mechanics

Charge Conservation

The observation that electric charge is conserved in any process of transformation of one group of particles into another.

Charge-Coupled Device (CCD)

A rectangular array of light sensitive pixels which forms the image detection sensor in comcoders and some telescopes. For telescopes, the number and distribution of the pixels matches the pixels on a PC monitor. The first diagram shows one CCD pixel. It consists of three tiny metal flags labelled, respectively, a, b and c. One row or horizontal line on the PC monitor. Three 'wires' connect all the a pixels together, all the *b* pixels together and all the *c* pixels together. The metal flags are mounted on a very thin layer of silicon dioxide which, in turn, covers a thin but stiff sheet of *p*-type semiconductor.

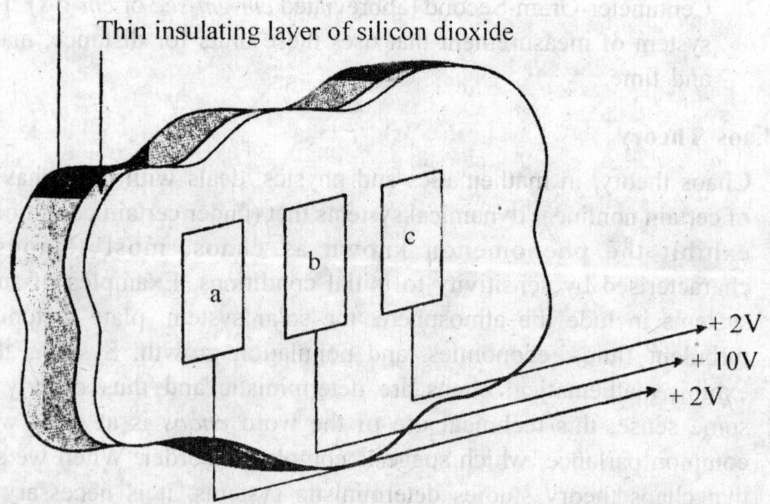

Charge

1. A numerical rating of how strongly an object participates in electrical forces.

2. A quantity carried by a particle that determines its participation in an *interactions* process. A particle with electric charge has electrical interactions; one with strong charge (or *color charge*) has strong interactions, etc.

3. A quantum number carried by a particle. Determines whether the particle can participate in an interaction process. A particle with electric charge has electrical interactions; one with h2 charge has h2 interactions, etc.

4. Charges are the basis of electricity. A charge exert forces on other charges. The smallest charge is the charge of an electron, 0.00000000000000016 Coulombs.

Charm Quark (c)

The fourth quark (in order of increasing mass), with electric charge +2/3.

Charm

One "*flavor*" of quarks. Also known as the C quark.

Chlorofluorocarbon (CFC)

A family of compounds of chlorine, fluorine, and carbon, entirely of industrial origin. CFCs include refrigerants, propellants for spray cans (this usage is banned in the U.S., although some other countries permit it) and for blowing plastic-foam insulation, styrofoam packaging, and solvents for cleaning electronic circuit boards. The compounds' lifetimes vary over a wide range, exceeding 100 years in some cases. CFCs' ability to destroy stratospheric ozone through catalytic cycles is contributing to the depletion of ozone worldwide. Because CFCs are such stable molecules, they do not react easily with other chemicals in the lower atmosphere. One of the few forces that can break up CFC molecules is ultraviolet radiation, however, the ozone layer protects the CFCs from ultraviolet radiation in the lower atmosphere. CFC molecules are then able to migrate intact into the stratosphere, where the molecules are bombarded by ultraviolet rays, causing the CFCs to break up and release their chlorine atoms. The released chlorine atoms participate in ozone destruction, with a single atom of chlorine able to destroy ozone molecules over and over again. International attention to CFCs resulted in a meeting of diplomats from around the world in Montreal in 1987. They forged a treaty that called for drastic reductions in the production of CFCs. In 1990, diplomats met in London and voted to significantly strengthen the Montreal Protocol by calling for a complete elimination of CFCs by the year 2000.

Chromosphere

The layer of the *solar atmosphere* that is located above the *photosphere* and beneath the transition region and the *corona*. The chromosphere is hotter than the photosphere but not as hot as the corona.

Chronic Dose

A Chronic dose means a person received a radiation dose over a long period of time.

Circadian Rhythm

The cyclical changes in physiological processes and functions that are related to the 24-hour *diurnal* cycle.

Circuit

1. An electrical device in which charge can come back to its starting point and be recycled rather than getting stuck in a dead end.
2. The complete path of an electric current; an assemblage of electronic elements; a means of two-way communication between two points—comprised of associated "go" and "return" channels.

Circularly Polarized RF

Radio frequency transmissions where the wave energy is divided equally between a vertically- and a horizontally-polarized components.

Clarke Belt

A belt 22,245 miles (35,800 kilometers) directly above the equator where a satellite orbits the Earth at the same speed the Earth is rotating. Science fiction writer and scientist Arthur C. Clarke wrote about this belt in 1945, hence the name.

Classical Mechanics

In physics, Classical mechanics is one of the two major sub-fields of study in the science of mechanics, which is concerned with the motions of bodies, and the forces that cause them. The other sub-field is quantum mechanics. Roughly speaking, classical mechanics was developed in the 400 years since the groundbreaking works of Brahe, Kepler, and Galilei, while quantum mechanics developed within the last 100 years, starting with similarly decisive discoveries by Planck, Einstein, and Bohr. The notion of "classical" may be somewhat confusing, insofar as this term usually refers to the era of classical antiquity in European history. While many discoveries within the mathematics of that period remain in full force today, and of the greatest use, the same cannot be said about its "science". This in no way belittles the many important developments, especially within technology, which took place in antiquity and during the Middle Ages in Europe and elsewhere. However, the emergence of classical mechanics was a decisive stage in the development of science, in the modern sense of the term. What characterizes it, above all, is its insistence on mathematics (rather than speculation), and its reliance on experiment (rather than observation). With classical mechanics it was established how to formulate quantitative predictions in theory, and how to test them by carefully designed measurement. The emerging globally cooperative endeavor increasingly provided for much closer scrutiny and testing, both of theory and experiment. This was, and remains, a key factor in establishing certain knowledge, and in bringing it to the service of society. History shows how closely the health and wealth of a society depends on nurturing this investigative and critical approach. The initial stage in the development of classical mechanics is often referred to as Newtonian mechanics and is characterized by the mathematical methods invented by Newton himself, in parallel with Leibniz and others. This is further described in the following sections. More abstract, and general methods include Lagrangean mechanics and Hamiltonian mechanics.

Classical Physics

1. Classical physics includes the traditional branches and topics that were recognized and fairly well developed before the beginning of the 20th cent.-*mechanics, sound, light, heat*, and

electricity and *magnetism*. Mechanics is concerned with bodies acted on by *forces* and bodies in *motion* and may be divided into *statics* (study of the forces on a body or bodies at rest), kinematics (study of motion without regard to its causes), and *dynamics* (study of motion and the forces that affect it); mechanics may also be divided into solid mechanics and fluid mechanics, the latter including such branches as hydrostatics, hydrodynamics, aerodynamics, and pneumatics. *Acoustics*, the study of sound, is often considered a branch of mechanics because sound is due to the motions of the particles of air or other medium through which sound waves can travel and thus can be explained in terms of the laws of mechanics. Among the important modern branches of acoustics is *ultrasonics*, the study of sound waves of very high frequency, beyond the range of human hearing. Optics, the study of light, is concerned not only with visible light but also with infrared and ultraviolet radiation, which exhibit all of the phenomena of visible light except visibility, e.g., *reflection, refraction, interference, diffraction, dispersion and polarization of light*. Heat is a form of energy, the internal energy possessed by the particles of which a substance is composed; *thermodynamics* deals with the relationships between heat and other forms of energy. Electricity and magnetism have been studied as a single branch of physics since the intimate connection between them was discovered in the early 19th cent.; an electric current gives rise to a magnetic field and a changing magnetic field induces an electric current. Electrostatics deals with electric charges at rest, electrodynamics with moving charges, and magnetostatics with magnetic poles at rest.

2. The physics developed before about 1900, before we knew about relativity and quantum mechanics.

Climate

The average weather condition in an area determined over a period of years.

Climatology

Science dealing with *climate* and climate phenomena.

Clinac

Varian trade name for a range of *linear accelerator* models used in cancer treatment and stereotactic radiosurgery.

Clone

A person or thing very much like another, e.g., a copy of another manufacturer's computer.

Closed System

A physical system on which no outside influences act; closed so that nothing gets in or out of the system and nothing from outside can influence the system's observable behavior or properties. Obviously we could never make measurements on a closed system unless we were in it†, for no information about it could get out of it! In practice we loosen up the condition a bit, and only insist that there be no interactions with the outside world which would affect those properties of the system which are being studied. † Besides, when the experimenter is a part of the system, all sorts of other problems arise. This is a dilemma physicists must deal with: the fact that if we take measurements, we are a part of the system, and must be very certain that we carry out experiments so that fact doesn't distort or prejudice the results.

Cloud Streets

Lines or rows of cumuliform clouds.

Cloudburst

Any sudden, heavy rain shower.

Clouds

A visible mass of liquid droplets suspended in the atmosphere above Earth's surface. Clouds form in areas where air rises and cools. The condensing water vapor forms small droplets of water (0.012 mm) that, when combined with billions of other droplets, form clouds. Clouds can form along warm and cold fronts, where air flows up the side of the mountain and cools as it rises higher into the atmosphere, and when warm air blows over a colder surface, such as a cool body of water. Clouds fall into two general categories: sheetlike or layer-looking stratus clouds (stratus means layer) and cumulus clouds (cumulus means piled up). These two cloud types are divided into four more groups that describe the clouds' altitude. High clouds form above 20,000 feet in the cold region of the troposphere, and are denoted by the prefix CIRRO or CIRRUS. At this altitude water almost always freezes so clouds are composed of ice crystals. The clouds tend to be wispy are often transparent, and include cirrus, cirrocumulus, and cirrostratus. Middle clouds form between 6,500 and 20,000 feet and are denoted by the preftx ALTO. They are made of water droplets and include altostratus and altocumulus. Low clouds are found up to 6,500 feet and include the stratocumulus and nimbostratus clouds. When stratus clouds contact the ground they are called fog. Vertical clouds, such as cumulus, rise far above their bases and can form at many heights. Cumulonimbus clouds, or thunderheads, can start near the ground and soar up to 75,000 feet.

Coastal Zone Color Scanner (CZCS)

The first spacecraft instrument devoted to measurement of ocean color. Although instruments on other satellites have sensed ocean color; their spectral bands, spatial resolution, and dynamic range were optimized for geographical or meteorological use. In the CZCS, every parameter is optimized for use over water to the exclusion of any other type of sensing. The CZCS flew on the Nimbus-7 spacecraft.

Coaxial Cable

A hollow copper cylinder; or other cylindrical conductor; surrounding a singlewire conductor having a common axis (hence coaxial). The space between the cylindrical shell and the inner conductor is filled with an insulator which may be plastic or mostly air; with supports separating the shell and the inner conductor every inch or so. The cable is used to carry radio frequency signals to or from antennas, etc.

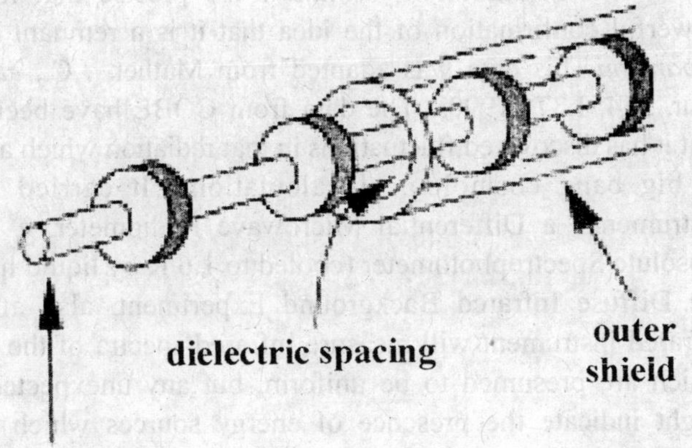

COBE Satellite

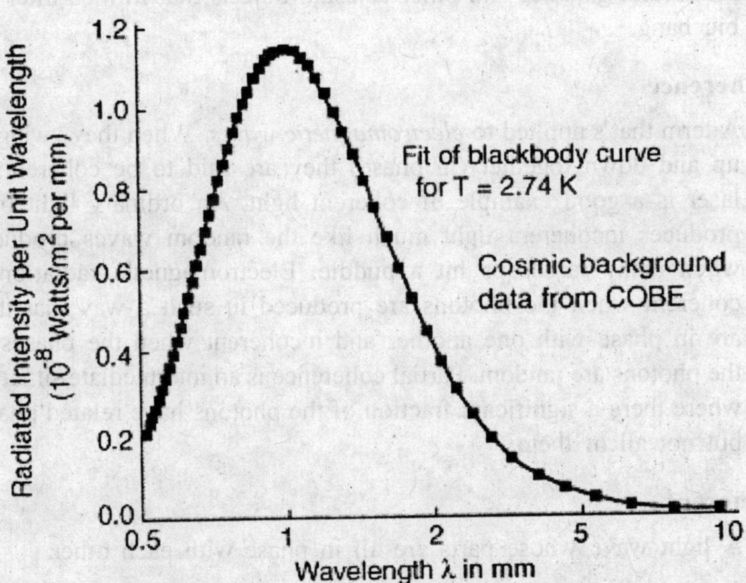

NASA's Cosmic Background Explorer satellite (COBE) was launched to explore the cosmic microwave *background* radiation. Data points are shown superimposed on the theoretical blackbody curve. The fit of the *Planck radiation formula* is so precise that it provides a powerful confirmation of the idea that it is a remnant of *big bang expansion.* This data was adapted from Mather, J.C., *et. al.,* Astro. Jour. 354, L37 (1990). The data from COBE have been so precise that it has discovered fluctuations in that radiation which are important to big bang cosmological calculations. It carried three main instruments, a Differential Microwave Radiometer, a Far-Infrared Absolute Spectrophotometer (cooled to 1.6 K by liquid helium) , and the Diffuse Infrared Background Experiment, also at 1.6K. The infrared instrument will measure infrared spectra of the background which are presumed to be uniform, but any unexpected variations might indicate the presence of energy sources which might have driven turbulence to trigger galaxy formation. The infrared instruments sensitivity is 100 times greater than that achievable from the Earth's surface. The Infrared Background Experiment will look at distant primordial galaxies and other celestial objects that formed after the big bang.

Coherence

A term that's applied to *electromagnetic waves.* When they "wiggle" up and down together (in phase) they are said to be coherent. A laser is a good example of coherent light. An ordinary light bulb produces incoherent light much like the random waves produced when many raindrops hit a puddle. Electromagnetic radiation is coherent when the photons are produced in such a way that they are in phase with one another and incoherent when the phases of the photons are random. Partial coherence is an intermediate situation where there a significant fraction of the photons have related phase, but not all of them.

Coherent

A light wave whose parts are all in phase with each other.

Coherent Light

Coherence is one of the unique properties of *laser* light. It arises from the *stimulated emission* process which provides the

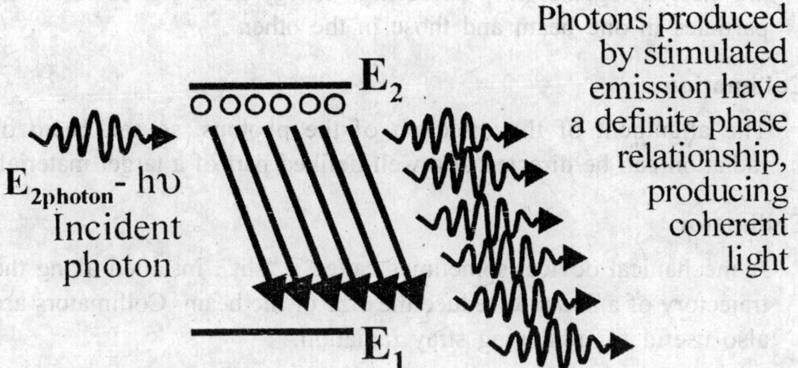

amplification. Since a common stimulus triggers the emission events which provide the amplified light, the emitted photons are "in step" and have a definite phase relation to each other. This coherence is described in terms of temporal coherence and spatial coherence, both of which are important in producing the interference which is used to produce holograms. Ordinary light is not coherent because it comes from independent atoms which emit on time scales of about 10^-8 seconds. There is a degree of coherence in sources like the mercury green line and some other useful spectral sources, but their coherence does not approach that of a laser.

Cohesion and Adhesion

Molecules liquid state experience strong intermolecular attractive forces. When those forces are between like molecules, they are referred to as cohesive forces. For example, the molecules of a water droplet are held together by cohesive forces, and the especially strong cohesive forces at the surface constitute *surface tension*. When the attractive forces are between unlike molecules, they are said to be adhesive forces. The adhesive forces between water molecules and the walls of a glass tube are stronger than the cohesive forces lead to an upward turning meniscus at the walls of the vessel and contribute to capillary action.

Collider

An accelerator in which two beams traveling in opposite directions are steered together to provide high-energy collisions between the particles in one beam and those in the other.

Collimation

The alignment of the direction of the photons, so the *beam* of radiation can be directed at a well-defined part of a target material.

Collimator

A mechanical device, sometimes called a "slit", installed along the trajectory of a *beam* to reduce the size of the beam. Collimators are also useful for removing stray radiation.

Collision

An interaction between moving objects that lasts for a certain time.

Color Charge

The *charge* associated with strong *interactions.* Quarks and *gluons* have color charge and consequently participate in strong interactions. *Leptons,* photons and W and Z *bosons* do not have color charge and therefore do not participate in strong interactions.

Color Force

A property of quarks labeled color is an essential part of the quark model. The force between quarks is called the color force. Since quarks make up the baryons, and the strong interaction takes place between baryons, you could say that the color force is the source of the strong interaction, or that the strong interaction is like a residual color force which extends beyond the proton or neutron to bind them together in a nucleus.

Colour Charge

The quantum number that determines participation in h2 interactions. Quarks and gluons carry nonzero colour charges.

Comets

Comets are small asteroid-like bodies when they are far from the

Sun, traveling in highly ellipical orbits about the Sun. When they sweep in close to the Sun, dramatic changes occur as they brighten and develop an extended tail. The nucleus is widely described as a "dirty snowball" composed of ice and some rocky debris. There is considerable vaporization as they approach the Sun and they develop ion tails and dust tails. The ion tails are almost straight streamers from the nucleus while the usually brighter dust tails are broad and diffuse and curve slightly, lagging behind the radial direction. The lighter ionized gas atoms of the ion tale cause it to point outward, directly away from the Sun, because the influence of the solar wind is dominant. I take the lower, more focused part of the tail in the image above to be the ion tail. The dust tail is made up of more massive particles and the role of gravity is important. If particles influenced by gravity are moved to an orbit further from the Sun, their radial direction falls behind that of the nucleus of the comet because their orbital period will be longer. The upper part of the Halley image would then appear to be the dust tail - you can see a slight curvature. In its most visible phase close to the Sun, the comet has a small solid nucleus and a ball of gas around it called the coma. Comas have been found to be on the order of 100,000 km in diameter at their maximum size, comparable to the largest planets. Most aof the light reflection is from the coma. Surrounding the coma and the visible tails is a hydrogen envelope which may extend millions of kilometers. The light from comets is purely reflected light; like the planets, the comets produce no light of their own. Current models of the nuclei of comets view them as balls of loosely packed ices, a cold mixture containing gas and dust. The dust is thought to be trapped in a mixture of methane, ammonia, and water ice. The smaller moons of the outer solar system are similar in constitution. Since they spend most of their time far from the Sun, their temperatures are thought to be a few tens of kelvins. Chaisson & McMillan suggest a core temperature of 200K and a surface temperature on the order of 350K for Halley when it made its close approach to the Sun. The short-period comets (less than 200 years) are thought to originate in a region of the solar system out past the orbit of Neptune called the Kuiper belt (30 to 100 AU).

Most of them are found to have prograde orbits (in the same orbital direction as the planets) and to be close to the ecliptic plane. The Kuiper belt is described as a region of asteroid-like comets, most of which travel in roughly circular orbits. It may be that occasional close encounters between comets or the cumulative gravitational pull of the outer planets brings one into the higly elliptical orbit which brings it close to the Sun. Other comets, characterized as "long-period comets", are found in random orientations with respect to the ecliptic plane. They are thought to originate in a large "cloud" of objects in a region perhaps 50,000 AU from the Sun called the Oort cloud.

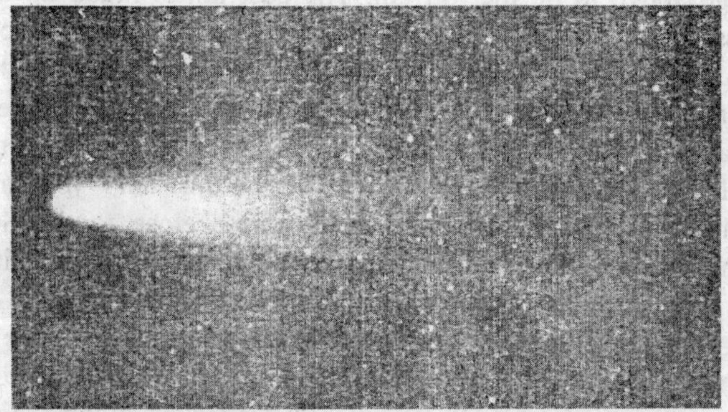

This image of comet Schwassmann-Wachmann 3 taken by Tim Puckett of Villa Rica, Ga. USA. It was obtained with a 12" Lx200 working at f/7. This is a 300 second exposure taken on 12-01-95.

While the most famous comet is Halley's Comet, some interesting recent encounters have been with comet Giacobini-Zinner and the impact of comet Shoemaker-Levy 9 on Jupiter.

Comma Cloud

Band of organized cumuliform clouds that look like a comma from a satellite's perspective. Comma clouds are indicators of heavy storms.

Common Business Oriented Language (COBOL)

A computer programming language written for business application.

Common Misconception

Many students think that kinetic energy is *defined* by $\frac{1}{2}mv^2$. It is not. That happens to be approximately the kinetic energy of objects moving slowly, at small fractions of the speed of light. If the body is moving at relativistic speeds, its kinetic energy is mc^2, which can be expressed as $\frac{1}{2}mv^2$ + *an infinite series of terms*. $^2 = 1/(1-(v/c)^2)$, where c is the speed of light in a vacuum.

Common Misuse

We often hear non-scientists say such things as 'The car was going at a high rate of speed.' This is redundant at best, since it merely means 'The car was moving at high speed.' It is the sort of mistake made by people who don't think while they talk.

Common Types of Radiation-Gamma Rays

Gamma rays are electromagnetic waves or photons emitted from the nucleus (center) of an atom.

Common Units - SI - International Standard

These are the common units used throughout the world in health physics.

Compact Disc Audio

Analog sound data is digitized by sampling at 44.1 kHz and coding as binary numbers in the pits on the compact disc. As the focused laser beam sweeps over the pits, it reproduces the binary numbers in the detection circuitry. The same function as the "pits" can be accomplished by magnetooptical recording. The digital signal is then reconverted to analog form by a D/A converter. The tracks on a compact disc are nominally spaced by 1.6 micrometers, close enough that they are able to separate reflected light into it's component colors like a diffraction grating.

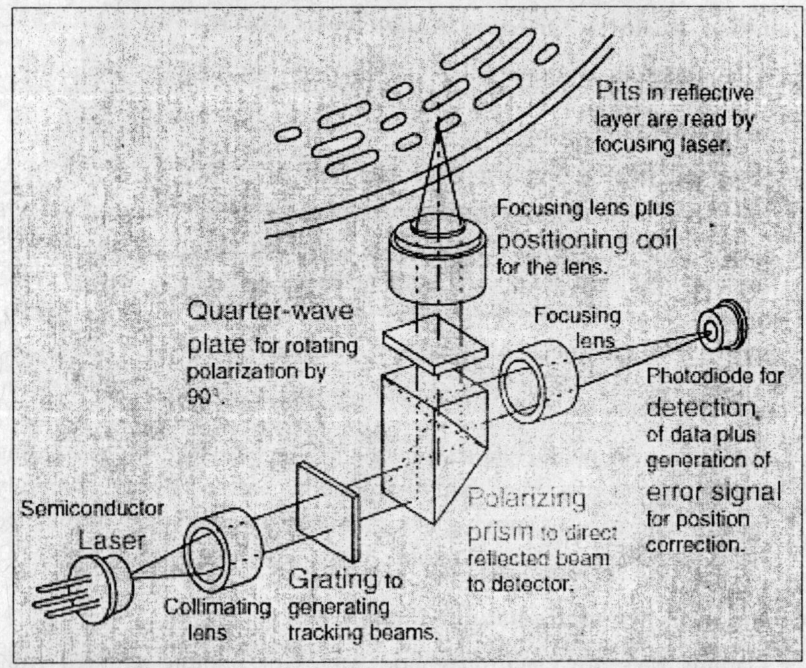

Compact Disk-Ready Only Memory

Type of computer memory that reads and uses information, but does not allow information to be added, changed, or erased. Digital information is read by laser. CD-ROM does not depend upon any proprietary hardware or software, making it an accessible vehicle for electronic publishing.

Component

The part of a velocity, acceleration, or force that is along one particular coordinate axis.

Compton Scattering

The scattering of photons from charged particles is called Compton scattering after Arthur Compton who was the first to measure photon-electron scattering in 1922.

Computer

Electronic machine capable of performing calculations and other manipulations of various types of data, under the control of a stored set of instructions. The machine itself is the hardware; the instructions are the program or software. Depending upon size computers are called mainframe, minicomputers, and microcomputers. Micro-computers include desk top and portable personal computers.

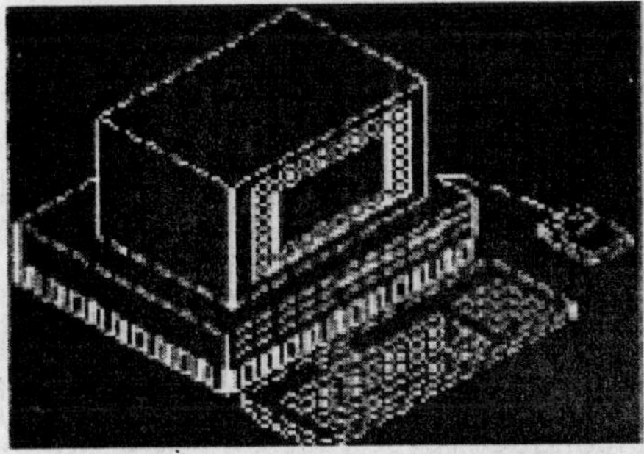

Concave

Describes a surface that is hollowed out like a cave.

Condensation

Change of a substance to a denser form, such as a gas to a liquid. The opposite of *evaporation.*

Condensed Matter Physics

Condensed matter physics is the field of physics that deals with the macroscopic physical properties of matter. In particular, it is concerned with the "condensed" phases that appear whenever the number of constituents in a system is extremely large and the interactions between the constituents are strong. The most familiar examples of condensed phases are solids and liquids, which arise from the electric force between atoms. More exotic condensed phases include the superfluid and the Bose-Einstein condensate found in certain atomic systems at very low temperatures, the superconducting phase exhibited by conduction electrons in certain materials, and the ferromagnetic and antiferromagnetic phases of spins on atomic lattices.

Conduction

1. The transfer of heat from one substance to another by direct contact. Denser substances are better conductors; the transfer is always from warmer to colder substances.

2. Conduction is the most common means of heat transfer in a solid. On a microscopic scale, conduction occurs as hot, rapidly moving or vibrating atoms and molecules interact with neighboring atoms and molecules, transferring some of their energy (heat) to these neighboring atoms. In insulators the heat current is carried almost entirely by phonon vibrations. The "electron fluid" of a conductive metallic solid conducts nearly all of the heat current through the solid. (Phonon currents are still there, but carry less than 1% of the energy.) Electrons also conduct electric current through conductive solids, and the thermal and electrical conductivities of most metals have about the same ratio. A good electrical conductor, such as copper, usually also conducts heat well. The Peltier-Seebeck effect exhibits the propensity of electrons to conduct heat through an electrically conductive solid. Thermoelectricity is caused by the relationship between electrons, heat currents and electrical currents. Convection is usually the dominant form of heat transfer in liquids and gases. This is a term used to characterize the combined effects of conduction and fluid flow. In convection, enthalpy transfer occurs by the movement of hot or cold portions of the fluid together with heat transfer by conduction. For example, when water is heated on a stove, hot water from the bottom of the pan rises, heating the water at the top of the pan. Two types of convection are commonly distinguished, *free convection*, in which gravity and buoyancy forces drive the fluid movement, and *forced convection*, where a fan, stirrer, or other means is used to move the fluid.

Buoyant convection is due to the effects of gravity, and absent in microgravity environments.

Conductors and Insulators

Conductors and Insulators have contrasting electrical properties. Conductors are materials in which three are innumerable electrons detached from atoms and forming what is known as an 'electron gas'. Should the conducting material be immersed in an electric field, the electrons will move to new position within the conductor so as to eliminate the electric field inside the conductor.

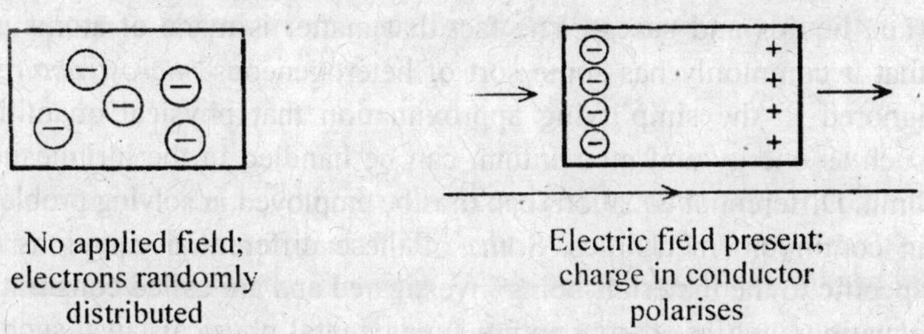

No applied field; electrons randomly distributed

Electric field present; charge in conductor polarises

Confinement

The property of the h_2 interaction that quarks or gluons are never found separately but only inside colour-neutral composite objects.

Conservation

When a quantity (e.g. electric charge, energy, or momentum) is conserved, it is the same after a reaction between particles as it was before.

Conserved

A quantity is said to be *conserved* if under specified conditions it's value does not change with time. *Example*: In a closed system, the charge, mass, total energy, linear momentum and angular momentum of the system are conserved. (Relativity theory allows that mass can be converted to energy and vice-versa, so we modify this to say that the mass-energy is conserved.)

Contamination

Radioactive material deposited or dispersed in materials or places where it is not wanted.

Continent

One of the large, continuous areas of the Earth into which the land surface is divided. The six geographically defined continents are politically defined as seven; Africa, Asia, Australia, Europe, North America, South America, and Antarctica.

Continuum Mechanics

Continuum mechanics is a branch of physics (specifically mechanics) that deals with continuous matter, including both solids and fluids (i.e., liquids and gases). The fact that matter is made of atoms and that it commonly has some sort of heterogeneous *microstructure* is ignored in the simplifying approximation that physical quantities, such as energy and momentum, can be handled in the infinitesimal limit. Differential equations can thus be employed in solving problems in continuum mechanics. Some of these differential equations are specific to the materials being investigated and are called constitutive equations, while others capture fundamental physical laws, such as conservation of mass or conservation of momentum. In fluids, the Knudsen number is used to assess to what extent the approximation of continuity can be made. The physical laws of solids and fluids do not depend on the coordinate system in which they are observed. Continuum mechanics thus uses tensors, which are mathematical objects that are independent of coordinate system. These tensors can be expressed in coordinate systems, for computational convenience.

Contrails

Condensation trails. Artificial clouds made by the exhaust of jet aircraft.

Conventional Current

Conventional current was defined early in the history of electrical science as a flow of positive charge. In solid metals, like wires, the positive charges are immobile, and only the negatively charged electrons flow in the direction opposite conventional current, but this is not the case in most non-metallic conductors. In other materials, charged particles flow in both directions at the same time. Electric currents in electrolytes are flows of electrically charged atoms (ions), which exist in both positive and negative varieties. For example, an electrochemical cell may be constructed with salt water (a solution of sodium chloride) on one side of a membrane and pure water on the other. The membrane lets the positive sodium ions pass, but not the negative chlorine ions, so a net current results. Electric currents in plasma are flows of electrons as well as positive and negative ions. In ice and in certain solid electrolytes, flowing protons constitute the electric current. To simplify this situation, the original definition of conventional current still stands.

Convection

1. The rising of warm air and the sinking of cool air. Heat mixes and moves air. When a layer of air receives enough heat from the Earth's surface, it expands and moves upward. Colder, heavier air flows under it which is then warmed, expands, and rises. The warm rising air cools as it reaches higher, cooler regions of the atmosphere and begins to sink. Convection causes local breezes, winds, and thunderstorms.
2. The physical upwelling of hot matter, thus transporting energy from a lower, hotter region to a higher, cooler region. A bubble of gas that is hotter than its surroundings expands and rises. When it has cooled by passing on its extra heat to its surroundings, the bubble sinks again. Convection can occur when there is a substantial decrease in temperature with height, such as in the Sun's *convection zone.*

Convection Zone

A layer in a star in which *convection* currents are the main mechanism by which energy is transported outward. In the Sun, a convection zone extends from just below the *photosphere* to about seventy percent of the solar radius.

Conversion Electron

An alternate process to *x-ray* emission during the de-excitation of an excited atom.

Convex

Describes a surface that bulges outward.

Coordinated Universal Time (UTC)

Local time at zero degrees longitude at the Greenwich Observatory England. UTC uses a 24-hour clock, i.e., 2:00 a.m. is 0200 hours, 2:00 p.m. is 1400 hours, midnight is 2400 or 0000 hundred hours.

Coriolis Force

The apparent tendency of a freely moving particle to swing to one side when its motion is referred to a set of axes that is itself rotating in space, such as Earth. The acceleration is perpendicular to the direction of the speed of the article relative to the Earth's surface and is directed to the right in the northern hemisphere. Winds are affected by rotation of the Earth so that instead of a wind blowing in the direction it starts, it turns to the right of that direction in the northern hemisphere; left in the southern hemisphere.

Corona

The outermost layer of the solar atmosphere. The corona consists of a highly rarefied gas with a low density and a temperature greater than one million degrees *Kelvin.* It is visible to the naked eye during a solar eclipse.

Cosmic Censorship

At the center of a mathematical description of a black hole discovered by the German mathematician, Karl Schwarzschild, is a singularity, a point at which the laws of physics break down and space becomes infinitely curved. Fortunately, the Schwarzschild spacetime has another feature, an event horizon, out through which nothing can pass (although things CAN pass IN through the horizon). Because the singularity is "clothed" by the event horizon it cannot affect the exterior spacetime and we do not need to worry about the presence of the singularity. If a *"naked" singularity* were to exist, one without a surrounding event horizon, it would present great difficulties for physicists. Cosmic censorship is a hypothesis which proposes that the natural laws do not permit a naked singularity to form, that these laws will always work to modestly clothe a singularity with an event horizon.

Cosmology

1. The study of the history of the universe.
2. *Cosmology* is the study of the large-scale structure and history of the universe. In particular, it deals with subjects regarding its origin and evolution. It is studied by astronomy, philosophy, and theology.

Cosmos

The universe regarded as an orderly, harmonious whole.

Coulomb (C)

The unit of electrical charge.

Contour Map

A map showing the intensity of radiation as a function of position. Each contour line corresponds to a specific intensity of radiation, with inner contours corresponding to higher intensities than outer contours. Therefore, a closed contour encircles a region where the intensity of the emitted radiation is greater than or equal to the intensity on the contour line. The contours outline the shape of the emitting source.

Coupled Equations

An example of a coupled equation may be found in the example of state and federal tax. The state (in our example) takes 20 percent of the part of your income (after the federal tax is deducted), and the federal government takes 10 percent of your income (after the state tax is deducted). Using the definitions. s = state tax, f = federal tax. i = your income. We can write the coupled equations:

$$s = 0.20 \times (i-f), \quad f = 0.10 \times (i-s).$$

Coupled, Hyperbolic-Elliptic, Nonlinear, Partial Differential Equation

This is name is a sequence of adjectives, each with a specific meaning, that tells us something about the type of *differential equation* we are trying to solve. The terms "Hyperbolic" or "Elliptic" describe the mathematical form of the differential equation. Hyperbolic equations describe the propagation of waves moving at some speed, such as water or gravitational waves. Hence a disturbance at one place (say a pebble falling into a pond, or two black holes colliding) is only felt at another place later in time, when the waves reach that point. Elliptic equations generally describe a function, like Newton's gravitational potential, whose effects are felt throughout space at one instant in time. This is the mathematical description of Newton's "action-at-a-distance." The term *coupled* tells us that we have a set of equations that must be solved all at once; the unknown quantities appear all mixed together in each equation. The term *nonlinear* means that the unknowns appear as squares or higher powers, and the solution is likely to be more difficult to solve. With nonlinear equations it will be more difficult to assess whether we have found a *general solution* or not. The Einstein equations have all of these properties.

Coupled System

Two or more processes that affect one another;

Cow

A radioisotope generator system.

Critical Circumference

This is the circumference below which an object of given mass would collapse to form a black hole. This circumference depends on the mass of the object in question. For example, a collapsing star equal to 10 suns will have a critical circumference of 198 kilometers or 118 miles.

Crop Calendar

The schedule of the maturing and harvesting of seasonal crops.

Cryosphere

One of the interrelated components of the Earth's system, the cryosphere is frozen water in the form of snow, permanently frozen ground (permafrost), floating ice, and glaciers. Fluctuations in the volume of the cryosphere cause changes in ocean sea-level, which directly impact the atmosphere and biosphere.

Crystalline

A regularly repeated crystal-like substructure.

CT or CAT Scan (also called Computed Tomography or CT)

Detailed pictures of areas of the body created by a computer linked to an *x-ray* machine. Also called computed tomography (CT) scan or computed axial tomography (CAT) scan.

Culmination

The point at which a satellite reaches its highest position or elevation in the sky relative to an observer (aka the closest point of approach).

Curie (Ci)

1. The basic unit used to describe the intensity of radioactivity in a sample of material. One curie equals thirty-seven billion disintegrations per second, or approximately the radioactivity of one gram of radium.
2. The curie is a unit used to measure a radioactivity. One curie is that quantity of a radioactive material that will have 37,000,000,000 transformations in one second. Often radioactivity is expressed in smaller units like: thousandths (mCi), one millionths (uCi) or even billionths (nCi) of a curie. The relationship between becquerels and curies is: 3.7×10^{10} Bq in one curie.

Current Law

The electric current in amperes which flows into any junction in an electric circuit is equal to the current which flows out. This can be seen to be just a statement of conservation of charge. Since you do not lose any charge during the flow process around the circuit, the total current in any cross-section of the circuit is the same. Along with the voltage law, this law is a powerful tool for the analysis of electric circuits.

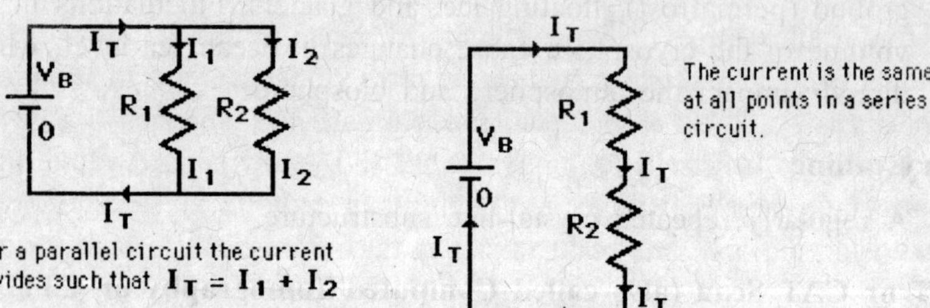

Current

1. Current is the rate of flow of electrical charge.
2. The rate at which charge crosses a certain boundary.
3. The time rate at which charge passes through a circuit element or through a fixed place in a conducting wire, I = dq/dt.
4. In electricity current is the rate of flow of charges, usually through a metal wire or some other electrical conductor. e symbol typically used for the amount of current (the amount of charge flowing per unit of time) is *I*, from the German word *Intensität*, which means 'intensity'. The SI unit of electrical current is the ampere. Electric current is therefore sometimes informally referred to as *amperage*, by analogy with the term *voltage*. Though this is a valid term, some engineers frown on it.

Cyclone

An area of low pressure where winds blow counterclockwise in the Northern Hemisphere and clockwise in the Southern Hemisphere.

Cygnus X-1

oppler studies of this blue supergiant in Cygnus indicate a period of 5.6 days in orbit around an unseen companion. The mass of the companion is calculated to be 8-10 solar masses, much too large to be a *neutron star*.

An x-ray source was discovered in the constellation Cygnus in 1972 (Cygnus X-1). X-ray sources are candidates for black holes because matter streaming into black holes will be ionized and greatly accelerated, producing x-rays.

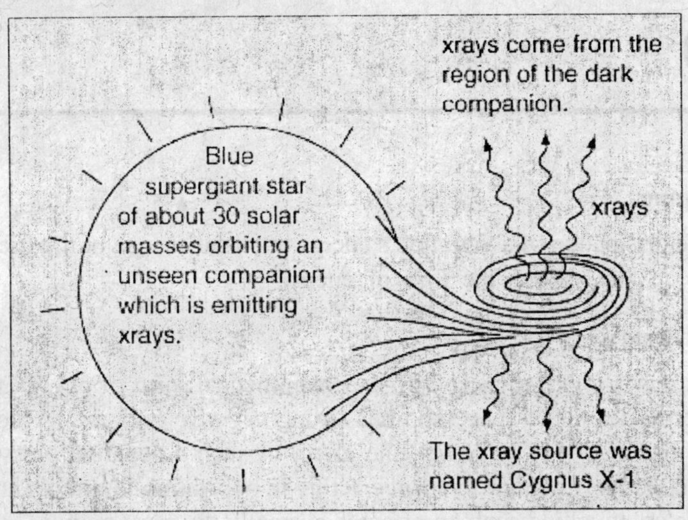

D

Damping

the dissipation of a vibration's energy into heat energy, or the frictional force that causes the loss of energy.

Dark Energy

In *cosmology* , dark energy is a hypothetical form of *energy* which permeates all of space and has strong *negative pressure*. According to the *theory of relativity*, the effect of such a negative pressure is qualitatively similar to a force acting in opposition to gravity at large scales. Invoking such an effect is currently the most popular method for explaining the observations of an accelerating universe as well as accounting for a significant portion of the *missing mass* in the universe.

Dark Matter

Matter that is in space but is not visible to us because it emits no radiation by which to observe it. The motion of stars around the centers of their galaxies implies that about 90% of the matter in a typical galaxy is dark. Physicists speculate that there is also dark matter between the galaxies but this is harder to verify.

Data Collection System (DCS)

DCS units are flown on both GOES and NOAA polar-orbiting spacecraft. They gather and relay data from both mobile and stationary platforms at various locations. DCS units on NOAA satellites can also determine the precise location of moving platforms at the time the data were acquired.

Data Rate

The amount of information transmitted per unit time.

Data

The word *data* is the plural of *datum*. Examples of correct usage: 'The data are reasonable, considering the..." The data were taken over a period of three days."How well do the data confirm the theory?'

Daughter

A nucleus formed by the radioactive decay of a different (parent) nuclide.

Davisson-Germer Experiment

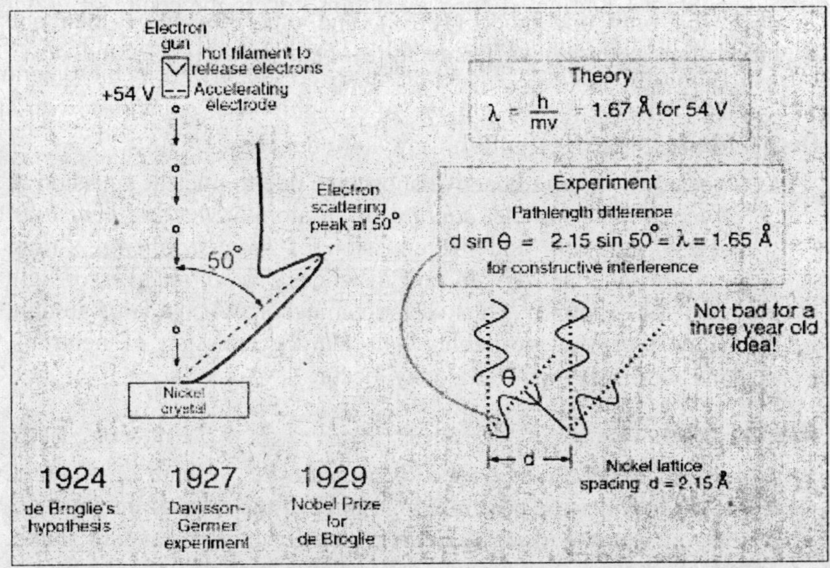

This experiment demonstrated the wave nature of the electron, confirming the earlier hypothesis of deBroglie. Putting wave-particle duality on a firm experimental footing, it represented a major step forward in the development of quantum mechanics. The *Bragg law* for diffraction had been applied to x-ray diffraction, but this was the first application to particle waves.

Decay (Radioactive)

1. The change of one radioactive nuclide into a different nuclide by the spontaneous emission of alpha, beta, or gamma rays, or by electron capture. The end product is a less energetic, more stable nucleus. Each decay process has a definite half-life.
2. A process in which a particle disappears and in its place different particles appear. The sum of the masses of the produced particles is always less than the mass of the original particle.
3. Any process in which a particle disappears and in its place two or more different particles appear.

Decay Branching %

The nuclide decay rate by a particular decay mode. Some nuclides decay by only one mode (100%), and others by more than one mode. For example, 187Pb decay by *beta* decay (98%) and *alpha* decay (2%).

Decay Mode

Disappearance of a radioactive substance due to nuclear emission of an *alpha* or *beta* particle, *capture* of an atomic *electron, neutrinos, spontaneous fission*, and the emission of *bremsstrahlung*, x-rays, and *conversion electrons*. In rare instances *proton, neutron*, or light element (for example ^{14}C) emission can occur. When a large amount of decay energy is available, *beta-delayed* emission of neutrons, protons, and other particles may occur.

Decibel (dB)

A tenth of a bel. A unit used to measure the volume of a sound, equal to ten times the common logarithm of the ratio of the intensity of the sound to the intensity of an arbitrarily chosen standard sound. The decibel also is used to measure relative strengths of antenna and amplified signals and always refers to a ratio or difference between two values.

Declination

The angular distance from the equator to the satellite, measured as positive north and negative south.

Decontamination

The removal of radioactive contaminants by cleaning and washing with chemicals.

Defense Meteorological Satellite Program (DMSP)

A U.S. Air Force-managed meteorological satellite program with satellites circling in sun-synchronous orbit. Imagery is collected in the visible- to near-infrared band (0.4 to 1.1 micrometers) and in the thermal-infrared band (about 8 to 13 micrometers) at a resolution of about three kilometers. DMSP data is available directly from the satellite for local use aboard ships and at military deployment locations, but is also usually available to civilian users.

Degree

A unit of angular measure represented by the symbol ^{o}. The circumference of a circle contains 360 degrees. When applied to the roughly spherical shape of the Earth for geographic and cartographic purposes, degrees are each divided into 60 minutes.

Delta

The fan-shaped area at the mouth or lower end of a river; formed by eroded material that has been carried downstream and dropped in quantities larger than can be carried off by tides or currents.

Demodulation

The process of retrieving information (data) from a modulated carrier wave, the reverse of modulation.

Density

1. That property of a substance which is expressed by the ratio of its mass to its volume.

2. The amount of mass or number of particles per unit volume. In *cgs* units mass density has units of $gm\ cm^{-3}$. Number density has units cm^{-3} (particles per cubic centimeter).

Department of the Interior (DOI)

Responsible for our nationally-owned public lands and natural resources, the DOI is chartered to foster the wisest use of our land and water resources, protect fish and wildlife, preserve the environmental and cultural values of national parks and historical places, and provide for the enjoyment of life through outdoor recreation. The department assesses energy and mineral resources and is responsible for assuring that their development is in the best interest of all citizens. The U.S. Geological Survey (USGS) is part of the DOI.

Dependent Variable

When two variables are related, we say that one depends on the other. This variable is called the dependent variable. The"other" variable is free to roam so it is called the independent variable. In a scientific experiment, the experimenter chooses values for the independent variable, runs the experiment, and *measures* the dependent variable. Ordered pairs of choosen and measured values (independent and dependent) are often plotted on a two dimensional graph for visualization. The dependent variable is traditionally plotted on the vertical axis.

Derive

To derive a result or conclusion is to show, using logic and mathematics, how a conclusion follows logically from certain given facts and principles.

Descending Node

The point in a satellite's orbit at which it crosses the equatorial plane from north to south.

Desert

A land area so dry that little or no plant or animal life can survive.

Desertification

The man-made or natural formation of desert from usable land.

Detector

1. A device in a radiometer that senses the presence and intensity of radiation. The incoming radiation is usually modified by filters or other optical components that restrict the radiation to a specific spectral band. The information can either be transmitted immediately or recorded for transmittal at a later time.
2. Any device used to sense the passage of a particle; also a collection of such devices designed so that each serves a particular purpose in allowing physicists to reconstruct particle *events*.

Deuterium-Tritium Fusion

The most promising of the hydrogen *fusion reactions* which make up the *deuterium cycle* is the fusion of deuterium and tritium. The reaction yields 17.6 MeV of energy but requires a temperature of approximately 40 million Kelvins to overcome the *coulomb barrier* and ignite it. The *deuterium fuel* is abundant, but tritium must be either *bred* from lithium or gotten in the operation of the deuterium cycle.

Dew

Atmospheric moisture that condenses after a warm day and appears during the night on cool surfaces as small drops. The cool surfaces cause the water vapor in the air to cool to the point where the water vapor condenses.

Dew Point

The temperature to which air must be cooled for saturation to occur; exclusive of air pressure or moisture content change. At that temperature dew begins to form, and water vapor condenses into liquid.

Diffraction

The behavior of a wave when it encounters an obstacle or a nonuniformity in its medium; in general, diffraction causes a wave to bend around obstacles and make patterns of strong and weak waves radiating out beyond the obstacle.

Diffraction of Sound

Diffraction: the bending of waves around small* obstacles and the spreading out of waves beyond small* openings. * small compared to the wavelength Important parts of our experience with sound involve diffraction. The fact that you can hear sounds around corners and around barriers involves both diffraction and reflection of sound. Diffraction in such cases helps the sound to "bend around" the obstacles. The fact that diffraction is more pronounced with longer wavelengths implies that you can hear low frequencies around obstacles better than high frequencies, as illustrated by the example of a *marching band* on the street. Another common example of diffraction is the contrast in sound from a close lightning strike and a distant one. The thunder from a close bolt of lightning will be experienced as a sharp crack, indicating the presence of a lot of high frequency sound. The thunder from a distant strike will be experienced as a low rumble since it is the long wavelengths which can bend around obstacles to get to you. There are other factors such as the higher air absorption of high frequencies involved, but diffraction plays a part in the experience.

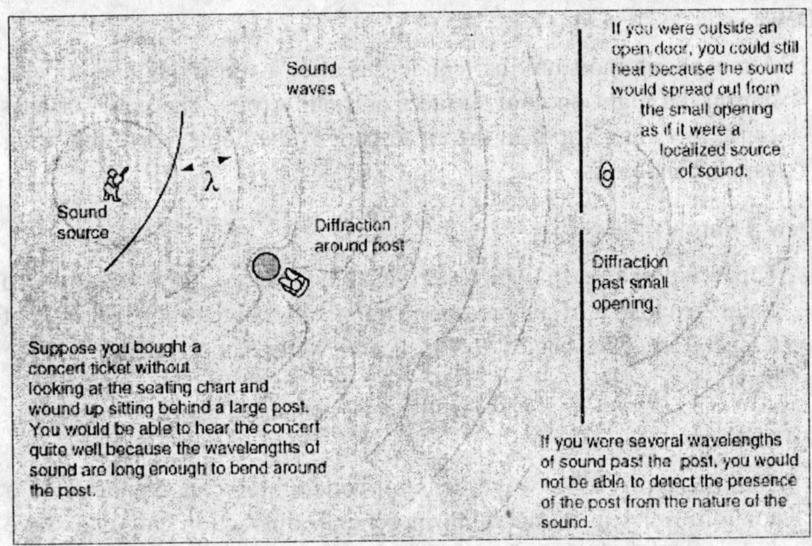

Differential Equation (Partial)

Solutions to algebraic equations, like x^2 = 2, are just numbers. The solutions to differential, or partial differential equations are functions. The term "differential" describes one aspect of the equation familiar from Calculus (invented by Newton to describe his theory of gravitation). This relates the slope of an unknown function to its value in some way. The most obvious equation of this type asks the question "what function has a slope equal to its value at each point" and the answer is y=e^x.

Diffuse Reflection

Reflection from a rough surface, in which a single ray of light is divided up into many weaker reflected rays going in many directions.

Diffusion

Diffusion refers to the process by which molecules intermingle as a result of their *kinetic energy* of random motion. Consider two containers of gas A and B separated by a partition. The molecules of both gases are in constant motion and make numerous collisions with the partition.

If the partition is removed as in the lower illustration, the gases will mix because of the *random velocities* of their molecules. In time a uniform mixture of A and B molecules will be produced in the container. The tendency toward diffusion is very strong even at room temperature because of the high molecular velocities associated with the *thermal energy* of the particles. (Fig. on next page)

Digital Image

An analog image converted to numerical form so that it can be stored and used in a computer. The image is divided into a matrix of small regions called picture elements or pixels. At sub-satellite point each pixel represents a specific amount of area. For example, in *APT* each pixel represents 4.1 kilometers. Each pixel has a numerical value or data number value, quantifying the radiance of the image at that spot. The data number value of each pixel usually

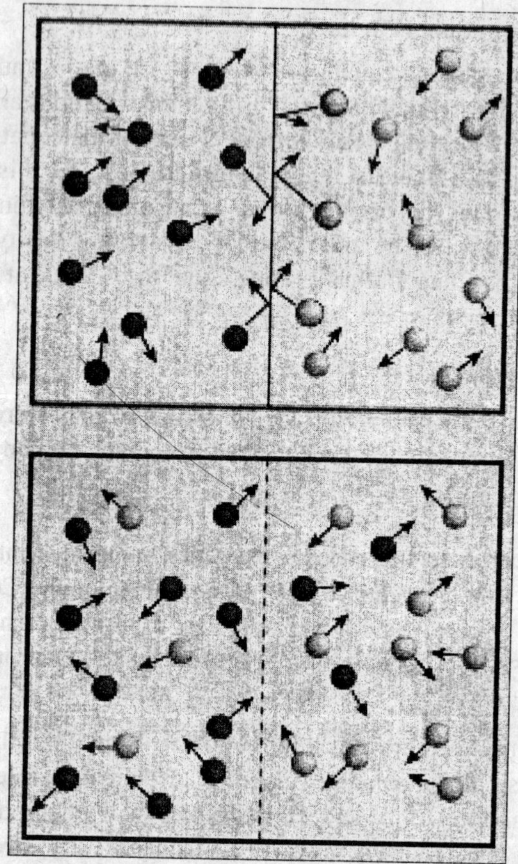

Diffusion

represents a value between black and white, *i.e.,* shades of gray. False color can be applied to the image by assigning a graduated color palette to represent the gray shades. The color is "false" because it represents an assigned, not actual, color.

Digital System

A system in which information is transmitted in a series of pulses. The source is periodically sampled, analyzed, and converted or coded into numerical values and transmitted. Digital transmissions typically use the binary coding used by computers so most data is in appropriate form, but verbal and visual communication must be converted. Many satellite transmissions use digital formats because noise will not interfere with the quality of the end product, producing clear and higher-resolution imagery.

Dimensions

The fundamental measurables of a unit system in physics-those which are defined through operational definitions. All other measurable quantities in physics are defined through mathematical relations to the fundamental quantities. Therefore any physical measurable may be expressed as a mathematical combination of the dimensions. *Example*: In the MKSA (meter-kilogram-second-ampere) system of units, length, mass, time and current are the fundamental measurables, symbolically represented by L, M, T, and I. Therefore we say that velocity has the dimensions LT^{-1}. Energy has the dimensions ML^2T^{-2}.

Dipole Magnet

Any magnet with one north and one south pole. In an accelerator the dipole magnets are used to steer a particle *beam* to the left or right by placing one pole above and the other below the beam pipe.

Direct Readout

The capability to acquire data directly from environmental satellites via an Earth station. Data can be acquired from NOAA and other nations' environmental satellites, which offer weather information from geostationary and polar-orbiting satellites.

Director

Parasitic element(s) of a VHF antenna located forward of the driven element.

DIS

Data and Information System.

Discrepancy

(1) Any deviation or departure from the expected.

(2) A difference between two measurements or results.

(3) A difference between an experimental determination of a quantity and its standard or accepted value, usually called the *experimental discrepancy.*

Displacement

1. A name for the symbol delta-x.
2. The displacement of an object is defined as the *vector* distance from some initial point to a final point. It is therefore distinctly different from the distance traveled except in the case of straight line motion. The distance traveled divided by the time is called the speed, while the displacement divided by the time defines the average velocity.

   If the *positions* of the initial and final points are known, then the *distance relationship* can be used to find the displacement.

Displacement Current

Displacement current is a pseudocurrent invented in 1865 by James Clerk Maxwell when formulating what are today known as Maxwell's equations. It is defined by the flux of the electric field through the surface:

$$I_D = \varepsilon \frac{d\Phi_E}{dt}$$

It is incorporated into Ampère's law, whose original form works only for surfaces that are well-defined (continuous and existing) in terms of current. A surface S_1 chosen to include only one plate of a capacitor should have the same current as a surface S_2 chosen to include both capacitor plates. However, because charge stops at the first plate, Ampère's Law concludes there is no charge enclosed by S_1. To compensate for this difference, Maxwell reasoned that charge is found in electric flux the change in the electric field; and while displacement current is not a current of electric charge it produces the same result of generating a magnetic field. A displacement current can be thought of as an elastic response of a material to an applied electric field. As an applied electric field is increased, the displacement current is stored in the material, and when the electric field is decreased, the material releases the displacement current. A perfect dielectric is a material that shows displacement current only, storing and returning electrical energy as if it were an ideal 'battery'.

Dispersion of Light

Dispersion of light is the separation of a beam of light into its component wavelengths or colours. This is possible because the speed of light in a substance such as glass changes slightly from one wavelength to the next. The dispersion of white light by glass is clearly demonstrated by the formation of a spectrum using a glass prism.

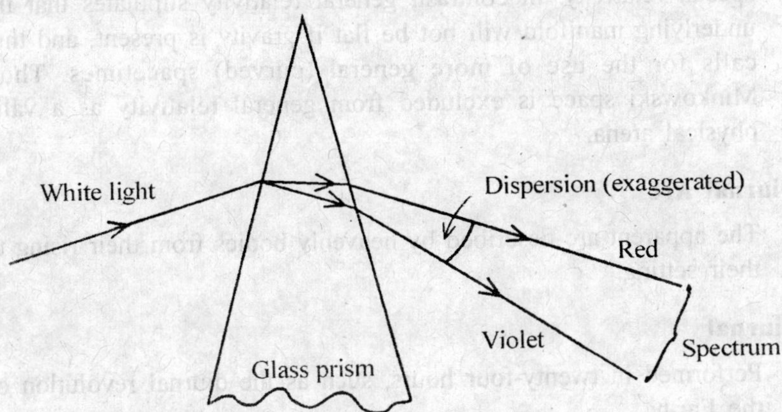

Distance

One of the undefined qualities of physics, it measures the separation of two points.

Distance in Spacetime

A *spacetime interval* between two events is the frame-invariant quantity analogous to distance in Euclidean space. The spacetime interval s along a curve (geodesic) is defined in special relativity by: $ds^2 = dx^2 + dy^2 + dz^2 - c^2dt^2$ where c is the speed of light and where Cartesian coordinates are used. A basic assumption of relativity is that coordinate transformations must leave spacetime intervals invariant. Intervals are invariant under Lorentz transformations. The spacetime intervals on a manifold define a pseudo-metric called the Lorentz metric. This metric is very similar to distance in Euclidean space. However, note that whereas distances are always positive in Euclidean space, intervals may be positive, zero, or negative in space-time. Events with a spacetime interval of zero are separated by the propagation of a light signal. Events with a positive spacetime interval are in each other's future or past, and the value of the interval defines the proper time measured by an observer travelling between them. One of the simplest interesting examples of a spacetime is R^4 with the spacetime interval defined above. This is known as Minkowski space, and is the usual geometric setting for special relativity. In contrast, general relativity stipulates that the underlying manifold will not be flat if gravity is present, and thus calls for the use of more general (curved) spacetimes. Thus, Minkowski space is excluded from general relativity as a valid physical arena.

Diurnal Arc

The apparent arc described by heavenly bodies from their rising to their setting.

Diurnal

Performed in twenty-four hours, such as the diurnal revolution of the Earth.

Dobson Unit (DU)

The standard way to express ozone amounts in the atmosphere. One DU is 2.7×10^{16} ozone molecules per square centimeter. One Dobson unit refers to a layer of ozone that would be 0.001 cm thick under conditions of standard temperature (0° C) and pressure (the average pressure at the surface of the Earth). For example, 300 Dobson units of ozone brought down to the surface of the Earth at 0° C would occupy a layer only 0.3 cm thick in a column. Dobson was a researcher at Oxford University who, in the 1920s, built the first instrument (no called the Dobson meter) to measure total ozone from the ground.

Doldrums

Region near the equator characterized by low pressure and light shifting winds.

Doppler Effect (Aka Doppler shift)

The apparent change in frequency of sound or light waves, varying with the relative velocity of the source and the observer. If the source and observer draw closer together; the frequency is increased. Named for Christian Doppler; Austrian mathematician and physicist (1803-1853).

Doppler Radar

The weather radar system that uses the Doppler shift of radio waves to detect air motion that can result in tornadoes and precipitation, as previously-developed weather radar systems do. It can also measure the speed and direction of rain and ice, as well as detect the formation of tornadoes sooner than older radars.

Dose *(Absorbed Dose)*

More specifically referred to as "absorbed dose", this is a measure of the energy deposited within a given mass of a patient. Absorbed dose is quantified by the unit called the "*rad*.

Dose Calibration

Determining the response of a dosimeter to a known radiation exposure or known absorbed dose. For a *beam* of radiation, this means determining the absorbed dose rate at a calibrated point in the beam under a specified set of conditions. Normally, such a determination is carried out with a number of beams under different specified conditions.

Dose Equivalent (DE)

Parameter used to express the risk of the deleterious effects of *ionization radiation* upon living organisms. For radiation protection purposes, the quantity of the effective irradiation incurred by exposed persons, measured on a common scale in sievert (SI) or rem (non-SI).

Dose Rate

A measure of the *dose* delivered per unit time.

Dose

A general term denoting the quantity of radiation or energy absorbed in a specific mass.

Dosimeter

A radiation sensitive device, e.g., film, monitor ion chamber, TLD, etc., with a known sensitivity that is placed in the *beam* path of radiation to *dose.*

Dosimetry

The calculations, measurements and other activities required for determining the radiation *dose* to be delivered.

Down Converter

Any radio frequency circuit that converts a higher frequency to a lower frequency This enables signal processing by a receiver. A typical down converter will feature one or more states of RE preamplification, a mixer where the frequency conversion occurs, a local oscillator chain, and often one or more intermediate frequency preamplifiers to minimize the effect of line losses between the converter and the receiver.

Down Quark (d)

The second flavour of quark (in order of increasing mass), with electric charge -1/3.

Drag (aka N1)

A retarding force caused by the Earth's atmosphere. Thus by definition, drag will act opposite to the vehicle's instantaneous velocity vector with respect to the atmosphere. The magnitude of the drag force is directly proportional to the product of the vehicle's cross-sectional area, its drag coefficient, its velocity, and the atmospheric density and inversely proportional to its mass. The effect of drag is to cause the orbit to decay or spiral downward. A satellite of very high mass and very low cross-sectional area, and in a very high orbit, may be very little affected by drag, whereas a large satellite of low mass, in a low altitude orbit may be affected very strongly by drag. Drag is the predominant force affecting satellite lifetime.

Driving Force

An external force that pumps energy into a vibrating system.

Dynamics

The study of the action of forces on bodies and the changes in motion they produce.

E

E = mc²

Where: e is energy, m is mass, and c is the speed of light. Einstein's famous equation describes how energy and mass are related. In our animated decays, mass is lost. That mass is converted into energy in the form of *electromagnetic waves*. Because the speed of light is so great, a little matter can transform into large amount of energy.

Earth Observing System (EOS)

A series of small- to intermediate-sized spacecraft that is the centerpiece of NASAs Mission to Planet Earth (MTPE). Planned for launch beginning in 1998, each of the EDS spacecraft will carry a suite of instruments designed to study global climate change. MTPE will use space-, aircraft-, and ground-based measurements to study our environment as an integrated system. Designing and implementing the MTPE is, of necessity, an international effort. The MTPE program involves the cooperation of the U.S., the European Space Agency (ESA), and the Japanese National Space Development Agency (NASDA). The MTPE program is part of the U.S. interagency effort, the Global Change Research Program.

Earth Observing System Data & Information System (EOSDIS)

The system that will manage a dataset of Earth science observations to be collected over a 15-year period. Existing data indicates that the Earth is changing, and that human activity increasingly contributes to this change. To monitor these changes, a baseline of "normal" performance characteristics must be obtained. For the Earth, these baseline characteristics must cover a global scale and a long enough

EOS Research

Spacecraft	Research purpose
EOS-AM1 (1998)	**characterization of tand and ocean surfaces sea-surface temperature terrestrial and ocean productivity clouds, aerosols, and radiative balance.**
EOS-COLOR (1998)	**ocean color and productivity**
EOS-AERO1 (2000)	**Atmospheric aerosols and ozone**
EOS-PM1 (2000)	**clouds, precipitation, radiative balance, terrestrial snow, and sea ice terrestrial and ocean productivity atmospheric temperature and moisture.**
EOS-ALT1 (2002)	**ocean circulation, ice sheet mass balance and land surface topography.**
EOS-CHEM1 (2002)	**almospheric chemical species and their transformations solar radiation**

Protected year of launch shown in parenmeses

period that the variation caused by seasonal changes and other cyclical or periodic events (*e.g.,* El Nino and the solar cycle) may be included in the analyses. The baseline characteristics also must enable scientists to quantify processes that govern the Earth's system. Functionally EOSDIS will provide computing and networking facilities supporting EOS research activities, including data interpretation and modeling; processing, distribution, and archiving of EDS data; and command and control of EGS observatories.

Earth Probes

Discipline-specific satellites and instruments that will be used by NASA to obtain observations before the launch of EDS spacecraft. Generally smaller than the EGS satellites and instruments, Earth Probes are planned to complement the broad environmental measurements from EOS with highly focused studies in areas such as tropical rainfall (TRMM), ocean productivity (SeaWi ES), atmospheric ozone (TOMS), and ocean surface winds (NSCAT).

Earth Radiation Budget Experiment (ERBE)

An experiment to obtain data to study the average radiation budget of the Earth and determine the energy transport gradient from the equator to the poles. Three satellites were flown in different orbits to obtain the data: the Earth Radiation Budget Satellite, ERBS (launched in October 1984), NOAA-9 (launched in December 1984), and NOAA-10 (launched in September 1986).

Earth Station (aka ground station)

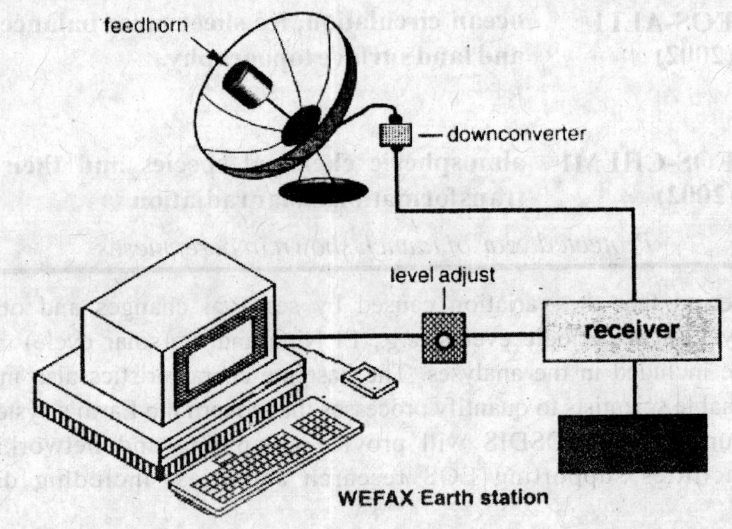

WEFAX Earth station

Hardware necessary to acquire data directly from environmental satellites. The WEFAX Earth station diagram illustrates a basic ground station configuration for obtaining direct readout data from geostationary environmental (weather) satellites.

Earth System

The earth regarded as a unified system of interacting components, including geosphere (land), atmosphere (air), hydrosphere (water and ice), and biosphere (life).

EARTH'S INTERACTING COMPONENTS	
Geosphere	Physical elements of the Earch's surface, crust, and interior. Proesses in the geosphere include continental drift. Volcanic eruptrons, and earthquakes.
Atmosphere	Thin layer of gas or air that surrounds the Earth. Processes in the atmosphere include winds, weather, and the exchange of gases with living organisms.
Hydrosphere	Water and ice on or near the Surface of the Earth. Incldes water vapor in clouds; ice caps and glaciers : and water in the oceans, rivers, takes, and aquifers. Processes in the hydrosphere include the flow of rivers, evaporation, and rain..
biosphere	The wealth and diversity of living organisms of the Earch. Processes in teh biosphere include life and death. evolution, and extinction.

Earth System Science

An integrated approach to the study of the Earth that stresses investigations of the interactions among the Earth's components in order to explain Earth dynamics, evolution, and global change.

Eccentricity

(aka ecce or EO or e) One of six Keplerian elements, it describes the shape of an orbit. in the Keplerian orbit model, the satellite orbit is an ellipse, with eccentricity defining the "shape" of the ellipse. When e = 0, the ellipse is a circle. When e is very near 1, the ellipse is very long and skinny.

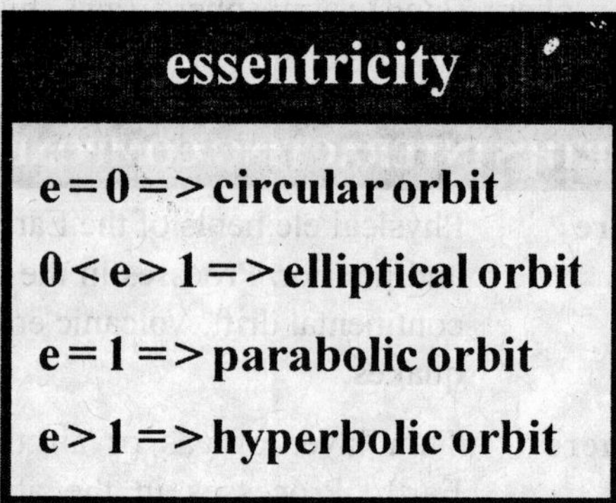

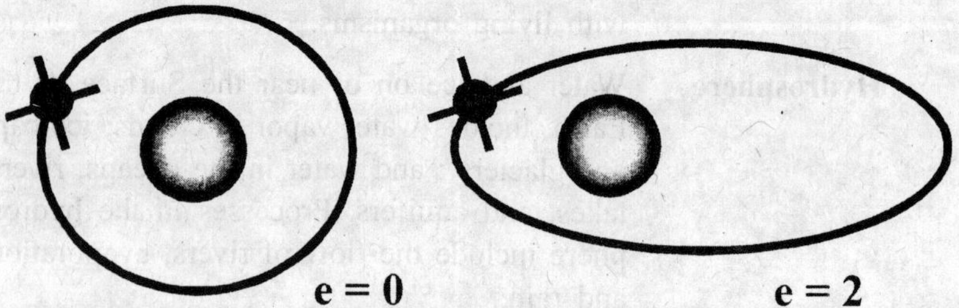

Eclipse Blindness

Focus-point type of vision loss caused by looking at the sun for too long a time, which can burn a hole in the retina of the eye.

Eclipse

The partial or total apparent darkening of the sun when the moon comes between the sun and the Earth (solar eclipse), or the darkening of the moon when the full moon is in the Earth's shadow (lunar eclipse).

Ecliptic Plane

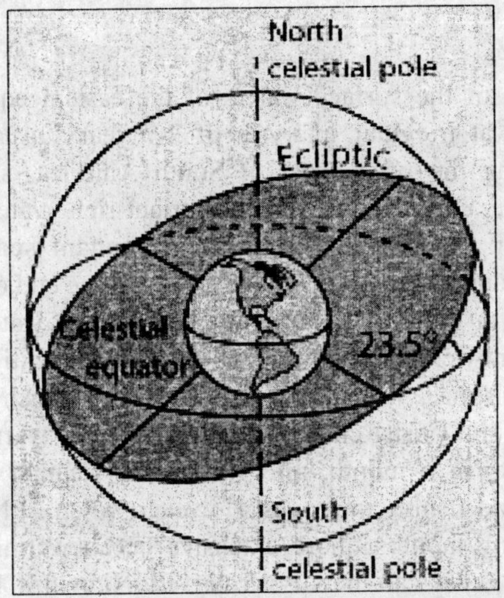

If the sun's path is observed from the Earth's reference frame, it appears to move around the Earth in a path which is tilted with respect to the spin axis at 23.5°. This path is called the *ecliptic*. It tells us that the Earth's spin axis is tilted with respect to the plane of the Earth's solar orbit by 23.5°. Observations show that the other planets, with the exception of *Pluto*, also orbit the sun in essentially the same plane. The ecliptic plane then contains most of the objects which are orbiting the sun. This suggests that the formation process of the solar system resulted in a disk of material out of which formed the sun and the planets. The 23.5° tilt of the Earth's spin axis gives the *seasonal variations* in the amount of sunlight received at the surface.

Ecology

Science dealing with the inter-relationships between living organisms and their environments.

Ecosystem

Any natural unit or entity including living and non-living parts that interact to produce a stable system through cyclic exchange of materials.

El Nino

A warming of the surface waters of the eastem equatorial Pacific that occurs at irregular intervals of 2-7 years, usually lasting 1-2 years. Along the west coast of South America, southerly winds promote the upwelling of cold, nutrient-rich water that sustains large fish populations, that sustain abundant sea birds, whose droppings support the fertilizer industry Near the end of each calendar year; a warm current of nutrient-poor tropical water replaces the cold, nutrient-rich surface water Because this condition often occurs around Christmas, it was named El Niho (Spanish for boy child, referring to the Christ child). In most years the warming lasts only a few weeks or a month, after which the weather patterns return to normal and fishing improves. However; when El Nino conditions last for many months, more extensive ocean warming occurs and economic results can be disastrous. El Nino has been linked to wetter; colder winters in the United States; drier; hotter summers in South America and Europe; and drought in Africa.

Electric Charge

1. The quantum number that determines participation in electromagnetic interactions.

2. Electric charge is a fundamental conserved property of some subatomic particles, which determines their electromagnetic interactions. Electrically charged matter is influenced by, and produces, electromagnetic fields. The interaction between charge and field is the source of one of the four fundamental forces the electromagnetic force. Electric charge is a quantum number. Electrons have a charge, by convention, of -1. Protons have the opposite charge of +1.Quarks have a fractional charge of -1/3 or +2/3. The antiparticle equivalents of these have the opposite charge. There are other charged particles. *Q* is a measurement of the charge held by an object. The SI unit of electric charge

is the coulomb, which represents approximately 6.24×10^{18} elementary charges (the charge on a single electron or proton). The coulomb is defined as the quantity of charge that has passed through the cross-section of a conductor carrying one ampere within one second. Electric charge can be directly measured with an electrometer. The discrete nature of electric charge was demonstrated by Robert Millikan in his oil-drop experiment. Formally, a measure of charge should be a multiple of the elementary charge *e*, but since it is an average, macroscopic quantity, many orders of magnitude larger than a single elementary charge, it can effectively take on any real value.

Electric Current

Electric current is the rate of *charge* flow past a given point in an electric circuit, measured in coulombs/second which is named amperes. In most DC electric circuits, it can be assumed that the resistance to current flow is a constant so that the current in the circuit is related to voltage and resistance by Ohm's law.

Electric Dipole

An object that has an imbalance between positive charge on one side and negative charge on the other; an object that will experience a torque in an electric field.

Electric Dipole Field

The *electric field* of an *electric dipole* can be constructed as a vector sum of the *point charge fields* of the two charges: Direction of electric dipole.

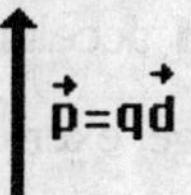

Electric Field

1. A force field which defines what acceleration an electric *charge* placed at rest at any point in space will feel. Electric charges cause electric fields around them, which then apply a force to any other electric charge placed in the field. The electric field E has both a magnitude and a direction at each point in space, and the magnitude and direction of the resulting force on a charge q at that point is given by F= qE. When you get a shock from a door handle after scuffing your feet on a carpet you feel the effect of an electric field accelerating *electrons*.

2. The force per unit charge exerted on a test charge at a given point in space.

3. In physics, an electric field or E-field is an effect produced by an electric charge that exerts a force on charged objects in its

vicinity. The units of the electric field are newtons per coulomb or volts per meter (both are equivalent). Electric fields are composed of photons and contain electrical energy with energy density proportional to the square of the field intensity. In the static case, an electric field is composed of virtual photons being exchanged by the charged particle(s) creating the field. In the dynamic case the electric field is accompanied by a magnetic field, by a flow of energy, and by real photons.

Electric Flux

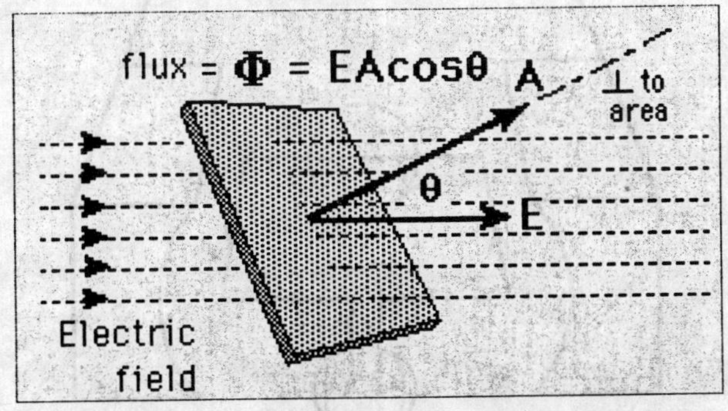

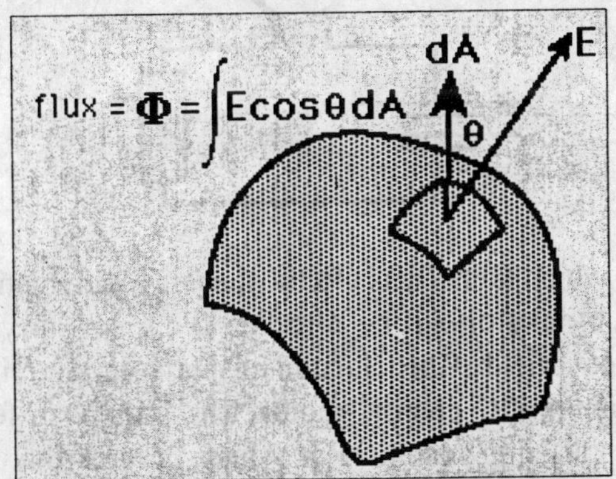

He concept of electric flux is useful in association with *Gauss' law*. The electric flux through a planar area is defined as the *electric field* times the component of the area perpendicular to the field. If the area is not planar, then the evaluation of the flux generally requires an *area integral* since the angle will be continually changing. When the area A is used in a vector operation like this, it is understood that the magnitude of the vector is equal to the area and the direction of the vector is perpendicular to the area.

Electrical Force

One of the fundamental forces of nature; a noncontact force that can be either repulsive or attractive.

Electric Motor

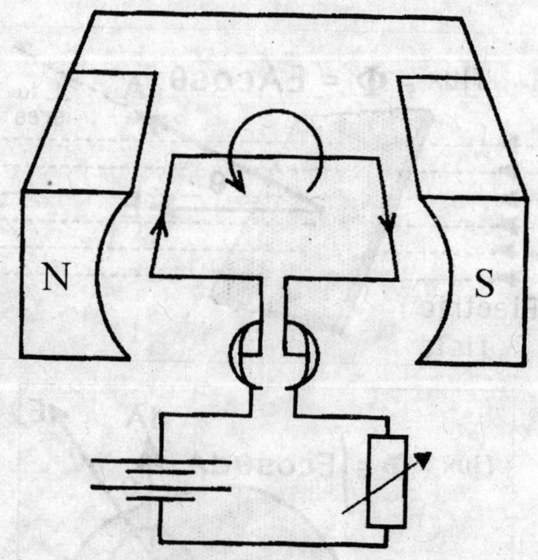

A machine which transforms electrical energy into rotational mechanical energy by exploiting the force on a current-carrying conductor in a magnetic field. The diagram displays in simplified from the principle of the DC electric motor. The current is fed to the coil through a split ring commutator which ensures that the left–hand half of tied coil always receives current and the right–hand half of the coil always returns it. A soft iron cylinder (not shown) between the curved poles of the magnet creates a radial field. This ensures that the plane of the coil is parallel to the magnetic field direction except when its inertia carries it through the vertical position. An application of *Fleming's left hand rule* to this situation shows that the force on the left–hand of the coil is always downwards and that the force on the right–hand half of the coil is always upwards. The coil is forced by the current and the magnetic field to rotate in an anti–clockwise sense and this can be used to drive machinery and do work.

Electric Potential

Electric potential is the potential energy per unit charge associated with a static (time-invariant) electric field, also called the electrostatic potential, typically measured in volts. Metaphorically, electric potential may be conceived of as "electric pressure that can push electric charges to different locations. Technically, it is the potential ö (a scalar field) associated with the conservative electric field E (E = –ö) that occurs when the magnetic field is time invariant. Like any potential function, only the potential difference (voltage) between two points is physically meaningful (neglecting quantum Aharonov Bohm effects), since any constant can be added to ö without affecting E.

Electrical Network

An electrical network or electrical circuit is an interconnection of analog electrical elements such as resistors, inductors, capacitors, diodes, switches and transistors. It can be as small as an integrated circuit on a silicon chip, or as large as an electricity distribution network. A circuit is a network that has a closed loop *i.e.* a return path. A network is a connection of 2 or more simple circuit elements, and may not be a circuit. The goal when designing electrical networks for signal processing is to apply a predefined operation on potential differences (measured in volts) or currents (measured in amperes). Typical functions for these electrical networks are amplification, oscillation and analog linear algorithmic operations such as addition, subtraction, multiplication, division, differentiation and integration. In the case of power distribution networks, engineers design the circuit to transport the energy as efficiently as possible while at the same time taking into account economic factors, network safety and redundancy. These networks use components such as power lines, cables, circuit breakers, switches and transformers. To design any electrical circuits, electrical engineers need to be able to predict the voltages and currents in the circuit. Linear circuits can be analysed to a certain extent by hand because complex number theory gives engineers the ability to treat all linear elements using a single mathematical representation.

Electrical Resistance

Electrical resistance is a measure of the degree to which an electrical component opposes the passage of current. It is the ratio of the potential difference (i.e. voltage) across an electric component to the current passing through that component:

$$R = \frac{V}{I}$$

where *R* is the resistance of the component. *V* the voltage across the component, measured in volts *I* is the current passing through the component, measured in amperes V can either be measured directly across the component or calculated from a subtraction of voltages relative to a reference point. The former method is simpler for a single component and is likely to be more accurate. The latter method is useful when analysing a larger circuit or if you want to work one handed with one lead clipped (which can be a useful safety precaution on systems using dangerous voltages. There may also be problems with the latter method if the system is AC and the two measurements from the reference point are not in phase with each other. Resistance is thus a measure of the component's opposition to the flow of electric charge. The SI unit of electrical resistance is the ohm. Its reciprocal quantity is electrical conductance measured in siemens. For a wide variety of materials and conditions, the electrical resistance does not depend on the amount of current flowing or the amount of applied voltage. This means that voltage is proportional to current and the proportionality constant is the electrical resistance. This case is described by Ohm's law and such materials are known as ohmic devices.

Electricity

1. This word names a branch or subdivision of physics, just as other subdivisions are named 'mechanics', 'thermodynamics', 'optics', etc.

2. *The article on electrical energy is located elsewhere.* Electricity is a property of certain subatomic particles (e.g. electrons/protons) which couples to electromagnetic fields and causes attractive and repulsive forces between them. Electricity gives rise to one of the four fundamental forces of nature, and is a conserved property of matter that can be quantified. In this sense, the phrase "quantity of electricity is used interchangeably with the phrases "charge of electricity and "quantity of charge." There are two types of electricity or charge: we call one kind of electricity positive and the other negative. Through experimentation, we find that like-charged objects repel and opposite-charged objects attract one another. The magnitude of the force of attraction or repulsion is given by Coulomb's law. Some electrical effects are discussed under electrical phenomenon and electromagnetism. The SI unit of electricity is the coulomb, which has the abbreviation "C". The symbol Q is used in equations to represent the quantity of electricity or charge. For example, "Q = 0.5 C" means "the quantity of electric charge is 0.5 coulomb."

Electromagnetic (em) Wave

Simplified EM spectrum, representing energy per photon, not total beam energy

Electromagnetic waves make up the electromagnetic spectrum. Visible light, ultraviolet, infrared, radio and TV signals are all examples of "everyday" em waves. X-rays, microwaves and high energy photons or *gamma rays* are also electromagnetic waves.

Electromagnetic Field

1. An electromagnetic field consists of energy oscillations associated with electric and magnetic fields initially caused by the motions of electric charges. The resulting waves propagate through space at the speed of light. The famous British Scientist, *James Clark Maxwell* (1871-79), formulated the mathematical laws governing the propagation of electromagnetic waves in space. These laws provided the underpinning for classical "electrodynamics." Certain problems in the manifestation of these laws prompted Einstein to formulate and publish his theory of Special Relativity in 1905.
2. This article needs to be cleaned up to conform to a higher standard of article quality. An electromagnetic field is composed of two related vectorial fields, the electric field and the magnetic field. This means that the vectors (E and B) that characterize the field each have a value defined at each point of space and time. If only E, the electric field, is nonzero and is constant in time, the field is said to be an electrostatic field. E and B are linked by Maxwell's equations. Electromagnetic fields can be explained with a quantum basis by quantum electrodynamics. Behaviour of the electromag-netic field (a hydrodynamic interpretation) The electric and magnetic vector fields can be thought of as being the velocities of a pair of fluids which permeate space. In the absence of charges these fluids would be at rest, so that their velocity fields would be zero. Electric charges act either as sources or sinks of the electric fluid. An electron is constantly absorbing electric fluid around it at some rate, call it å. Protons are the reverse: they constantly pour electric "liquid" towards the surrounding space at rate å, so liquid moves away from the proton with speed

$$V = \frac{\varepsilon}{4\pi r^2}$$

(where r is distance of the fluid away from the proton) so that the total flux of liquid going through any (imaginary) sphere which contains that proton is the area of the sphere times the speed of the fluid flowing through it:

$$4\pi r^2 \, . \, v = \varepsilon$$

Magnetic liquid, on the other hand, has no sources or sinks: nothing can pour out or suck up magnetic fluid. Magnetic fluid is incompressible, which means that its density does not change: it is not possible to compress a lot of magnetic fluid into a smaller space, or to squash it out of a given volume. (Electric fluid is also incompressible, but it has sources and sinks.) If magnetic fluid is standing still, it can be stirred up, making it move in closed circles and closed loops.

Electromagnetic Interaction

1. The interaction due to electric charge; this includes magnetic interactions.
2. The *interaction* due to electric *charge*; this includes magnetic effects that have to do with moving electric charges.

Electromagnetic Induction

1. It is the setting of an electromotive force (*e.m.f.*) in a conductor by a change of *magnetic flux linkage.* Two typical situations, illustrated in the diagram below, are a magnet falling through a coil and a conductor moving at right angles to a magnetic field. They represent two general cases : a fixed conductor in a changing magnetic field and a moving conductor in a fixed magnetic field. In neither case need the conductor be part of a complete circuit. If the circuit is incomplete there will be no current, the *e.m.f.* is unaffected (this is the same situation as a battery in a circuit as a battery in a circuit with the switch left open). With a coil left f represent the instantaneous flux threading the coil and let N represent the number of turns in the coil. The number of *flux linkages* for the coil is Nf. If Nf is changing, then the induced *e.m.f.* E equals the rate of change of flux linkages :

$$E = -\frac{d}{dt}(N\phi) \quad = -NA\frac{dB}{dt}$$

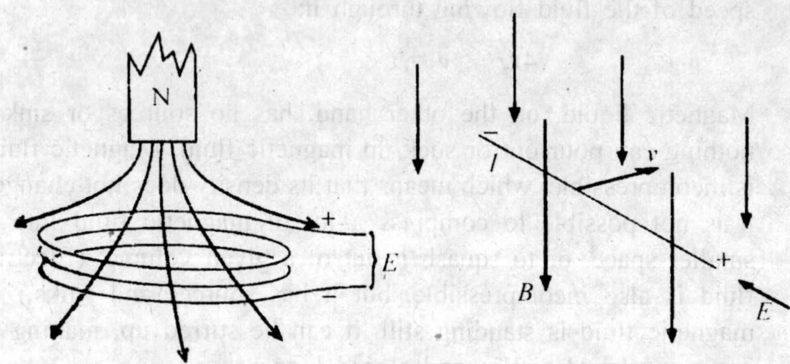

2. Electromagnetic induction is the production of an electrical potential difference (or voltage) across a conductor situated in a changing magnetic flux. Michael Faraday discovered in 1831 the induction phenomenon. He found that the electromotive force (EMF) produced along a closed path is proportional to the rate of change of the magnetic flux through any surface bounded by that path. In practice, this means that an electrical current will flow in any closed conductor, when the magnetic flux through a surface bounded by the conductor changes. This applies whether the field itself changes in strength or the conductor is moved through it. Electromagnetic induction underlies the operation of generators, induction motors, transformers, and most other electrical machines.

Electromagnetic Radiation

1. Energy propagated as time-varying electric and magnetic fields. These two fields are inextricably linked as a single entity since time-varying electric fields produce time-varying magnetic fields and vice versa. Light and radar are examples of electromagnetic radiation differing only in their wavelengths (or frequency). Electric and magnetic fields propagate through space at the speed of light. Energy propagated as time-varying electric and magnetic

fields. These two fields are inextricably linked as a single entity since timevarying electric fields produce time-varying magnetic fields and vice versa. Light and radar are examples of electromagnetic radiation differing only in their wavelengths (or frequency). Electric and magnetic fields propagate through space at the speed of light. Electromagnetic Radiation Radiation consisting of electric and magnetic waves that travel at the speed of light. *Examples:* light, radio waves, gamma rays, x-rays.

2. Although initially created by moving charges, electromagnetic radiation *electromagnetic fields* propogates freely through the vacuum requiring no further influence from matter to sustain it. Such radiation is both generated by and indicative of a wide range of phenomena in the universe (since visible light, radio, x-ray and infrared are all manifestations of electromagnetic radiation). By measuring the intensity and *wavelength* of such radiation, scientists can gain insights into the underlying chemistry and physics.

3. Radiation consisting of electric and magnetic waves that travel at the speed of light. Examples: light, radio waves, *gamma rays, x-rays.*

4. Radiation that travels through vacuous space at the speed of light and propagates by the interplay of oscillating electric and *magnetic fields*. This radiation has a *wavelength* and a *frequency*.

5. Electromagnetic radiation or EM radiation is a combination (cross product) of oscillating electric and magnetic fields perpendicular to each other, moving through space as a wave, effectively transporting energy and momentum. EM radiation is quantized as particles called photons. EM radiation with a wavelength between 400 nm and 700 nm is detected by the human eye and perceived as visible light. The physics of electromagnetic radiation is electro-dynamics, a subfield of electromagnetism. Electromagnetic waves were predicted by Maxwell's equations and subsequently discovered by Heinrich Hertz. Any electric charge which accelerates, or any changing magnetic field, produces

electromagnetic radiation. Electromagnetic information about the charge travels at the speed of light. Accurate treatment thus incorporates a concept known as retarded time (as opposed to advanced time, which is unphysical in light of causality), which adds to the expressions for the electrodynamic electric field and magnetic field. These extra terms are responsible for electromagnetic radiation. When any wire (or other conducting object such as an antenna) conducts alternating current, electromagnetic radiation is propagated at the same frequency as the electric current. Depending on the circumstances, it may behave as waves or as particles. As a wave, it is characterized by a velocity (the speed of light), wavelength, and frequency. When considered as particles, they are known as photons, and each has an energy related to the frequency of the wave given by Planck's relation $E = hí$, where E is the energy of the photon, $h = 6.626 \times 10^{-34}$ J·s is Planck's constant, and $í$ is the frequency of the wave. Generally, EM radiation is classified by wavelength into electrical energy, radio microwave, infrared, the visible region (we perceive as light,) ultraviolet, X-rays and gamma rays. The details of this classification are contained in the article on the electromagnetic spectrum. The behavior of EM radiation depends on its wavelength. Higher frequencies have shorter wavelengths, and longer wavelengths have lower frequencies. When EM radiation interacts with single atoms and molecules, its behavior depends on the amount of energy per quantum it carries. One rule is always obeyed, regardless of the circumstances. EM radiation in a vacuum always travels at the speed of light, *relative to the observer*, regardless of the observer's velocity. (This observation led to Albert Einstein's development of the theory of special relativity).

Electromagnetic Spectrum

1. The entire range of radiant energies or wave frequencies from the longest to the shortest wavelengths—the categorization of solar radiation. Satellite sensors collect this energy, but what the detectors capture is only a small protion of the entire electro-

magnetic spectrum. The spectrum usually is divided into seven sections: radio, microwave, infrared, visible, ultra-violet, x-ray, and gamma-ray radiation.

Electromagnetic Spectrum

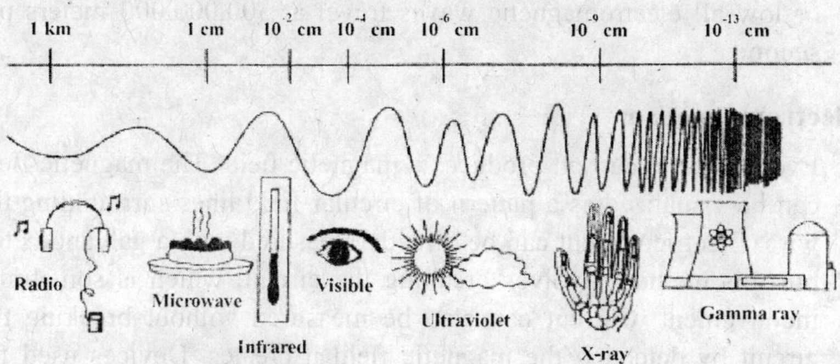

2. The entire range of all the various kinds or *wavelengths* of *electro-magnetic radiation*, including (from short to long wavelengths) *gamma rays, x-rays*, ultraviolet, optical (visible), infrared, and radiowaves.

Electromagnetic Wave

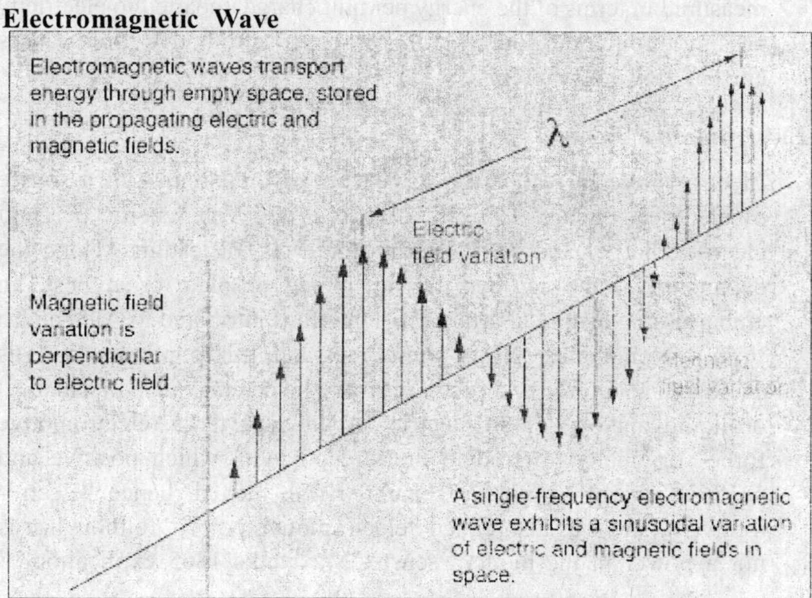

Method of travel for radiant energy (all energy is both particles and waves), so called because radiant energy has both magnetic and electrical properties. Electromagnetic waves are produced when electric charges change their motion. Whether the frequency is high or low all electromagnetic waves travel at 300,000,000 meters per second.

Electromagnetism

Every electric current produces a magnetic field The magnetic field can be visualized as a pattern of circular field lines surrounding the wire. Electric current can be directly measured with a galvanometer, but this method involves breaking the circuit, which is sometimes inconvenient. Current can also be measured without breaking the circuit by detecting the magnetic field it creates. Devices used for this include Hall effect sensors current clamps and Rogowski coils.

Electromotive

Producing an electric current through differences in potential. electromotive force The force that can alter the motion of electricity, measured in terms of the energy per unit charge imparted to electricity passing through the source of this force. Electromotive force causes current flow in a circuit.

Electromotive Force

Electromotive force (*emf*) is a measure of the strength of a source of electrical energy. The unit of emf is the Volt (energy per unit electric charge) and so the term 'force' is misleading. Thus, the expansion of the acronym is considered obsolete or at best, an embarrassing historical artifact. (The term is attributed to Alessandro Volta). Nonetheless, it is sometimes helpful to picture emf as analogous to a force or a pressure such as when making a mechanical or liquid analogy of an electric circuit. The term "electromotive force" originally referred to the strength with which positive and negative charges could be separated (*i.e.* moved, hence "electromotive"), and was also called "electromotive power" (although it is not a power in the modern sense). Maxwell's 1865 explication of what are now called Maxwell's equations used the term "electromotive force" for what is now called the electric field. Commonly, emf is generated by chemical reaction (*e.g.*, a battery or a fuel cell), absorption of radiant or thermal energy (*e.g.*, a solar cell or a thermocouple), or electromagnetic induction (*e.g.*, a generator or an alternator). Electromagnetic induction is a means of converting mechanical energy, *i.e.*, energy of motion into electrical energy. The emf generated in this way is often referred to as *motional emf.*

Electron (e)

The least massive electrically-charged particle, hence absolutely stable. It is the most common lepton, with electric charge -1.

Electron Accelerator

Electrons carry electrical *charge* and successful manipulation of *electrons* allows electronic devices to function. The picture and text on the video terminal in front of you is caused by electrons being accelerated and focused onto the inside of the screen, where a *phosphor* absorbs the electrons and light is produced. A television screen is a simple, low-energy example of an electron accelerator. A typical medical electron accelerator used in medical *radiation therapy* is about 1000 times more powerful than a color television set, while the electron accelerator at SLAC is about 2,000,000 times more powerful than a color TV. One example of an electron accelerator used in radiotherapy is the *Clinac.*

Electron Beam Therapy

Treatment by electrons accelerated to high energies in a *linear accelerator.* Primarily used for lesions situated at or near the surface.

Electron Beam

The stream of electrons generated by the electron gun and accelerated by the accelerator guide.

Electron Capture

A radioactive decay process in which an orbital electron is captured by and merges with the nucleus. The mass number is unchanged, but the atomic number is decreased by one.

Electron Degeneracy

Electron degeneracy is a stellar application of the Pauli Exclusion Principle, as is neutron degeneracy. No two electrons can occupy identical states, even under the pressure of a collapsing star of several solar masses. For stellar masses less than about 1.44 solar masses, the energy from the gravitational collapse is not sufficient to produce the neutrons of a neutron star, so the collapse is halted by electron degeneracy to form white dwarfs. This maximum mass for a white dwarf is called the Chandrasekhar limit. As the star contracts, all the lowest electron energy levels are filled and the electrons are forced into higher and higher energy levels, filling the lowest unoccupied energy levels. This creates an effective pressure which prevents further gravitational collapse.

Electron

1. An elementary particle with a unit electrical charge and a mass 1/1837 that of the proton. Electrons surround the atom's positively charged nucleus and determine the atom's chemical properties.
2. The least massive electrically *charged* particle, therefore absolutely stable. It is the most common *lepton* with charge -1. An electron is one of the *fundamental particles* in nature. Fundamental means that, as far as we know, an electron cannot be broken down into smaller particles. (This concept is one of the things SLAC physicists always challenge by looking for other particles.) Electrons are responsible for many of the phenomena that we observe in everyday life. Mutual repulsion between electrons in the atoms of the floor and those within your shoes keeps you from sinking and disappearing into the floor!!! Electrons carry electrical current and successful manipulation of electrons allows electronic devices, such as the one you are using, to function.
3. Thomson's name for the particles of which a cathode ray was made; a subatomic particle.

4. A negatively charged elementary particle that normally resides outside (but is bound to) the *nucleus* of an atom.
5. An elementary particle with a unit electrical charge and a mass 1/1837 of the *proton*. Electrons surround the atom's positively charged *nucleus* and determine the atom's chemical properties.

Electron Flux

The rate of flow of electrons through a reference surface. In *cgs* units, measured in electrons s^{-1}, or simply s^{-1}.

Electron Volt

Abbreviated *eV*. A unit of energy used to describe the total energy carried by a particle or *photon*. The energy acquired by an *electron* when it accelerates through a potential difference of 1 volt in a vacuum. 1 eV = 1.6×10^{-12} *erg*.

Electronic Structure

The distribution of electrons in the material and the energies related to changes in this distribution.

Electroweak Interaction

In the Standard Model, electromagnetic and weak interactions are related (unified); physicists use the term electroweak to encompass both of them.

Element Set (Aka Keplerian Elements, Classical Elements, Satellite Elements)

Specific information used to define and locate a particular satellite. The set includes the catalog number; epoch year; day and fraction of day; period decay rate; argument of perigee; inclination; eccentricity; right ascension of the ascending node; mean anomaly; mean motion; revolution number at epoch; and element set number. This data is contained in the two-line orbital elements provided by NASA in the NASA Prediction Bulletin.

Elementary Particles

The name given to protons, neutrons and electrons before it was discovered that protons and neutrons had substructure (quarks). Today we use the term "fundamental" for the six types of quarks and the six *leptons* and their *antiparticles* , which have no known substructure. *Gluons*, photons and W and Z *bosons* are also fundamental particles. All other particles are composite, that is made from combinations of *fundamental particles*.

Elevation

The angle at which an antenna must be pointed above the horizon for optimal reception from a spacecraft.

Elliptical Orbits

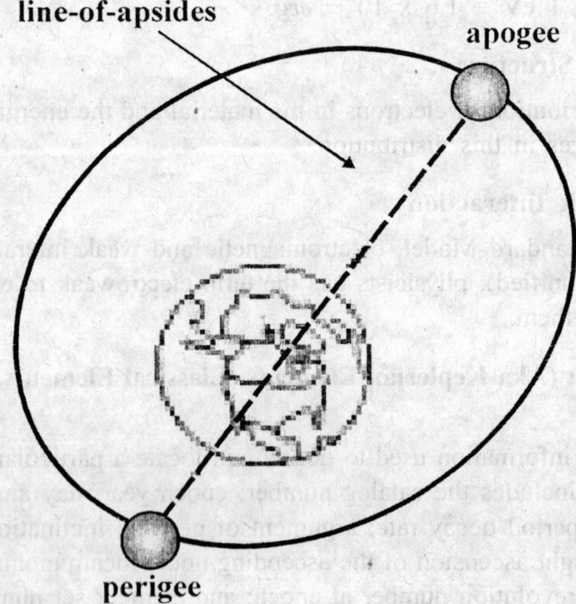

Bodies in space orbit in elliptical rather than circular orbits because of factors such as gravity and drag. The point where the orbiting satellite is closest to Earth is the perigee, sometimes called per-apsis or perifocus. The point where the satellite is farthest from Earth is called apogee apoapsis, or api-focus. A line drawn from perigee to apogee is the line-of-apsides, sometimes called the major-axis of the ellipse. It's simply a line drawn through the ellipse the long way.

ELT

Emergency Locator Transmitter.

Eluant

Washing solution (The solution that is introduced into the cow).

Eluate

The washings obtained by elution (the solution that comes out of the cow).

Elute

To separate by washing (to milk).

Empirical Law

A law strictly based on experiment, which may lack theoretical foundation.

Energy (E)

The energy of an object increases when work is done on it. An increase in energy may heat up the object, speed it up, lift it up, or all of the above.

Energy

1. A numerical scale used to measure the heat, motion, or other properties that would require fuel or physical effort to put into an object; a scalar quantity with units of joules (J).
2. Energy is a property of a body, not a material substance. When bodies interact, the energy of one may increase at the expense of the other, and this is sometimes called a *transfer of energy*. This does *not* mean that we could intercept this energy in transit and bottle some of it. After the transfer one of the bodies may have higher energy than before, and we speak of it as having

'stored energy'. But that doesn't mean that the energy is 'contained in it' in the same sense as water in a bucket.

Energy Budget

A quantitative description of the energy exchange for a physical or ecological system. The budget includes terms for radiation, conduction, convection, latent heat, and for sources and sinks of energy.

Energy Flux

The rate of flow of energy through a reference surface. In *cgs* units, measured in *erg* s^{-1}. Also measured in watts, where 1 watt = 1×10^7 erg s^{-1}. *Flux density*, the flux measured per unit area, is also often referred to as "flux".

Energy Scale

The energy scale used by most nuclear scientists is electron volts (eV), thousands of electron volts (keV), and millions of electron volts (MeV). An electron volt is the energy acquired when an *electron* falls through a potential difference of 1 volt. 1 eV=$1.602*10^{12}$ergs. Masses are also given by their "mass-equivalent" energy ($E=mc^2$). The mass of the *proton* is 938.27231 MeV.

Energy–separation Curve

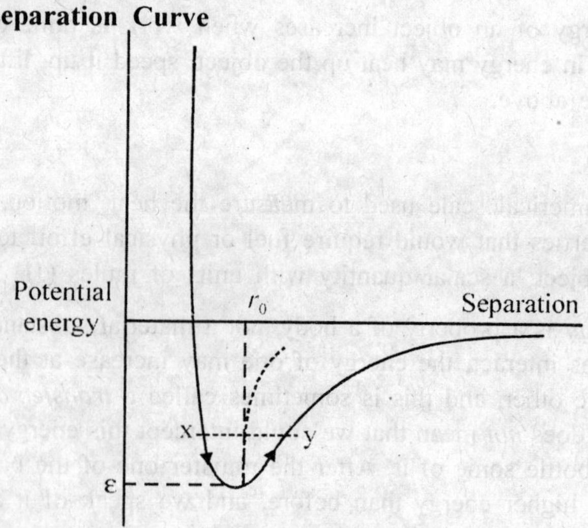

A graph which shows how the energy stored in the bond between two atoms varies with their separation. The diagram shows a typical example. The part of the curve where the bond energy is negative corresponds to the stable states; energy is needed to separate the atoms from one another.

ENSDF

The Evaluated Nuclear Structure Data File is evaluated by an international collaboration of nuclear scientists. ENSDF is a database of nuclear structure and decay data.

Ensembles

The language of density matrices is also used to describe quantum ensembles, or a collection of identical quantum systems. Consider a "black-box" apparatus that spits electrons towards an observer. The electrons' Hilbert spaces are identical. The apparatus might produce electrons that are all in the same state; in this case, the electrons received by the observer are then called a *pure ensemble*. However, the apparatus could produce electrons in different states. For example, it could produce two populations of electrons: one with state |**z**+> (spins aligned in the positive **z** direction), and the other with state |**y**-> (spins aligned in the negative **y** direction.) Generally, there can be any number of populations, each corresponding to a different state. This is a *mixed ensemble*. We can describe an ensemble as a collection of populations with weights w_i and corresponding states |$á_i$>.

ENSO (El Nino-Southern Oscillation)

Interacting parts of a single global system of climate fluctuations. ENSO is the most prominent known source of interannual variability in weather and climate around the world, though not all areas are affected. The Southern Oscillation (SO) is a global-scale seesaw in atmospheric pressure between Indonesia /North Australia, and the southeast Pacific. In major warm events El Nino warming extends over much of the tropical Pacific and becomes clearly linked to the SO pattern. Many of the countries most affected by ENSO events are developing countries with economies that are largely dependent upon their agricultural and fishery sectors as a major source of food supply employment, and foreign exchange. New capabilities to predict the onset of ENSO events can have a global impact. While ENSO is a natural part of the Earth's climate, whether its intensity or frequency may change as a result of global warming is an important concern.

Entropy

For other uses of the term entropy, The thermodynamic entropy *S*, often simply called the entropy in the context of thermodynamics, is a measure of the amount of energy in a physical system that cannot be used to do work. It is also a measure of the disorder present in a system. The SI unit of entropy is $J \cdot K^{-1}$ (joule per kelvin), which is the same unit as heat capacity. Thermodynamic entropy is closely related to information entropy. Entropy : S Two more parameters can be considered for Open system : Number of particles N, of each component of the system Chemical potential, ì of each component of the system. The mechanical parameters listed above can be described in terms of fundamental classical or quantum physics, while the statistical parameters can only be understood in terms of statistical mechanics. In most applications of thermodynamics, one or more of these parameters will be held constant, while one or more of the remaining parameters are allowed to vary. Mathematically, this means the system can be described as a point in n-dimensional space, where n is the number of parameters not held fixed. Using statistical mechanics, combined with the laws of classical or quantum physics, equations of state can be derived which express each of these parameters in terms of the others. The simplest and most important of these equations of state are the ideal gas law and its derived equations: $pV = NkT$ where k is the Boltzmann constant.

Environment

The complex of physical, chemical, and biological factors in which a living organism or community exists.

EPA (Environmental Protection Agency)

U.S. agency that ensures: Federal environmental laws are implemented and enforced effectively; U.S. policy—both foreign and domestic—fosters the integration of economic development and environmental protection so that economic growth can be sustained over the long term; public and private decisions affecting energy, transportation, agriculture, industry, international trade, and natural resources fully integrate considerations of environmental quality; national effort sto reduce environmental risk are based on the bvest available scientific information communicated clearly to the public; everyone in our society recognizes the value of preventing pollution before it is created; people have the information and incentives they need to make environmentally-responsible choices in their daily lives; and schools and community institutions promote environmental stewardship as a national ethic.

Ephemeris

A tabulation of a series of points that define the position and motion of a satellite.

EPIRB

Emergency Position Indicating Radio Beacon.

Epoch (Aka Epoch Time Or TO)

Epoch specifies the time of a particular description of a satellite orbit.

Equalization

One of the most powerful tools for increasing the potential acoustic gain of a sound amplification system is the production of a "mirror image" filter to level the frequency response of the system. The process of equalizing an auditorium also improves the fidelity of the sound. The process of leveling out the frequency response of the sound system removes the peaks which will ring the system before sufficient gain is achieved.

Equator

An imaginary circle around the Earth that is everywhere equally distant (900) from the North Pole and the South Pole. The equator is a great circle and defines latitude 00.

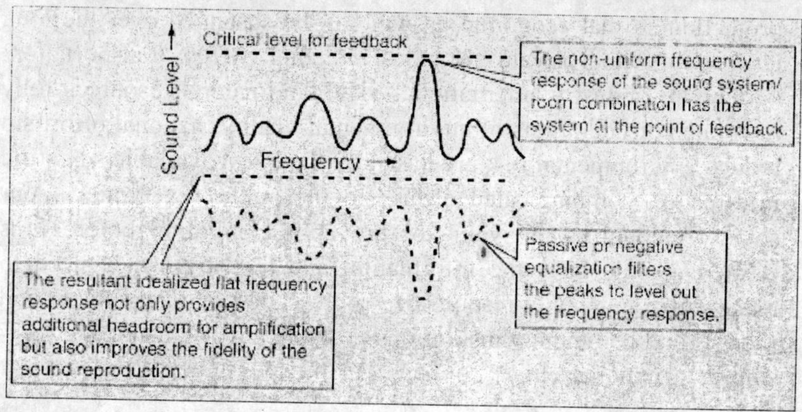

Equilibrium

A state in which an object's momentum and angular momentum are constant.

Equilibrant

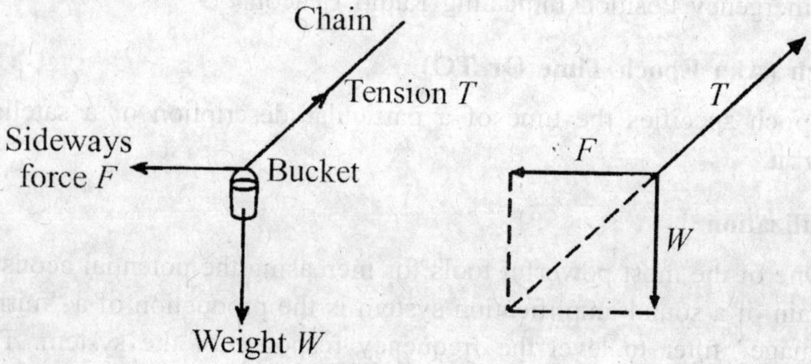

Equilibrant is the name given to the single vector which is of equal magnitude to the resultant of two or more vectors, but which points in the opposite direction. It balances the results. For example, the tension T in the chain shown in the diagram on the next page is the equilibrant of the weight *W* of the bucket and the sideways force *F*. In fact, when three or more forces are in equilibrium, any one force is the equilibrant of the remaining forces.

Equivalence Principle

In relativity the *equivalence principle* is applied to several related concepts dealing with frames of reference. In particular, the several forms of the principle describe the uniformity of physical measurements in inertial frames of reference.

Erg

A *cgs* unit of energy equal to work done by a force of 1 dyne acting over a distance of 1 cm. 10^7 (ten million) erg s^{-1} (ergs per second) = 1 watt. Also, 1 Calorie = 4.2×10^{10} (42 billion) ergs.

Error Analysis

The mathematical analysis done to show quantitatively how uncertainties in data produce uncertainty in calculated results, and to find the sizes of the uncertainty in the results. [In mathematics the word *analysis* is synonymous with *calculus*, or 'a method for mathe-matical calculation.' Calculus courses used to be named *Analysis*.]

ESA

European Space Agency.

Escape Velocity = 618 km/s

Being a gaseous body, the Sun does not have a single period of rotation like a rigid body. The sunspots provide a convenient reference for the measurement of the rotation period at different latitudes. The period of rotation averages 25.4 days, varying from 34.4 days at the poles to 25.1 days at the equator (Chaisson). Its axis is tilted 7.25° relative to the ecliptic.

The visible surface of the Sun (the photosphere) has a granular appearance with a typical dimension of a granule being 1000 kilometers. The image at right is from the NASA Solar Physics website and is credited to G. Scharmer and the Swedish Vacuum Solar Telescope. The granules are described as convection cells which transport heat from the interior of the Sun to the surface.

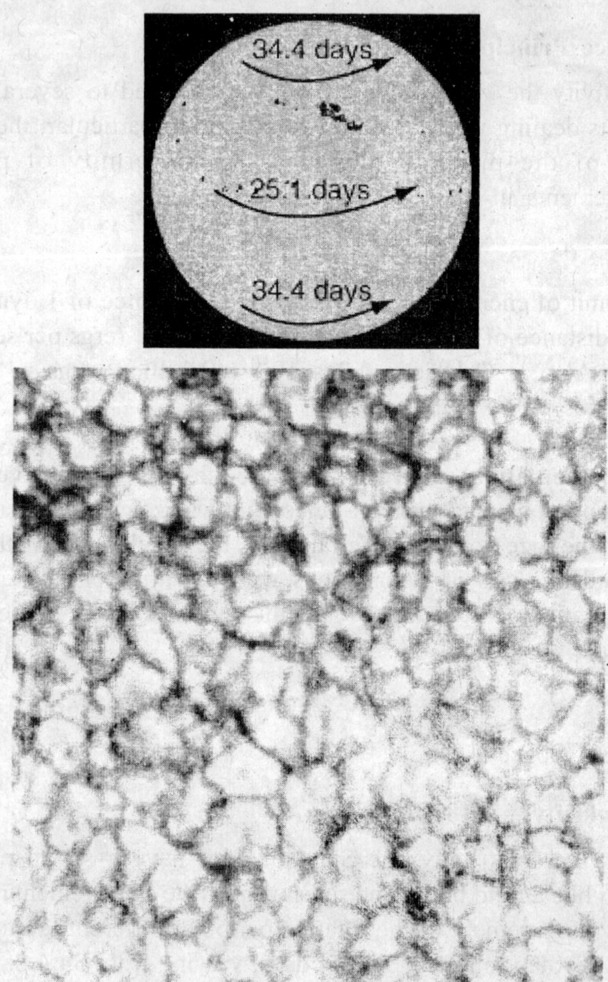

The Sun's apparent magnitude is -26.8 and its absolute magnitude is +4.8. Its greatest angular diameter as seen from the Earth is 32.5'. The sun is 7.7 kpc or 25,000 light years from the center of the galaxy and orbits the galaxy in about 200 million years. This corresponds to an orbital velocity of 230 km/s.

Eutrophication

The process whereby a body of water becomes rich in dissolved nutrients through natural or man-made processes. This often results in a deficiency of dissolved oxygen, producing an environment that favors plant over animal life.

eV (electronvolt)

The basic *unit* of energy used in high energy physics. It is the energy gained by one *electron* when it moves through a potential difference of one volt. By definition an eV is equivalent to 1.6×10^{-19} joules. This is a very small amount of energy and the more commonly used multiples are *MeV* (million eV), *GeV* (billion eV or giga-electronvolt) and *TeV* (trillion eV).

Evacuated

Sealed and pumped down to a pressure very much below atmospheric pressure. Typical pressures inside accelerators or waveguides are about 10^{-12} times atmospheric pressure.

Evaporation

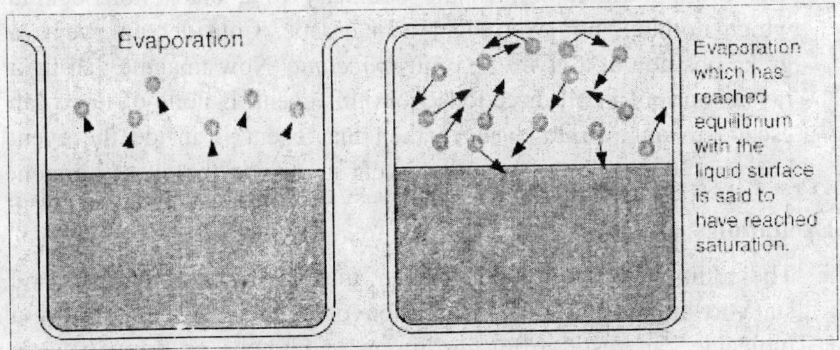

1. Change from a liquid (more dense) to a vapor or gas (less dense) form. When water is heated it becomes a vapor that increases humidity Evaporation is the opposite of condensation.
2. Ordinary evaporation is a surface phenomenon - some molecules have enough kinetic energy to escape. If the container is closed.

an equilibrium is reached where an equal number of molecules return to the surface. The pressure of this equilibrium is called the saturation vapor pressure.

In order to evaporate, a mass of water must collect the large heat of vaporization, so evaporation is a potent cooling mechanism. Evaporation heat loss is a major climatic factor and is crucial in the cooling of the human body.

Event

1. An event occurs when two particles collide or a single particle *decay*. Particle theories predict the probabilities of various events occurring when many similar collisions or decays are studied. They cannot predict the outcome for a single collision or decay.
2. What occurs when two particles collide or a single particle decays. Particle theories predict the probabilities of various possible events occurring when many similar collisions or decays are studied. They cannot predict the outcome for any single event.

Event Horizon

The event horizon defines the boundary of a black hole behind which nothing, not even light, can escape. Consider an event (a given position at a given time) in *spacetime*. Now imagine that light rays shoot out in all directions from this event. If none of them can escape to an infinite distance then that event is inside the event horizon. If any can escape, that event is outside the event horizon.

Excitation

The addition of energy to a system, transferring it from its ground state to an excited state. Excitation of a nucleus, an atom, or a molecule can result from absorption of photons or from inelastic collisions with other particles.

Excited State

The state of an atom or nucleus when it possesses more than its normal energy. The excess energy is usually released eventually as a gamma ray.

Exosphere

The uppermost layer of the atmosphere, its lower boundary is estimated at 500 km to 1000 km above the Earth's surface. It is only from the exosphere that atmospheric gases can, to any appreciable extent, escape into outer space.

Expanding Universe

The galaxies we see in all directions are moving away from the Earth, as evidenced by their red shifts. Hubble's law describes this expansion.

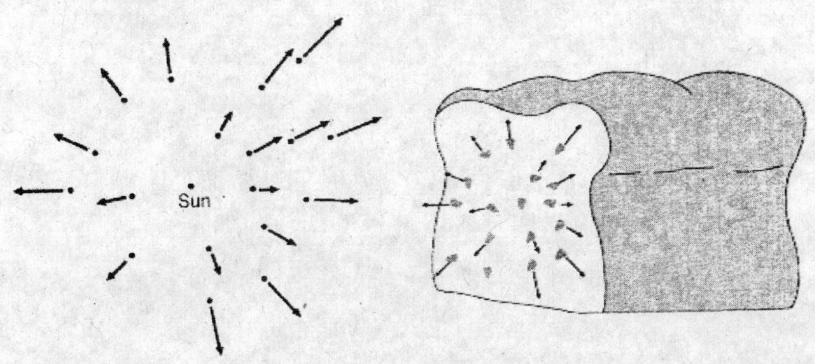

The fact that we see all stars moving away from us does not imply that we are the center of the universe! All stars will see all other stars moving away from them in an expanding universe. A rising loaf of raisin bread is a good visual model: each raisin will see all other raisins moving away from it as the loaf expands.

Experimental Eror

The uncertainty in the value of a quantity. This may be found from (1) statistical analysis of the scatter of data, or (2) mathematical analysis showing how data uncertainties affect the uncertainty of calculated results.

Extensive Poperty

A measurable property of a thermodynamic system is extensive if, when two identical systems are combined into one, the value of that property of the combined system is double its original value in each system. Examples: mass, volume, number of moles.

External Forcing

Influence on the Earth system (or one of its components) by an external agent such as solar radiation or the impact of extraterrestrial bodies such as meteorites.

F

FAA

Federal Aviation Administration.

Facsimile (FAX)

A process by which graphic or photographic information is transmitted or recorded by electronic means.

Factor

One of several things multiplied together.

Fahrenheit

Temperature scale designed by the German scientist Gabriel Fahrenheit in 1709, based upon water freezing at 32°F and water boiling at 212°F under standard atmospheric pressure.

Far Infrared

Electromagnetic radiation, longer than the thermal infrared, with wavelengths between about 25 and 1000 micrometers.

Feedhorn

A metallic cylinder closed at one end, used to obtain and direct radio frequency (RF) energy reflected from a satellite dish. It acts as a wave guide at microwave frequencies. RE energy inside the horn is picked up by a small probe; once inside the horn, the wavelength (energy) of the microwave radiation changes to a guided wave.

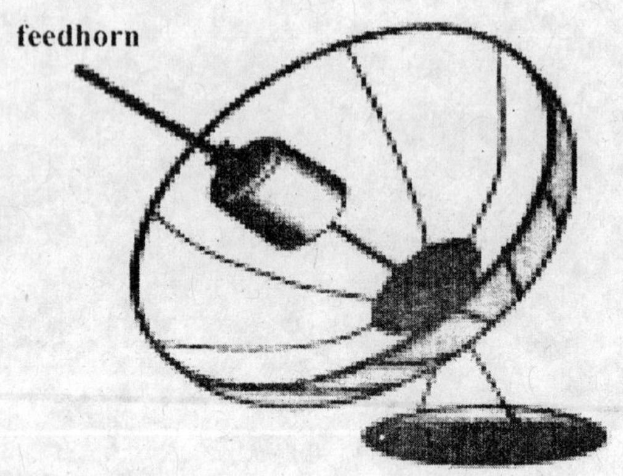

FEMA

U.S. Federal Emergency Management Agency.

Feng Yun

Chinese geostationary environmental satellite that was destroyed by an explosion before launch in April 1994. The name Feng Yun, meaning Wind and Cloud, was originally applied to the Chinese polar-orbiting environmental satellite launched in September 1991 (Feng Yun 1-2), which offered direct readout services. The Chinese polar-orbiter program has since been abandoned.

Fermion

1. Any particle that has odd-half-integer (1/2, 3/2, ...) intrinsic angular momentum (spin), measured in units of h-bar. As a consequence of this peculiar angular momentum, fermions obey a rule called the Pauli Exclusion Principle, which states that no two fermions can exist in the same state at the same time. Many of the properties of ordinary matter arise because of this rule. Electrons, protons, and neutrons are all fermions, as are all the fundamental matter particles, both quarks and leptons.

2. General name for a particle that is a matter constituent, characterized by spin in odd half integer quantum units (1/2,3/2,5/2...). Named for Italian physicist Enrico Fermi. Quarks, *leptons* and *baryons* are all fermions.

Fermions and Bosons

The choice of symmetry or antisymmetry is determined by the species of particle. For example, we must always use symmetric states when describing photons or helium-4 atoms, and antisymmetric states when describing electrons or protons. Particles which exhibit symmetric states are called bosons. As we will see, the nature of symmetric states has important consequences for the statistical properties of systems composed of many identical bosons. These statistical properties are described as Bose-Einstein statistics. Particles which exhibit antisymmetric states are called fermions. As we have seen, antisymmetry gives rise to the Pauli exclusion principle, which forbids identical fermions from sharing the same quantum state. Systems of many identical fermions are described by Fermi-Dirac statistics.

Field of View

The range of angles that are scanned or sensed by a system or instrument, measured in degrees of arc.

Field

1. A property of a point in space describing the forces that would be exerted on a particle if it was there.
2. The set of influences (electricity magnetism, gravity) that extend throughout space.

Filter

Device that while selectively passing desired frequencies removes undesired ones.

Fission

1. The radioactive decay of a nucleus by splitting into two parts.
2. The splitting of a heavy nucleus into two roughly equal parts (which are nuclei of lighter elements), accompanied by the release of a relatively large amount of energy in the form of kinetic energy of the two parts and in the form of emission of neutrons and gamma rays.

Fission Products

Nuclei formed by the fission of heavy elements. They are of medium atomic weight and almost all are radioactive.

Fixed-Target Experiment

An experiment in which the beam of particles from an accelerator is directed at a stationary (or nearly stationary) target. The target may be a solid, a tank containing liquid or gas, or a gas jet.

Flare (Solar)

Rapid release of energy from a localized region on the Sun in the form of *electromagnetic radiation*, energetic particles, and mass motions.

Flare Star

A member of a class of stars that show occasional, sudden, unpredicted increases in light. The total energy released in a flare on a flare star can be much greater that the energy released in a *solar flare*.

Flavor

The name used for the different quark types and the different *lepton* types. The six flavors of quarks are *up, down, strange, charm, bottom, top*, in increasing order of mass. The flavors of *charged* leptons are *electrons, muon* and *tau*, again in increasing order of mass. For each charged lepton flavor there is a corresponding *neutrino* flavor.

Flavour

The name used for the different quarks types (up, down, strange, charm, bottom, top) and for the different lepton types (electron, muon, tau). For each charged lepton flavour there is a corresponding neutrino flavour. In other words, flavour is the quantum number that distin-guishes the different quark/lepton types. Each flavour of quark and charged lepton has a different mass. For neutrinos we do not yet know if they have a mass or what the masses are.

Fleming's Left Hand Rule

Its gives the direction of the force on a current–carrying conductor in a magnetic field. It states that if the thumb and first two fingers of the left hand are held at right angles to each other, and if the first finger points in the direction of the magnetic field while the second finger points in the direction of the current in a conductor immersed in the magnetic field, them the thumb will be pointing in the direction of the force on the conductor, as in the diagram.

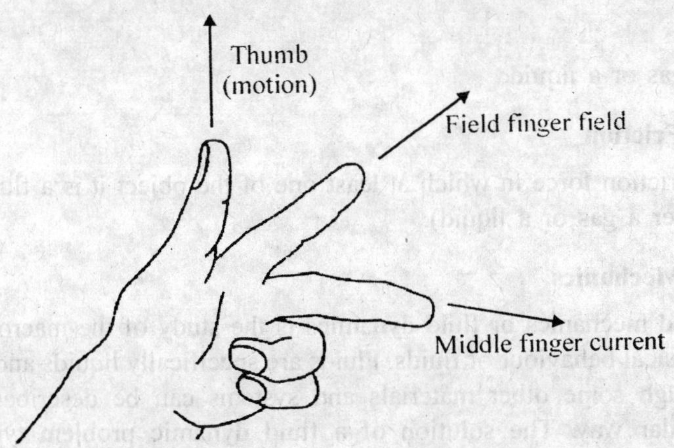

Fleming's Right Hand Rule

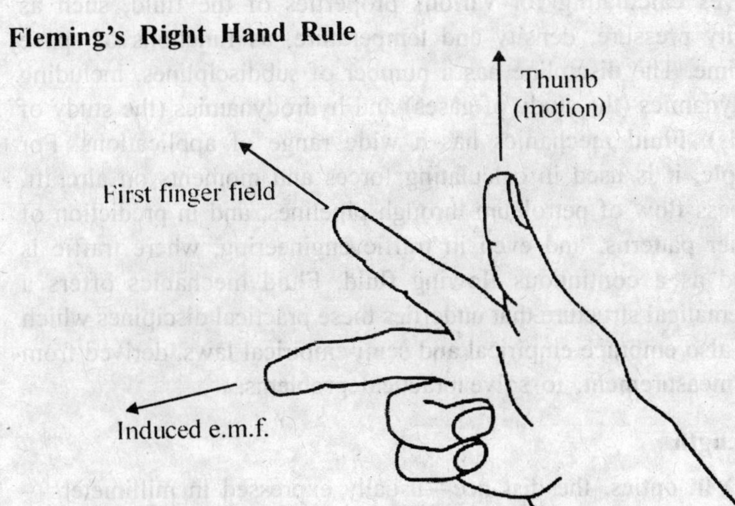

Its gives the direction of the induced *e.m.f.* in a conductor. It states that if the first two fingers and the thumb and the thumb of the right hand are held are held at right angles to one another. It the first finger points in the direction of the magnetic field and the thumb point in the direction of the motion of the conductor, then the second finger will be pointing in the direction of the induced *e.m.f.* the diagram is above.

Fluid

A gas or a liquid.

Fluid Friction

A friction force in which at least one of the object it is a fluid (i.e. either a gas or a liquid).

Fluid Mechanics

Fluid mechanics or fluid dynamics is the study of the macroscopic physical behaviour of fluids. Fluids are specifically liquids and gases though some other materials and systems can be described in a similar way. The solution of a fluid dynamic problem typically involves calculating for various properties of the fluid, such as velocity pressure, density and temperature, as functions of space and time. The discipline has a number of subdisciplines, including aerodynamics (the study of gases) and hydrodynamics (the study of liquids). Fluid mechanics has a wide range of applications. For example, it is used in calculating forces and moments on aircraft, the mass flow of petroleum through pipelines, and in prediction of weather patterns, and even in traffic engineering, where traffic is treated as a continuous flowing fluid. Fluid mechanics offers a mathematical structure that underlies these practical discipines which often also embrace empirical and semi-empirical laws, derived from flow measurement, to solve practical problems.

Focal Length

1. (1) In optics, the distance—usually expressed in millimeters—from the principal point of a lens or concave mirror to its focal

point. (2) The distance, measured from the center of the surface of a parabolic or spherical reflector (e.g., satellite dish) where RF energy is brought to essential point focus.

2. A property of a lens or mirror, equal to the distance from the lens or mirror to the image it forms of an object that is infinitely far away.

Focal Point

1. The area where weak signals collected by a satellite dish, concentrated into a smaller receiving area, converge.

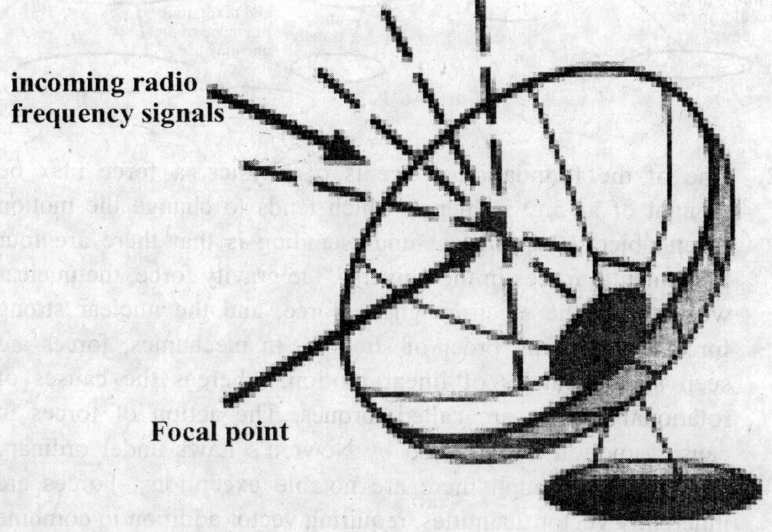

2. The focal point of a lens is defined by considering a parallel bundle or beam of light incident upon the lens, parallel to the optic (symmetry) axis of the lens. The focal point is that point to which the rays converge or from which they diverge. The first case is that of a *converging* (positive) lens. The second case is that of a *diverging* (negative) lens. It's easy to tell which kind of lens you have, for converging lenses are thicker at their center than at the edges, and diverging lenses are thinner at the center than at the edges.

Fog

A cloud on the ground.

Footpoint

The intersection of *magnetic* loops with the *photosphere.*

Force (F)

1. The strength of a force is defined by the rate at which it can speed up one kilogram of mass.

2. One of the foundation concepts of physics, a force may be thought of as any influence which tends to change the motion of an object. Our present understanding is that there are four fundamental forces in the universe, the gravity force, the nuclear weak force, the electromagnetic force, and the nuclear strong force in ascending order of strength. In mechanics, forces are seen as the causes of linear motion, whereas the causes of rotational motion are called torques. The action of forces in causing motion is described by Newton's Laws under ordinary conditions, although there are notable exceptions. Forces are inherently vector quantities, requiring vector addition to combine them. The SI unit for force is the Newton, which is defined by Newton = kg m/s^2 as may be seen from Newton's second law.

Force on Moving Charge

A phrase which relates to the force on a free charge, usually an electron, moving through a magnetic field at an angle to the field. The force on an electron in an electric field equals *Ee* whether or not the electron is moving. The force *F* on charge *q* moving through a uniform magnetic field of strength B with velocity v along a path at right angles to the field direction is given by the formula : $F = Bqv$. The direction of the force is given by *Fleming's left hand rule.* Remember when applying the rule that the current direction is opposite to the direction of motion for negative charge. The diagram illustrates an important special case. A beam of electrons, all travelling with speed v at right angles to an extensive uniform magnetic field of strength B is pulled by the force F into a circular path of radius r. The centripetal force equal Bev.

$$Bev = \frac{mv^2}{r}, \qquad r = \frac{mv}{Be}.$$

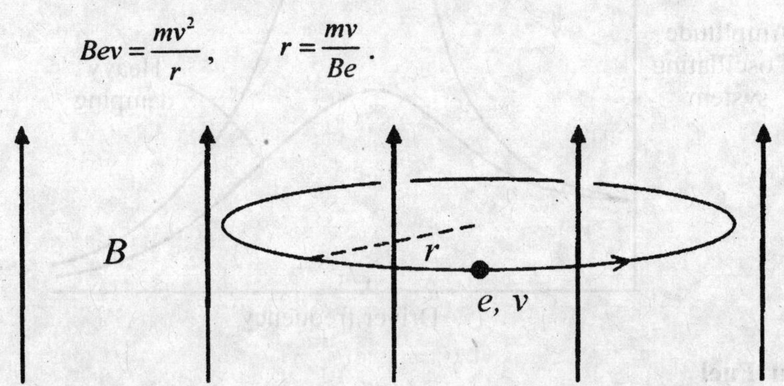

Forced Oscillations

Forced Oscillations are the oscillations of a damped of a damped system acted on by a periodic driving force. Were the system not damped, the oscillations would soon go wildly out of control. If the system is damped, an equilibrium situation is reached when per cycle, the energy transferred to the system from driver equals the energy transferred from the system through the damping mechanism. The amplitude of this equilibrium state for a given driver amplitude depends on the frequency of the driver. It is a maximum when the driver frequency is the same as the natural frequency f_0 of the undamped system. This is called *resonance.* The link between resonance and physical damage is connected to this maximum amplitude in an obvious manner : the vibrating system is subject to maximum distortion. Less obvious, but just as important, is the fact that the rate of energy flow from the driver, through the system into the damping mechanism is also at its maximum value. This is the principle behind tuning circuits, musical instrument design and many other engineering applications.

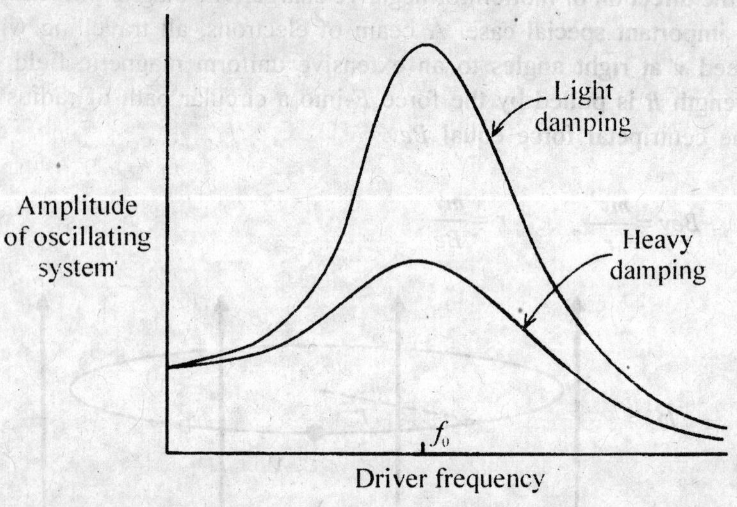

Fossil Fuel

Any hydrocarbon deposit that can be burned for heat or power, such as petroleum, coal, and natural gas.

Fossil

Hardened remains or traces of plant or animal life from a previous geological period preserved in the Earth's crust.

FPS

The system of units based on the fundamental units of the 'English system': foot, pound and second.

Frame

A single image or picture. A single complete vertical scan of the cathode ray tube (CRT).

Frame of Reference

A frame of reference in physics is a set of axes which enable an observer to measure the aspect, position and motion of all points in a system relative to the reference frame. Two observers may choose to use different frames of reference to investigate a common system. The measurements that an observer makes about a system generally depend on the observer's frame of reference. This definition applies to "classical" physics, *i.e.* before the general theory of relativity. In relativity, a larger refrerence frame (one that far exceeds the bounds of the local speed of light) cannot be described in terms of a rigid reference frame, but rather must be considered as an *inertial* reference frame, wherein a classical physical space can be mapped to a larger *relative* spacetime (via the Lorentz transformations). The principle of relativity states that, even though a set of measurements may depend on an observer's *particular* frame of reference, the observed physical events still must follow the same physical laws in all *inertial frames of reference.* Nomenclature and notation When working a problem involving one or more frames of reference it is common to designate an inertial frame of reference. An accelerated frame of reference is often delineated as being the "primed" frame, and all variables that are dependent on that frame are notated with primes, e.g. x', y' , a'. The vector from the origin of an inertial reference frame to the origin of an accelerated reference frame is commonly notated as R. Given a point of interest that exists in both frames, the vector from the inertial origin to the point is called **r**, and the vector from the accelerated origin to the point is called r'. From the geometry of the situation, we get

$$\vec{r} = \vec{R} + \vec{r}'$$

Taking the first and second derivatives of this, we obtain

$$\vec{v} = \vec{V} + \vec{v}'$$

$$\vec{a} = \vec{A} + \vec{a}'$$

where V and A are the velocity and acceleration of the accelerated system with respect to the inertial system and v and a are the velocity and acceleration of the point of interest with respect to the inertial frame.

Free Electron

An *electron* that has broken free of it's atomic bond and is therefore not bound to an atom.

Free-Electron Laser

The radiation from a free-electron laser is produced from free electrons which are forced to oscillate in a regular fashion by an applied field. They are therefore more like synchrotron light sources or microwave tubes than like other lasers. They are able to produce highly coherent, collimated radiation over a wide range of frequencies. The magnetic field arrangement which produces the alternating field is commonly called a "wiggler" magnet.

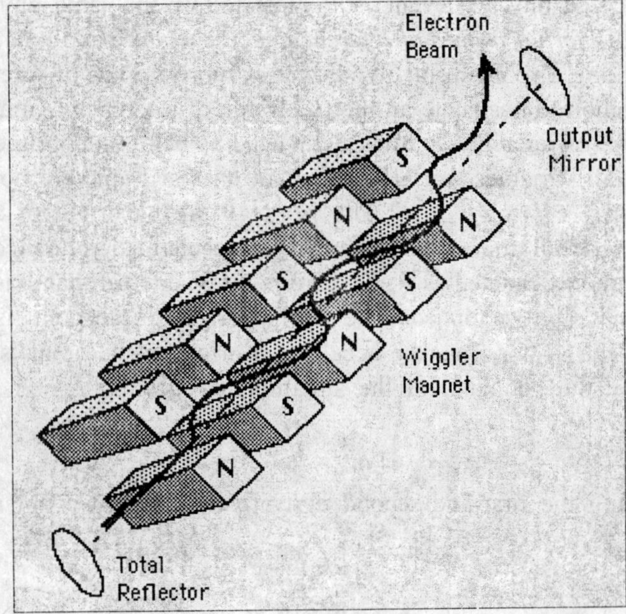

Free Energy

In thermodynamics *free energy* is a measure of the amount of work that can be extracted from a system. In this sense, it measures not the energy content of the system, but the "useful energy" content. In different situations, *free energy* is related to internal energy in different ways. For instance, in chemistry one is usually concerned with fluid systems at constant temperature, undergoing chemical reaction in closed containers (at constant volume) or in open containers (at constant pressure). In the first case one would use the *Helmholtz free energy* as a measure of available work, and in the second one would use the *Gibbs free energy.* Physicists carry out more abstract analysis of systems which are not necessarily fluid. In that context, the *Helmholtz free energy* is more common, and it is also directly related to the partition function of a canonical ensemble in statistical mechanics. In the isothermal-isobaric ensemble, the partition function is related to the Gibbs free energy. For this reason, there is some confusion in terminology, which we disentangle below. The usage in physics and chemistry is as follows: In physics, *free energy* denotes the thermodynamic potential $F=U-TS$, also called Helmholtz free energy. In chemistry, this quantity is called the Helmholtz function or the work content and is denoted A after the German word *Arbeit*, meaning *work*. In chemistry , *free energy* denotes the thermodynamic potential $F=U-TS+PV$, also called the Gibbs function. In physics, this quantity is called the Gibbs free energy and denoted G. Note that it is common usage in chemistry to denote internal energy by E instead of U Recently a compromise notation has become common, using A for the Helmholtz function, G for the Gibbs function and avoiding F entirely. The functions are then referred to as the *Helmholtz free energy* and *Gibbs free energy.*

Free Radicals

Atomic or molecular species with unpaired electrons or an otherwise open shell configuration, usually very reactive. Specific to atmospheric chemistry free radicals are: short-lived, highly reactive, intermediate species produced by dissociation of the source molecules by solar ultraviolet radiation or by reactions with other stratospheric constituents. Free radicals are the key to intermediate species in many important stratospheric chain reactions in which an ozone molecule is destroyed and the radical is regenerated.

Frequency (F)

1. Number of cycles and parts of cycles completed per second. F=1/T, where T is the length of one cycle in seconds.
2. The number of repetitions per unit time of the oscillations of an *electromagnetic* wave (or other wave). The higher the frequency, the greater the energy of the radiation and the smaller the *wavelength.* Frequency is measured in *Hertz.*

Frequency Division Multiplexing

The combining of a number of signals to share a medium by dividing it into different frequency bands for each signal.

Frequency Modulation (FM)

The instantaneous variation of the frequency of a carrier wave in response to changes in the amplitude of a modulating signal. As applied to APT the radio signal from the satellite is broadcast on an FM transmitter and received on the ground on an FM radio receiver.

Frequency

The number of cycles per second, the inverse of the period (q.v.).

Front

A boundary between two different air masses. The difference between two air masses sometimes is unnoticeable. But when the colliding air masses have very different temperatures and amounts of water in them, turbulent weather can erupt. A cold front occurs when a cold air mass moves into an area occupied by a warmer air mass. Moving at an average speed of about 20 mph, the heavier cold air moves in a wedge shape along the ground. Cold fronts bring lower temperatures and can create narrow bands of violent thunderstorms. In North America, cold fronts form on the eastern edges of high pressure systems. A warm front occurs when a warm air mass moves into an area occupied by a colder air mass. The warm air is lighter; 50 it flows up the slope of the cold air below it. Warm fronts usually form on the eastern sides of low pressure systems, create wide areas of clouds and rain, and move at an average speed of 15 mph. When a cold front follows and then overtakes a warm front (warm fronts move more slowly than cold fronts) lifting the warm air off the ground, an occluded front forms.

Frost

Water condensation occurring on surfaces below freezing. Condensing water turns to ice.

Fundamental Interaction

1. In the Standard Model the fundamental interactions are the electromagnetic, weak, strong and gravitational interactions. There is at least one more fundamental interaction in the theory that is responsible for fundamental particle masses. Five interaction types are all that are needed to explain all observed physical phenomena.

2. The known fundamental *interactions* are the *strong, electromagnetic, weak* and *gravitational interaction.* These interactions explain all observed physical processes but do not explain particle masses. Any force between two objects is due to one or another of these interactions. All known particle *decays* can be understood in terms of these strong, electromagnetic or weak interactions.

Fundamental Particle

1. A particle with no internal substructure. In the Standard Model the quarks, leptons, photons, gluons, W+ and W- bosons, and the Z bosons are fundamental. All other objects are made from these.

2. A particle with no internal substructure. In the *Standard Model*, the quarks, *leptons*, photons, *gluons*,W-*boson* and Z-bosons are fundamental. All other objects are made from these particles.

Fusion

A nuclear reaction in which two nuclei stick together to form one bigger nucleus.

G

Gain

The increase in signal power produced by an amplifier; usually expressed in decibels as the ratio of the output to the input. A measure of the effectiveness of a directional antenna as compared to a non-directional antenna.

Gala Hypothesis

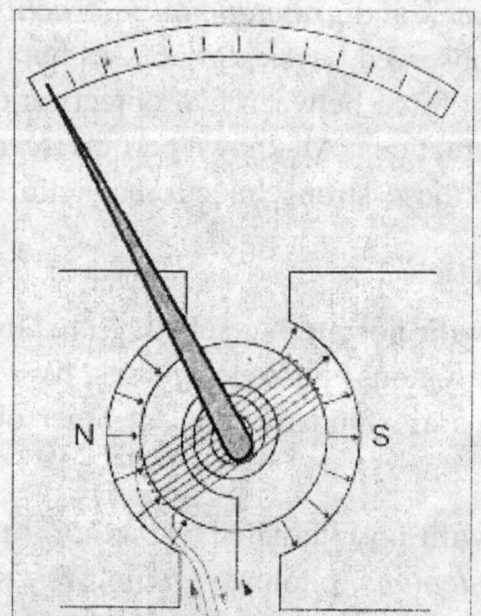

The hypothesis that the Earth's atmosphere, biosphere, and its living organisms behave as a single system striving to maintain a stability that is conducive to the existence of life.

Galaxy Geometry

There are many geometries of galaxies including the spiral galaxy characteristic of our own Milky Way. In the remarkable deep space photograph made by the Hubble Space Telescope, every visible object except for the one obvious foreground star seems to be another galaxy.

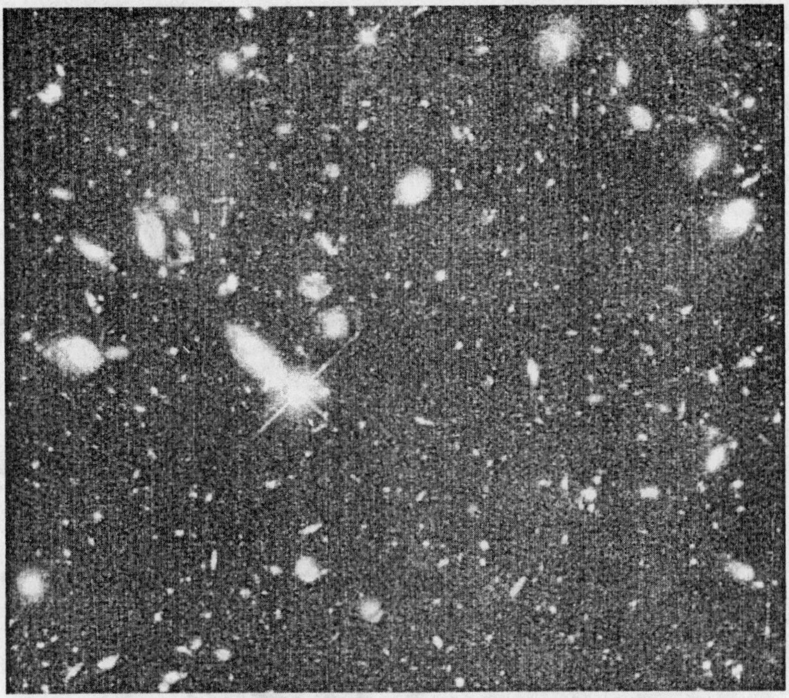

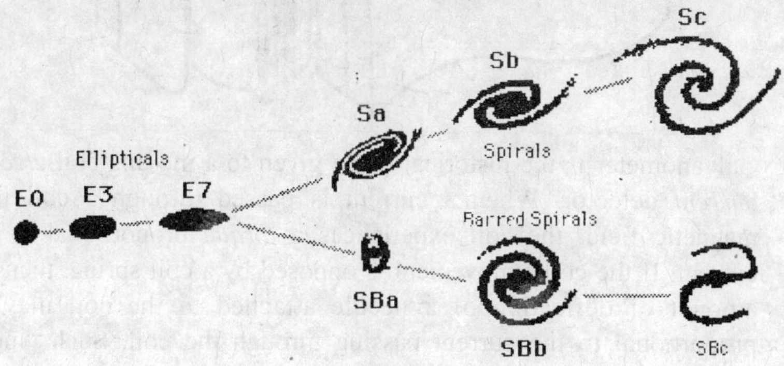

Elliptical galaxies usually have very little gas or dust and hence little evidence of new star formation. The spiral galaxies may have an abundance of gas and dust and show evidence of star formation in the form of lots of hot blue stars.

Galvanometer

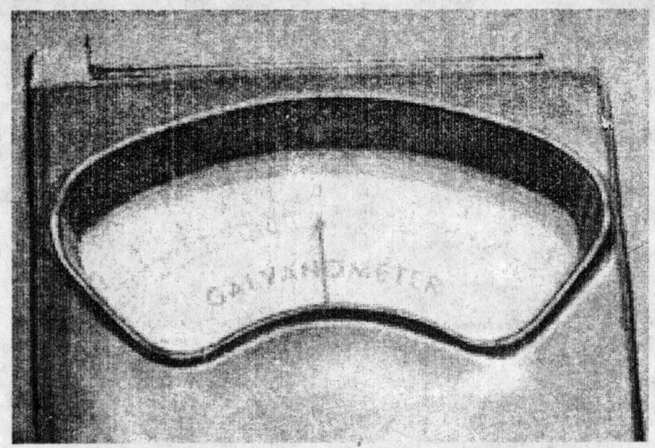

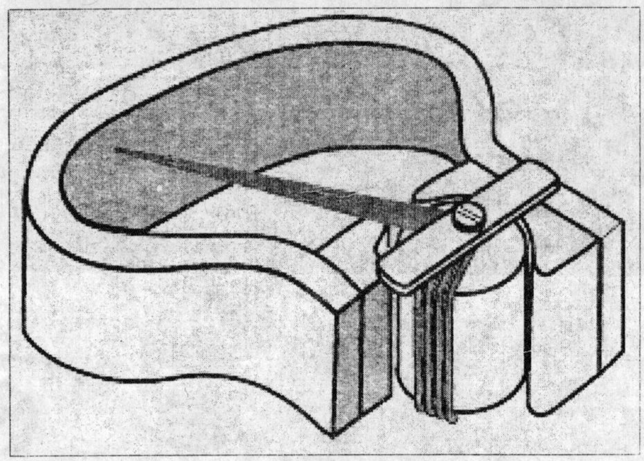

Galvanometer is the historical name given to a moving coil *electric current* detector. When a current is passed through a coil in a magnetic field, the coil experiences a *torque* proportional to the current. If the coil's movement is opposed by a coil spring, then the amount of deflection of a needle attached to the coil may be proportional to the current passing through the coil. Such "meter movements" were at the heart of the *moving coil meters* such as *voltmeters* and *ammeters* until they were largely replaced with solid state meters. The accuracy of moving coil meters is dependent upon having a uniform and constant magnetic field. The illustration shows one configuration of permanent magnet which was widely used in such meters.

Gamma Camera

A device for generating an image of an extended gamma radiation source. The essential features are shown in the diagram. The source, typically a radioactive tracer isotope such as ^{99m}Tc, is attached to a carrier compound which takes it to the organ under test. Gamma radiation which gets through a lead collimator reaches a sodium

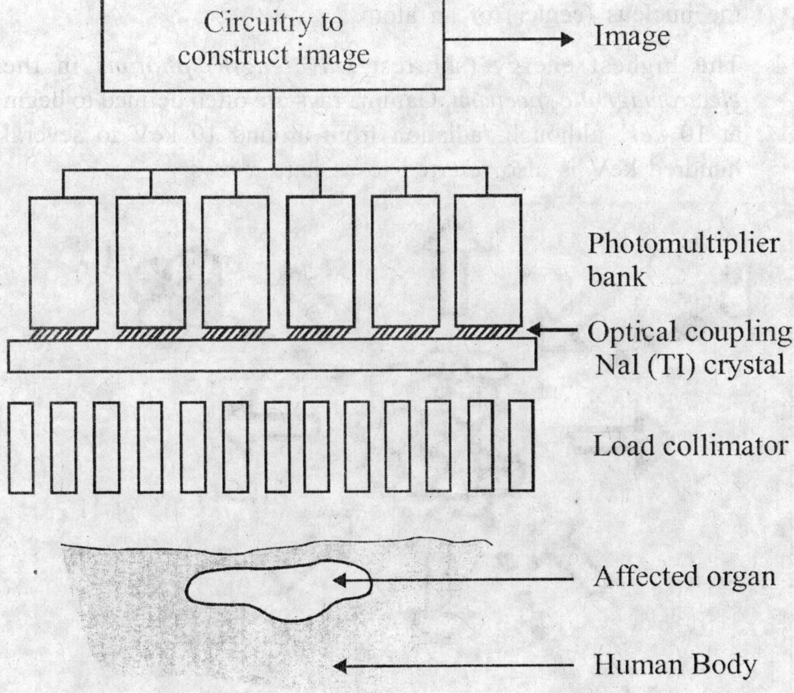

iodide scintillator crystal and is absorbed. A proportion of the absorbed energy re–appears in the form of visible photons. The different readings from the bank of photomultipliers enables the precise position of the gamma emission event to be calculated. The computer uses the information from tens of thousands of such events to build up an accurate picture of the gamma source, from which an image of the organ can be displayed on a screen.

Gamma Ray

1. A highly penetrating type of nuclear radiation, similar to x-radiation, except that it comes from within the nucleus of an atom, and, in general, has a shorter wavelength.
2. Aform of radioactivity consisting of a very high-frequency form of light.
3. Gamma rays are *electromagnetic waves* or photons emitted from the nucleus (center) of an atom.
4. The highest energy (shortest *wavelength*) *photons* in the *electromagnetic spectrum*. Gamma rays are often defined to begin at 10 *keV*, although radiation from around 10 keV to several hundred keV is also referred to as hard *x-rays*.

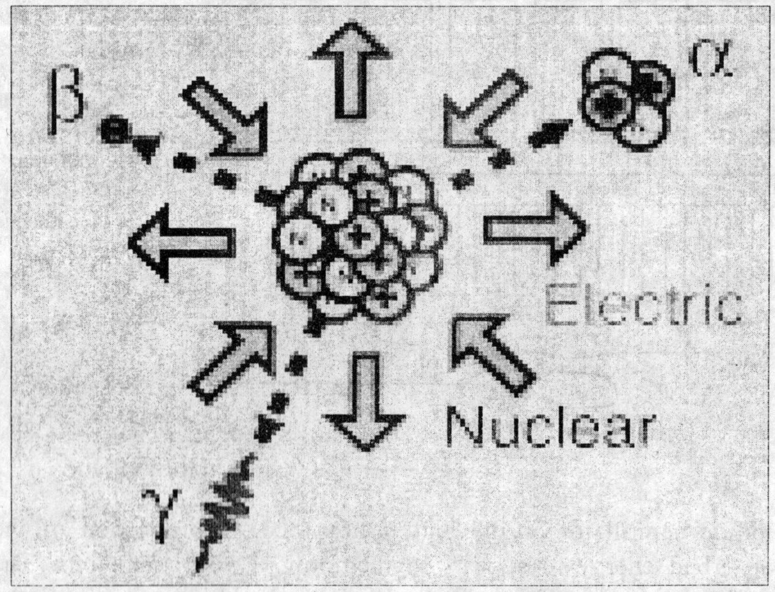

Gamma Radioactivity

Gamma *radioactivity* is composed of *electromagnetic rays*. It is distinguished from *x-rays* only by the fact that it comes from the nucleus. Most gamma rays are somewhat higher in energy than x-rays and therefore are very *penetrating*. It is the most useful type of radiation for medical purposes, but at the same time it is the most dangerous because of its ability to penetrate large thicknesses of material.

Geiger Counter

1. A device that detects the passage of *charged* particles via the *ionization* of gas that they cause as they pass through a region. Used to detect the particles produced in certain forms of radio-activity.

2. A Geiger-Müller detector and measuring instrument. It contains a gas-filled tube which discharges electrically when ionizing radiation passes through it and a device that records the events.

3. A radiation detector consisting of two electrodes with a low-pressure gas in between. A voltage is maintains such that if radiation passing through the counter ionizes the gas, an avalanche of *electrons* will occur. Geiger counters can count radiation but cannot distinguish either the energy or kind of radiation.

General Relativity

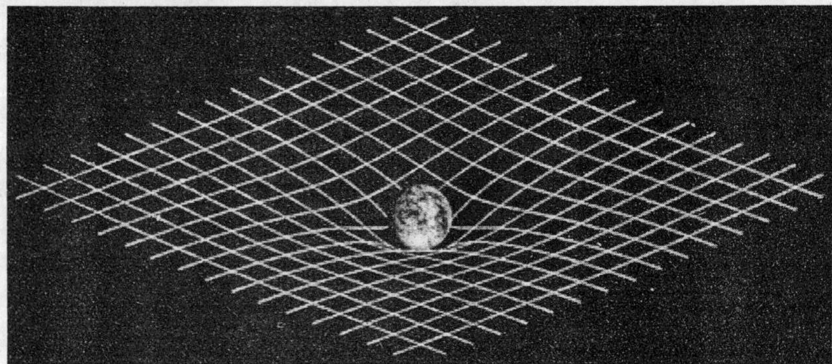

Two-dimensional visualisation of space-time distortion. The presence of matter changes the geometry of spacetime, this (curved) geometry being interpreted as gravity. *General relativity (GR)* or *general relativity theory (GRT)* is a fundamental physical theory of gravitation which corrects and extends Newtonian gravitation, especially at the macroscopic level of stars or planets. General relativity may be regarded as an extension of special relativity, this latter theory correcting Newtonian mechanics at high velocities. General relativity has a unique role amongst physical theories in the sense that it interprets the gravitational field as a geometric phenomena. More specifically, it assumes that any object possessing mass curves the 'space' in which it exists, this curvature being equated to gravity. It deals with the motion of bodies in such 'curved spaces' and has survived every experimental test performed on it since its formulation by Albert Einstein in 1915. General relativity forms the basis for modern studies in fields such as astronomy, cosmology and astrophysics. It describes with great accuracy and precision many phenomena where classical physics fails, such as the perihelion motion of planets (classical physics cannot fully account for the perihelion shift of Mercury, for example) and the bending of starlight by the Sun (again, classical physics can only account for half the experimentally observed bending). It also predicts phenomena such as the existence of gravitational waves, black holes and the expansion of the universe. In fact, even Einstein himself initially believed that the universe cannot be expanding, but experimental observations of distant galaxies by Edwin Hubble finally forced Einstein to concede.

Generation

A set of one of each charge type of quark and lepton, grouped by mass. The first generation contains the up and down quarks, the electron and the electron neutrino.

Generator

1. A cow-a system containing a parent-daughter set of radioisotopes in which the parent decays through a daughter to a stable isotope. The daughter is a different element from that of the parent, and, hence, can be separated from the parent by elution (milking).
2. A device for transforming mechanical energy into electrical energy by means of a coil rotating in a magnetic field. The principle should be clear from the diagram on page 116. A coil constructed from many turns of fine wire is mounted on a spindle by a turning force applied to the pulley. The coil rotates in the magnetic field and the *e.m.f.* induced in the coil is picked up from the two commutator rings. The output voltage is the familiar sine wave alternating voltage the frequency of which is decided by the angular speed of the coil. The magnitude is decided by the strength of the magnetic field, the number of turns in the coil, the area of the coil and (again) the angular speed. The angular speed of the coil governs the output frequency because output voltage goes through one complete cycle each complete turn of the coil. The angular speed influences the output voltage amplitude because an increasing angular speed causes the magnetic flux linkages to change at a faster rate. Also, if the angular speed is zero, the output voltage is zero. All this should be clear form the equation for the output voltage :

$$V = \omega NAB \sin \omega t = V_0 \sin \omega t = V_0 \sin 2\pi f t .$$

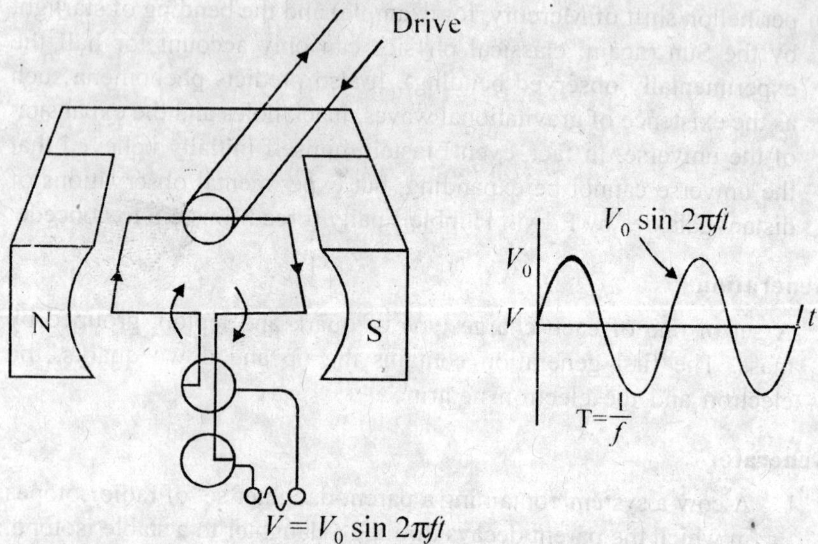

Genetic Effects

Genetic effects are effects from some agent, that are seen in the offspring of the individual who received the agent. The agent must be encountered pre-conception.

Geodesy

A branch of applied mathematics concerned with measuring the shape of the Earth and describing variations in the Earth's gravity field.

Geomagnetic Storm

A worldwide disturbance of the Earth's *magnetic field*, associated with solar activity.

Geographic Information System (GIS)

A system for archiving, retrieving, and manipulating data that has been stored and indexed according to the geographic coordinates of its elements. The system generally can utilize a variety of data types, such as imagery maps, tables, etc.

Geold

A surface of constant gravitational potential around the Earth-an averaged surface per pendicular to the force of gravity.

Geosphere

The physical elements of the Earth's surface, crust, and interior.

Geostationary Meteorological Satellite (GMS)

Japan's geostationary weather satellite.

Geostationary Operational Environmental Satellite (GOES)

NASA-developed, NOAA-operated series of satellites that: provide continuous day and night weather observations; monitor severe weather events such as hurricanes, thunderstorms, and flash floods; relay environmental data from surface collection platforms to a processing center; perform facsimile transmissions of processed weather data to low-cost receiving stations; monitor the Earths' magnetic field, the energetic particle flux in the satellite's vicinity and x-ray emissions from the sun; detect distress signals from downed aircraft and ships. GOES observes the U.S. and adjacent ocean areas from vantage points 35,790 km (22,240 miles) above the equator at 75° west and 35° west. GOES satellites have an equatorial, Earth-synchronous orbit with a 24-hour period, a resolution of 8 km, an IR resolution of 4 km, and a scan rate of 1864 statute miles in about three minutes. See geostadonary The transmission of processed weather data (both visible and infrared) by GOES is called weather facsimile (WEFAX). GOES WEFAX transmits at 1691 + MHz and is accessible via a ground station with a satellite dish antenna. GOES carries the following five major sensor systems: (1) The imager is a multispectral instrument capable of sweeping simultaneously one visible and four infrared channels in a north-to-south swath across an east-to-west path, providing full disk imagery once every thirty minutes. (2) The sounder has more spectral bands than the imager for producing high-quality atmospheric profiles of temperature and moisture. It is capable of stepping one visible and eighteen infrared channels in a north-to-south swath across an east-to-west path. (3) The Space Environment Monitor (SEM) measures the condition of the Earth's magnetic field, the solar activity and radiation around the spacecraft, and transmits these data to a central processing facility. (4) The Data Collection System (DCS) receives transmitted meteorological data from remotelylocated platforms and relays the data to the end users. (5) The Search and Rescue Transponder can relay distress signals at all times, but cannot locate them. While only the polar-orbiting satellite can locate distress signals, the two types of satellites work together to create a comprehensive search and rescue system.

Geostationary

Describes an orbit in which a satellite is always in the same position (appears stationary) with respect to the rotating Earth. The satellite travels around the Earth in the same direction, at an altitude of approximately 35,790km (22,240 statute miles) because that produces an orbital period equal to the period of rotation of the Earth (actually 23 hours, 56 minutes, 04.09 seconds). A worldwide network of operational geostationary meteorological satellites provides visible and infrared images of Earths' surface and atmosphere. The satellite systems include the U.S. GOES, METEOSAT (launched by the European Space Agency and operated by the European Weather Satellite Organization-EUMETSAT), the Japanese GMS, and most commercial, telecommunications satellites.

Geosynchronous (Aka GEO)

Synchronous with respect to the rotation of the Earth.

Geosynchronous Orbit

The orbit of a satellite that travels above the Earth's equator from west to east so that it has a speed matching that of the Earth's rotation and remains stationary in relation to the Earth (also called geostationary). Such an orbit has an altitude of about 35,900 *km* (22,300 miles).

GeV (Giga Electron Volt)

Unit of energy equal to that acquired by a particle with one electronic *charge* in passing through a potential difference of one billion volts.

Glacier

A multi-year surplus accumulation of snowfall in excess of snowmelt on land and resulting in a mass of ice at least 0.1 km2 in area that shows some evidence of movement in response to gravity. A glacier may terminate on land or in water. Glacier ice is the largest reservoir of fresh water on Earth, and second only to the oceans as the largest reservoir of total water Glaciers are found on every continent except Australia.

Global Change Research Program (GCRP)

The USGCRP is a government-wide program whose goal is "to establish a scientific basis for national and international policy-making relating to natural and human-induced changes in the global Earth system." Mission to Planet Earth is NASAs central contribution to the U.S. Global Change Research Program. The Global Change Research Program coordinates and guides the efforts of federal agencies. The program examines such questions as, is the Earth experiencing global warming? Is the depletion of the ozone layer expanding? How do we determine and understand the causes of global climate changes? Are they reversible? What are the implications for human needs and activities?

Global Measurement

All of the activities required to specify a global variable, such as ozone. These activities range from data acquisition to the generation of a data-analysis product, and include estimates of the uncertainties in that product. A global measurement often will consist of a combination of observations from a spacecraft instrument (required for global coverage) and measurements in situ (needed to provide reference points for long-term accuracy).

Global Positioning Satellites

In June of 1993 the last of the 24 satellites of the *Global Positioning System* was placed into orbit, completing a satellite network capable providing position data to locate you anywhere on Earth within 30 meters. The satellites carry up to four cesium and rubidium *atomic clocks*–which are periodically updated from a ground station in Colorado. The satellites transmit timing signals and position data. A *GPS receiver*, which can be a small, hand-held device, decodes the timing signals from several of the satellites, interpreting the arrival times in terms of latitude, longitude, and altitute with an uncertainty which may be as small as 10 meters. In differential mode, accuracies of less than a centimeter can be obtained for distances of hundreds of kilometers. *Hand-held units* read out longitude and latitude to a thousandth of an arcminute and change in the last decimal place in a couple of paces while walking. The satellites are in *orbits* much lower than the *syncom satellites*, orbiting around 17.7 million meters (11000 mi) above the earth, with orbit periods on the order of 10 hours.

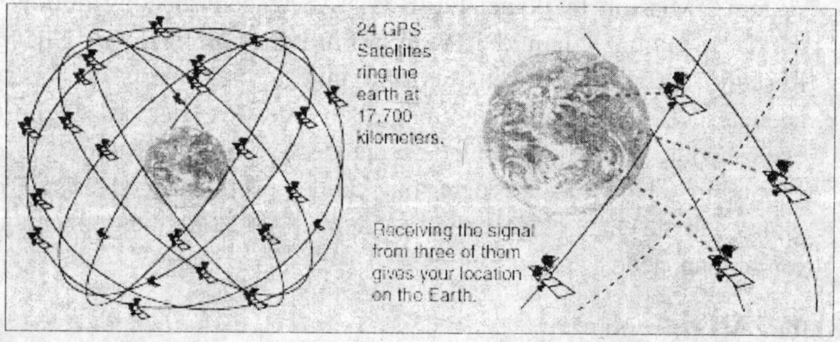

Global Variables

Functions of space and time that describe the large scale state and evolution of the Earth system. The Earth system's geosphere, hydrosphere, atmosphere, and biosphere and their components are, or potentially are, global variables.

Gluon (g)

The carrier particle of the strong interactions.

GOES 1/GOES 8

NOAA geostationary satellite launched in April 1, 994 (alphabetical designators are used while on the ground and before geostationary orbit, after it achieves geostationary orbit it became GOES 8). GOES

8 is the first in a series of five new geostationary satellites that will ensure dual-satellite coverage of the U.S. into the next century and will provide better advanced warnings of thunderstorms, flash floods, hurricanes, and other severe weather. GOES 8 will also contribute important information to a new flood and water management system which will assist decision-makers with the allocation of precious western water resources.

GOES NEXT

The next generation of NOAA geostationary satellites, scheduled for launch beginning sometime after 2003. Currently in the planning phase, these satellites will follow the series of five geostationary satellites which are being launched beginning in 1994.

Gravitational Field

The force per unit mass exerted on a test mass at a given point in space.

Gravitational Interaction

1. An attractive force between any two objects or particles. The "charge" that determines the strength of the gravitational interaction is energy. For a static object it is mass-energy but in fact all forms of energy both cause and feel gravitational effects.

2. The interaction of particles due to their massenergy.

Gravitational Lensing

Consider the example given under *spacetime curvature* where we describe two dimensional creatures living in a bedsheet. In that example we saw how parallel light rays could be caused to meet by passing on either side of a massive object. This tells us that the curvature of spacetime can focus light rays. In effect, the curvature of spacetime acts on light somewhat like a giant convex lens extending around the massive object. Because of this effect, sensitive devices sometimes see two images of an astronomical object. A real image, formed by light rays travelling without significant deflection to Earth, and another image which is formed by light rays that pass near a massive object and then toward the earth. Although this effect is very small, its significance is magnified by the great distances involved.

Gravitational Radiation

The energy that is emitted by strong sources of *gravitational waves*, for example, certain collapsing or colliding stars.

Gravitational Waves

Think about the example described under *spacetime curvature* in which we have two dimensional creatures living on the surface of a bedsheet. Now imagine that a physics professor grabs one end of the bedsheet and begins to shake it violently up and down. This will cause ripples to travel through the fabric. The imaginary creatures within the bedsheet will not be able to see what is happening, but they they will be able to measure the time variation in the geometry of their space. The wave travelling on the bedsheet is analagous to a gravitational wave in our universe, the difference being that our universe exhibits three spatial dimensions not two! Gravitational waves have never been observed directly, but scientists hope to detect them soon with extremely sensitive instruments now under construction.

Graviton

1. The carrier particle of the gravitational interactions; not yet directly observed.
2. The graviton is the exchange particle for the gravity force. Although it has not been directly observed, a number of its properties can be implied from the nature of the force. Since gravity is an inverse square force of apparently infinite range, it can be implied that the rest mass of the graviton is zero.

Gravity

A general term for the phenomenon of attraction between things having mass. The attraction between our planet and a human-sized object causes the object to fall.

Gray (Gy)

1. The SI unit for absorbed dose equal to an energy absorbed of one joule/kilogram in the stated medium.

2. The gray is a unit used to measure a quantity called absorbed dose. This relates to the amount of energy actually absorbed in some material, and is used for any type of radiation and any material. One gray is equal to one joule of energy deposited in one kg of a material. The unit gray can be used for any type of radiation, but it does not't describe the biological effects of the different radiations. Absorbed dose is often expressed in terms of hundredths of a gray, or centi-grays. One gray is equivalent to 100 rads.

Grayscale

Environmental satellite scanners, rather than photographing a scene, scan a scene line-byline measuring light or heat levels and transmitting this information as a video image via an amplitude modulated (AM) subcarrier contained in the satellite's FM signal.

Greenhouse Effect

Process by which significant changes in the chemistry of Earth's atmosphere may enhance the natural process that warms our planet and elevates temperatures. If the effect is intensified and Earth's average temperatures change, a number of plant and animal species could be threatened with extinction. Certain gaseous components of the atmosphere, called greenhouse gases, transmit the visible portion of solar radiation but absorb specific spectral bands of thermal radiation emitted by the Earth. The theory is that terrain absorbs radiation, heats up, and emits longer wavelength thermal radiation that is prevented from escaping into space by the blanket of carbon dioxide and other greenhouse gases in the atmosphere. As a result, the climate warms. Because atmospheric and oceanic circulations play a central role in the climate of the Earth, improving our knowledge about their interaction becomes essential.

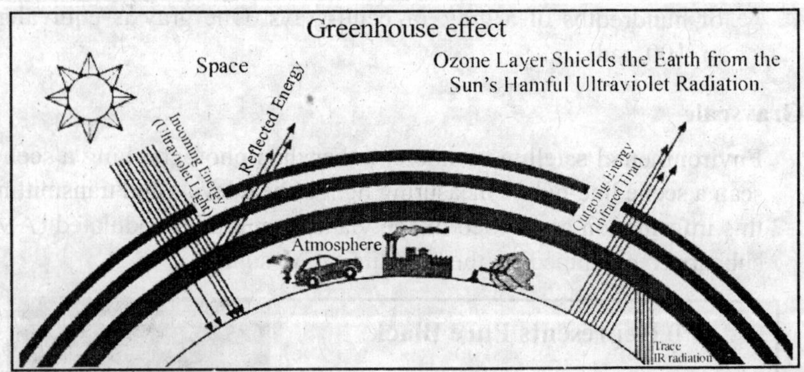

Greenhouse Gas

A gaseous component of the atmosphere contributing to the greenhouse effect. Greenhouse gases are transparent to certain wavlengths of the suns' radiant energy allowing them to penetrate deep into the atmosphere or all the way into the Earths' surface.

KEY GREENHOUSE GASES

Water Vapor

Ozone

Carbon dioxide

Methane

Nitrous oxide

Cholofluorocarbons (CFCs)

and clouds prevent some of infrared radiation from escaping, trapping the heat near the Earths' surface where it warms the lower atmosphere. Alteration of this natural barrier of atmospheric gases can raise or lower the mean global temperature of the Earth. Greenhouse gases include carbon dioxide. methane, nitrous oxide, chloro fluorocarbons, and water vapor Carbon dioxide, methane, and nitrous oxide have significant natural and human sources while only industries produce chlorofluorocarbons. Water vapor has the largest greenhouse effect, but its concentration in the troposphere is determined within the climate system. Water vapor will increase in response to global warming, which in turn may further enhance global warming.

Gross Feature Map

Map that displays geographic characteristics rather than political boundaries.

Ground Control (Points)

Identifiable points on the ground whose locations on the surface of the Earth are accurately known for use as geodetic references in mapping, charting, and other related mensuration applications.

Ground State

A lowest energy state of the *nucleus*.

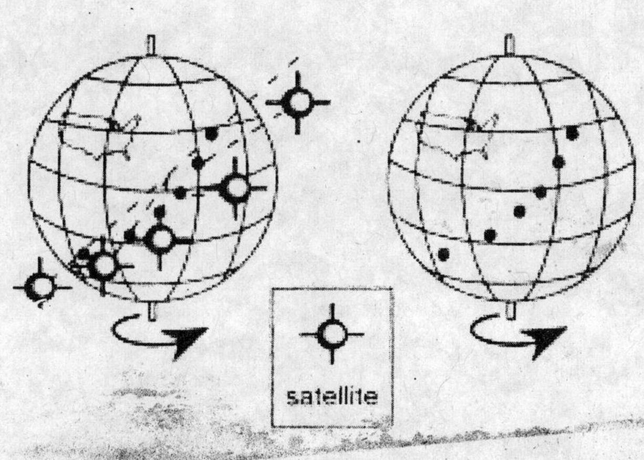

Ground Track

The inclination of a satellite, together with its orbital altitude and the period of its orbit, creates a track defined by an imaginary line connecting the satellite and the Earth's center; The intersection on the line with the Earth's surface is the subsatellite point. As the Earth turns on its axis and the satellite orbits overhead, a line is created by the satellite's apparent path over the ground (the series of subsatellite points connected). A geostationary satellite has an inclination of essentially zero, and, because its orbital period exactly matches the Earth's rotation, its ground track is reduced to an apparent stationary point on the equator.

Guided Wave

Electromagnetic or acoustic wave that is constrained within certain boundaries, as in a wave guide (transmission line).

Gulf Stream

A warm, swift ocean current that flows along the coast of the Eastern United States and makes Ireland, Great Britain, and the Scandinavian countries warmer than they would be otherwise.

Gulf

A large arm of an ocean or sea extending into a land mass.

H

Habitat

The area or region where a particular type of plant or animal lives and grows.

Hadron

1. A particle made of strongly-interacting constituents (quarks and/or gluons). These include the mesons and baryons. Such particles participate in residual strong interactions.

2. Any particle made of quarks and *gluons*, i.e. a meson or a *baryon*. All such particles have no strong charge (i.e. are strong charge neutral objects) but participate in residual strong *interactions* due to the strong charges of their constituents.

3. Particles that interact by the *strong interaction* are called hadrons. This general classification includes *mesons* and *baryons* but specifically excludes *leptons,* which do not interact by the strong force. The *weak interaction* acts on both hadrons and leptons. Hadrons are viewed as being composed of *quarks,* either as quark-antiquark pairs (mesons) or as three quarks (baryons). There is much more to the picture than this, however, because the constituent quarks are surrounded by a cloud of *gluons*, the exchange particles for the *color force*.

Hail

Precipitation composed of balls or irregular lumps of ice. Hail is produced when large frozen raindrops, or almost any particles, in cumulonimbus clouds act as embryos that grow by accumulating supercooled liquid droplets. Violent updrafts in the cloud carry the particles in freezing air; allowing the frozen core to accumulate more ice. When the piece of hail becomes too heavy to be carried by upsurging air currents it falls to the ground.

Half-life

1. The amount of time that a radioactive atom has a probability of 1/2 of surviving without decaying.
2. The time in which half the atoms of a particular radioactive nuclide disintegrate. The half-life is a characteristic property of each radio active isotope.
3. The time required for half the nuclei in a sample of a specific isotopic species to undergo radioactive decay.
4. Used to measure the rate of radioactive decay of disintegration. The time lapse during which a radioactive mass loses one half of its radioactivity.

Hall Effect

A difference in potential V_H set up in a conductor at right angles to the current flow direction by a magnetic field *B*. The diagram shows a thin rectangular conductor with the current *I* flowing parallel to the

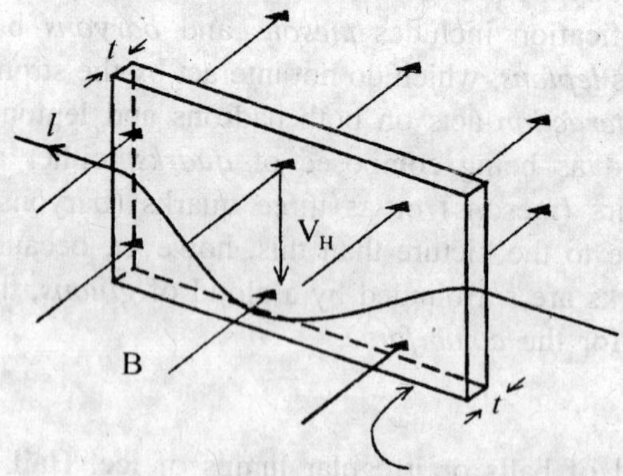

Specimen with n carriers per limit volume
(V_H is correct for electrons as carries)

long side. A magnetic field of strength B is switched on at right angles to the face of the specimen with the largest area. The force on the current in the magnetic field causes it to be deflected downwards; check this with *Fleming's left hand rule.* The concentration of charge on the lower edge sets up a voltage which opposes further deflection of the moving charges. Equilibrium is reached when :

$$V_{\mathrm{H}} = \frac{Bl}{nqt}$$

where n is the number of charge carriers per unit volume of conductor, q is the charge per carrier and t is the thickness of the specimen in the direction of the magnetic field.

Hamiltonian (Quantum Mechanics)

The *Hamiltonian*, denoted H, has two distinct but closely related meanings. In classical mechanics, it is a function that describes the state of a mechanical system in terms of position and momentum variables (i.e. symplectic variables), which is the basis for a re-formulation of classical mechanics known as Hamiltonian mechanics. In quantum mechanics, the Hamiltonian is the observable corresponding to the total energy of a system.

Hamiltonian Mechanics

Hamiltonian mechanics is a re-formulation of classical mechanics that was invented in 1833 by William Rowan Hamilton. It arose from Lagrangian mechanics, another re-formulation of classical mechanics, introduced by Joseph Louis Lagrange in 1788. It can however be formulated without recourse to Lagrangian mechanics, using symplectic spaces. See the section on its mathematical formulation for this. Starting with Lagrangian mechanics, the equations of motion are based on generalized coordinates and matching generalized velocities Abusing the notation, we write the Lagrangian as : $L(q_j, q_j, t)$, with the subscripted variables understood to represent all N variables of that type. Hamiltonian mechanics aims to replace the generalized velocity variables with generalized momentum variables, also known as *conjugate momenta*. By doing so, it is possible to handle certain systems, such as aspects of quantum mechanics that would otherwise be even more complicated. For each generalized velocity, there is one corresponding conjugate momentum, defined as:

$$p_j = \frac{\partial L}{\partial q_j}.$$

In Cartesian coordinates the generalized momenta are precisely the physical linear momenta. In circular polar coordinates, the generalized momentum corresponding to the angular velocity is the physical angular momentum. For an arbitrary choice of generalized coordinates, it may not be possible to obtain an intuitive interpretation of the conjugate momenta.

Hardware

The electrical and mechanical components of a system, as opposed to software.

Haze

Fine dry or wet particles of dust, salt, or other impurities that can concentrate in a layer next to the Earth when air is stable.

Health Physics

1. That science devoted to recognition, evaluation, and control of all health hazards from ionizing radiation.
2. Health Physics is an interdisciplinary science and its application, for the radiation protection of humans and the environment. Health Physics combines the elements of physics, biology, chemistry, statistics and electronic instrumentation to provide information that can be used to protect individuals from the effects of radiation.

Heat

1. Heat, like work, is a measure of the amount of energy *transferred* from one body to another because of the temperature difference

between those bodies. Heat is *not* energy *possessed* by a body. We should *not* speak of the 'heat *in* a body.' The energy a body possesses due to its temperature is a different thing, called *internal thermal energy*. The misuse of this word probably dates back to the 18th century when it was still thought that bodies undergoing thermal processes exchanged a substance, called *caloric* or *phlogiston*, a substance later called *heat*. We now know that heat is not a substance.

2. The energy that an object has because of its temperature. Heat is different from temperature (q.v.) because an object with twice as much mass requires twice as much heat to increase its temperature by the same amount. There is a further distinction in the terminology, not emphasized in this book, between heat and thermal energy. See the entry under thermal energy for a discussion of this distinction.

3. *Heat* (abbreviated *Q*, also called heat change) is the transfer of thermal energy between two bodies which are at different temperatures. The SI unit for heat is the joule. The relationship between *heat* and energy is similar to that between work and energy. Heat flows between regions that are not in thermal equilibrium; in particular, it flows from areas of high temperature to areas of low temperature. All objects (matter) have a certain amount of internal energy that is related to the random motion of their atoms or molecules. This internal energy is directly proportional to the temperature of the object. When two bodies of different temperature come into thermal contact, they will exchange internal energy until the temperature is equalized. The amount of energy transferred is the amount of heat exchanged. It is a common misconception to confuse heat with internal energy, but there is a difference: heat is related to the change in internal energy and the work performed by the system. The term heat is used to describe the *flow* of energy; while the term internal energy is used to describe the energy itself. Understanding this difference is a necessary part of understanding the first law of thermodynamics. Infrared radiation is often linked to heat,

since objects at room temperature or above will emit radiation mostly concentrated in the mid-infrared band

Heat Balance

The equilibrium existing between the radiation received and emitted by a planetary system.

Heat Capacity Mapping Mission (HCMM)

A two-channel radiometer launched by NASA to measure the thermal properties of the terrestrial surface. It had an application to identify and locate rocks and minerals. One radiometer channel was in the visible to near infrared (0.5-1.1 micrometers), and the other in the thermal infrared (10.5-12.5 micrometers). The instantaneous field of view (IEOV) was about 600 meters.

Heat Dissipation

In cold climates, houses with their heating systems form dissipative systems. In spite of efforts to insulate such houses, to reduce heat losses to their exteriors, considerable heat is lost, or dissipated, from them which would make their interiors uncomfortably cool or cold. The house is an open system inasmuch as it is incapable of preventing heat from escaping. Furthermore, the interior of the house must be maintained out of thermal equilibrium with its exterior for the sake of its inhabitants. In such a house, a thermostat is a device capable of starting the heating system when the house's interior falls to a set temperature, and of stopping that same system when another set temperature has been achieved. Thus the thermostat controls the flow of energy into the house, that energy eventually being dissipated to the exterior.

Heat Engine

A *heat engine* performs the conversion of heat energy to work by exploiting the temperature gradient between a hot "source and a cold "sink. Heat is transferred to the sink from the source, and in this process some of the heat is converted into work. The theoretical maximum efficiency of any heat engine is defined by the Carnot

Cycle. The carnot heat engine (the ideal imaginary heat engine) has an efficiency equal to $(T_1 - T_2)/T_1$ where T_1 is the absolute temperature of the hot source and T_2 that of the cold sink. Examples of everyday heat engines include: the steam engine the diesel engine and the gasoline (petrol) engine in an automobile. All of these familiar heat engines are powered by the expansion of heated gases. The general surroundings are the heat sink, providing relatively cool gases which when heated, expand rapidly to drive the mechanical motion of the engine.

Heat Transfer

The transfer of heat is normally from a high temperature object to a lower temperature object. Heat transfer changes the internal energy of both systems involved according to the First Law of Thermodynamics.

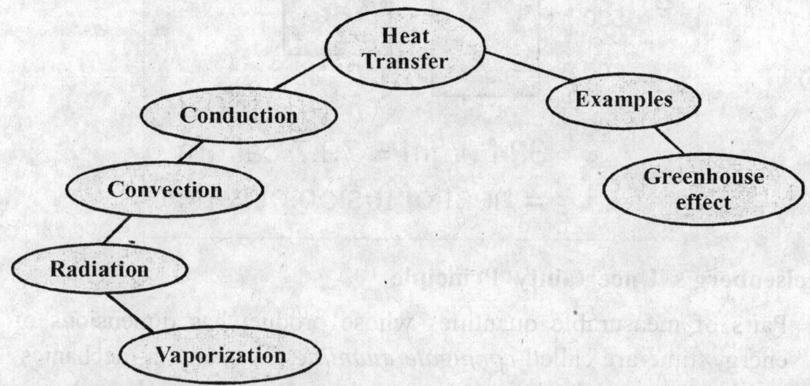

Heat Transfer Mechanisms

As mentioned previously, heat tends to move from a high temperature region to a low temperature region. This heat transfer may occur by the mechanisms conduction and radiation. The term convection is used to describe the combined effects of conduction and fluid flow. In the past, this has been regarded as a third mechanism of heat transfer, but, logically, it is not a mechanism of its own.

Heat of Fusion

The energy required to change a gram of a substance from the solid to the liquid state without changing its temperature is commonly called it's "heat of fusion". This energy breaks down the solid bonds, but leaves a significant amount of energy associated with the intermolecular forces of the liquid state.

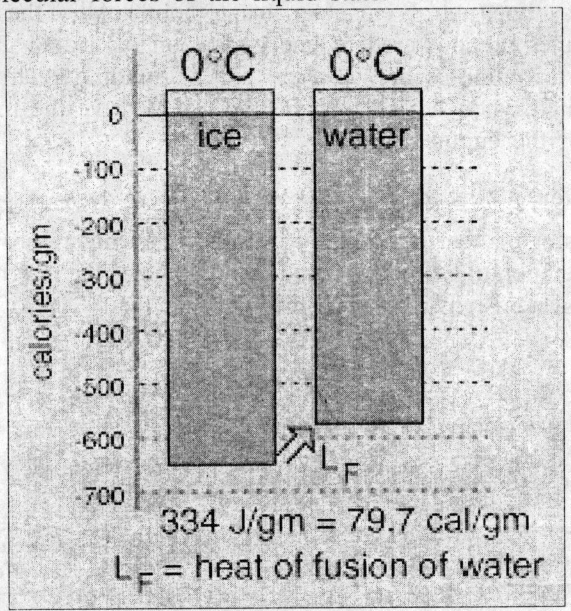

Heisenberg's Uncertainty Principle

Pairs of measurable quantities whose product has dimensions of energy×time are called *conjugate quantities* in quantum mechanics, and have a special relation to each other, expressed in Heisenberg's uncertainty principle. It says that the product of the uncertainties of the two quantities is no smaller than h/2. Thus if you improve the measurement precision of one quantity the precision of the other gets worse.

Helical Spring

A spring formed by winding a wire round and round a uniform cylinder. The usefulness of a helical spring is linked to its main property : its extension is proportional to the load applied. This relationship is illustrated by the graph in the diagram. The proportionality of extension to load is called *Hooke's low*. It can be stated in algebraic form : $F = kx$ Where k is called the spring constant, F is the stretching force and x is the extension. A typical spring in a school laboratory will be stretched by about 5 cm by the weight of a 100 g mass. The weight of a 100 g mass is about one newton and 5 cm is a twentieth of a metre, so the resulting spring constant is in the region of 20 N m^{-1}. It can be checked by suspending the spring from a rigid support and measuring carefully the extensions corresponding to different loads. The gradient of the load/extension graph is the spring constant k.

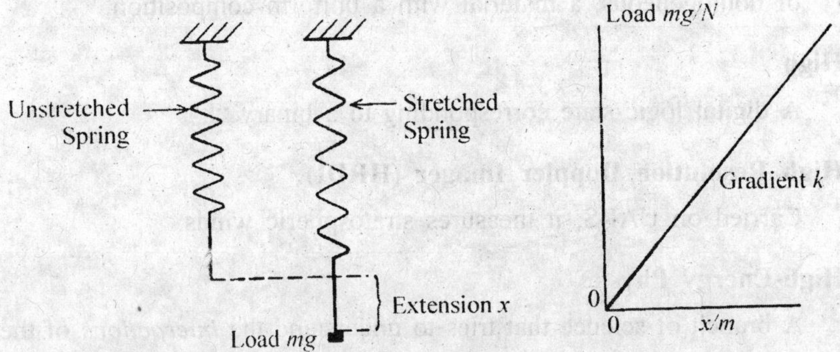

Helium Burning

When temperature in the core of a star reaches 100 million degrees, three colliding helium nuclei fuse to form a carbon nucleus. This process occurs when the star is a red giant.

Hemisphere

Half of the Earth, usually conceived as resulting from the division of the globe into two equal parts, north and south or east and west.

Hertz

1. The international unit of frequency equal to one cycle per second. Radio frequencies are usually expressed in kilohertz/kHz (1,000 cycles per second) or megahertz/MHz (1,000,000 cycles per second).

2. Abbreviated *Hz*. A unit of frequency equal to one cycle per second. One kHz = 1000 Hz. One MHz = 10^6 (one million) Hz. One GHz = 10^9 Hz.

Hertzian Waves

Radio waves or other electromagnetic radiation resulting from the oscillations of electricity in a conductor.

Heterogeneous

Materials made from more than one type of substructure. Opposite of homogeneous, a material with a uniform composition.

High

A digital logic state corresponding to a binary "1."

High Resolution Doppler Imager (HRDI)

Carried on UARS, it measures stratospheric winds.

High-Energy Physics

A branch of science that tries to understand the *interactions* of the *fundamental particles*, such as *electrons,* photons, neutrons and protons (and many others than can be created). These particles are the basic building blocks of everyday matter, making up the human body as well as the entire universe. This type of physics is called *high-energy* because very powerful machines, such as the Two-Mile Accelerator at SLAC, are created to make these particles go very fast so that they can probe deeply into other particles and try to understand what they are made of.

High-Resolution Infrared Radiation Sounder (HIRS)

Instrument carried by NOAA polar-orbiting satellites that detects and measures energy emitted by the atmosphere to construct a vertical temperature profile from the Earth's surface to an altitude of about 40 km. Measurements are made in 20 spectral regions in the infrared band.

High-Resolution Picture Transmission (HRPT)

Real-time, 1.1-kilometer resolution, digital images provided by NOAAs polar-orbiting environmental satellites, containing all five spectral channels and telemetry data transmitted as high-speed digital transmissions. The Advanced Very High Resolution Radiometer (AVHRR) provides the primary imaging system for APT and HRPT.

Horse Latitudes

The subtropical latitudes (30-35 degrees), where winds are light and weather is hot and dry. According to legend, ships traveling to the New World often stagnated in this region and had to throw dead horses overboard or eat them to survive, hence the name horse latitudes.

Hubble Constant

The proportionality between recession velocity and distance in the Hubble Law is called the Hubble constant, but the value of this proportionality is a source of disagreement. The values proposed by different experimental and theoretical groups range from 50 to 100 km/sec per megaparsec. The lower extreme would point to an age of about 20 billion years, while the higher value indicates about half of that age. The recession velocities of distant galaxies are known from the red shift, but the distances are much more uncertain. Distance measurement to nearby galaxies uses Cepheid variables as the main standard candle, but more distant galaxies must be examined to determine the Hubble constant since the direct Cepheid distances are all within the range of the gravitational pull of the local cluster. Use of the Hubble Space Telescope has permitted the detection of Cepheid variables in the Virgo cluster and may contribute to settling the distance scale dispute. The Particle Data Group documents quote a "best modern value" of the Hubble constant as 72 km/s per megaparsec (+/- 10%). This value comes from the use of type Ia supernovae (which give relative distances to about 5%) along with data from Cepheid variables gathered by the Hubble Space Telescope. The value from the WMAP survey is 71 km/s per megaparsec.

Humidity

The amount of water vapor in the air. The higher the temperature, the greater the number of water molecules the air can hold. For example: at 60 ºF (15 ºC), a cube of air one yard on each side can hold up to 4.48 ounces of water. At 104 ºF (40 ºC), the same cube of air can hold up to 17.9 ounces of water. Relative humidity describes the amount of water in the air compared with how much the air can hold at the current temperature. Example: 50% relative humidity means the air holds half the water vapor that it is capable of holding; 100% relative humidity means the air holds all the water vapor it can. At 100% humidity no more evaporation can occur until the temperature rises, or until the water vapor leaves the air through condensation. Absolute humidity is the ratio of the mass of water vapor present in a system of moist air to the volume occupied by the mixture, that is, the density of water vapor.

Hurricanes

Severe tropical storms whose winds exceed 74 mph. Hurricanes originate over the tropical and subtropical North Atlantic and North Pacific oceans, where there is high humidity and light wind. These conditions prevail mostly in the summer and early tall. Since hurricanes can take days or even weeks to form, time is usually available for preventive or protective measures. From space, hurricanes look like giant pinwheels, their winds circulating around an eye that is between 5 and 25 miles in diameter. The eye remains calm with light winds and often a clear sky. Hurricanes may move as fast as 50 mph, and can become incredibly destructive when they hit land. Although hurricanes lose power rapidly as soon as they leave the ocean, they can cause high waves and tides up to 25 feet above normal. Waves and heavy flooding cause the most deaths during a hurricane. The strongest hurricanes can cause tornadoes.

Hydrochloroflourocarbon (HCFC)

One of a class of compounds used primarily as a CFC substitute. Work on CFC alternatives began in the late 1970s after the first warnings of CFC damage to stratospheric ozone. By adding hydrogen to the chemical formulation, chemists made CFCs less stable in the lower atmosphere enabling them to break down before reaching the ozone layer. However, HCFCs do release chlorine and have contributed more to atmospheric chlorine buildup than originally predicted. Development of non-chlorine based chemical compounds as a substitute for CFCs and HCFCs continues.

Hydrogen-Helium Abundance

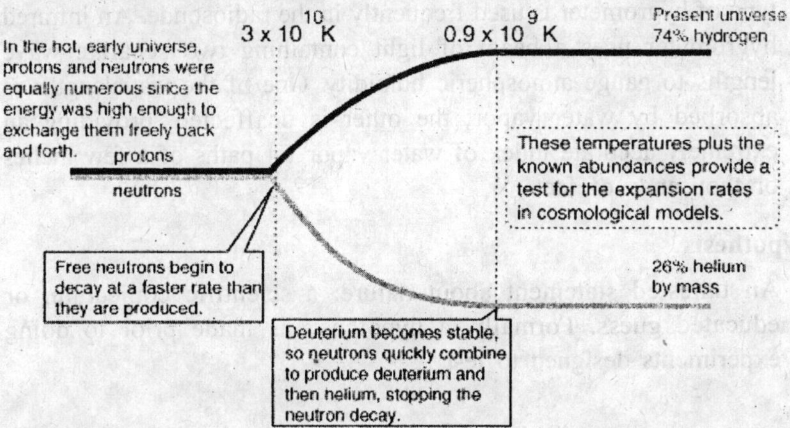

Hydrologic Cycle

The pathways through which water is cycled in the terrestrial biosphere.

Hydromagnetic Wave

A wave in which both the *plasma* and *magnetic field* oscillate.

Hydrosphere

The totality of water encompassing the Earth, comprising all the bodies of water; ice, and water vapor in the atmosphere.

Hygrometer

Instrument that measures water vapor content in the air and communicates changes in humidity visibly and immediately through a graph or a dial. There are three types of hygrometers: The hair hygrometer uses a human hair as the sensing instrument. The hair lengthens when the air is moist and contracts when the air is dry but remains unaffected by air temperature. However; the hair hygrometer cannot respond to rapid fluctuations in humidity. An electric hygrometer uses a plate coated with carbon. Electrical resistance of the carbon coating changes as the moisture content of the air changes-changes that translate into relative humidity. This type of hygrometer is used frequently in the radiosonde. An infrared hygrometer uses a beam of light containing two separate wave lengths to gauge atmospheric humidity. One of the wavelengths is absorbed by water vapor; the other is unaffected, providing an extremely accurate index of water vapor for paths of a few inches or thousands of feet.

Hypothesis

An untested statement about nature; a scientific conjecture, or educated guess. Formally, a hypothesis is made prior to doing experiments designed to test it.

I

Ice Shelf

A thick mass of ice extending from a polar shore. The seaward edge is afloat and sometimes extends hundreds of miles into the sea.

Ideal-lens Quation

$1/p + 1/q = 1/f$, where p is the distance from object to lens, q is the distance from lens to image, and f is the focal length of the lens. This equation has important limitations, being only valid for *thin* lenses, and for *paraxial rays.* Thin lenses have thickness small compared to p, q, and f. Paraxial rays are those which make angles small enough with the optic axis that the approximation *(angle in radian measure) = sin(angle)* may be used.

Identical Particles

Identical particles, or indistinguishable particles, are particles that cannot be distinguished from one another, even in principle. Species of identical particles include elementary particles such as electrons, as well as composite microscopic particles such as atoms. There are two main categories of identical particles: bosons, which can share quantum states, and fermions, which are forbidden from sharing quantum states (this property of fermions is known as the Pauli exclusion principle.) Examples of bosons are photons, gluons, phonons, and helium-4 atoms. Examples of fermions are electrons, neutrinos, quarks, protons and neutrons, and helium-3 atoms. The fact that particles can be identical has important consequences in statistical mechanics. Calculations in statistical mechanics rely on probabilistic arguments, which are sensitive to whether or not the objects being studied are identical. As a result, identical particles exhibit markedly different statistical behavior from distinguishable particles.

IFOV

Instantaneous Field of View.

Image

(Optics) A surprising number of physics glossaries omit a definition of this! No wonder. It's difficult to put in a few words, and still be comprehensive in scope. Try this. Image: A point mapping of luminous points of an object located in one region of space to points in another region of space, formed by refraction or reflection of light in a manner which causes light from each point of the object to converge to or diverge from a point somewhere else (on the image). The images which are useful generally have the character that adjacent points of the object map to adjacent points of the image without discontinuity, and is a recognizable (though perhaps somewhat distorted) mapping of the object.

Image

1. A place where an object appears to be, because the rays diffusely reflected from any given point on the object have been bent so that they come back together and then spread out again from the image point, or spread apart as if they had originated from the image.

2. Pictorial representation of data acquired by satellite systems, such as direct readout images from environmental satellites. An image is not a photograph. An image is composed of two-dimensional grids of individual picture elements (pixels). Each pixel has a numeric value that corresponds to the radiance or temperature of the specific ground area it depicts.

Image-Res

The area represented by each pixel of a satellite image. The smaller the area represented by a pixel, the more accurate and detailed the image. For example, if a U.S. map and a world map are printed on identically sized sheets of paper, one square inch on the U.S. map will represent far less area and provide for more detail than one square inch on the world map. In this example the U.S. map has higher resolution. APT has a resolution of 4 km, HRPT has a resolution of 1.1 km and WEFAX resolution is 8 km.

Imager

A satellite instrument that measures and maps the Earth and its atmosphere. Imager data are converted by computer into pictures.

Impedance

In electrical engineering, *impedance* is a measure for the manner and degree a component resists the flow of electrical current if a given voltage is applied. It is denoted by the symbol *Z* and is measured in ohms: Impedance differs from simple *resistance* in that it takes into account possible phase offset. The concept of impedance also has significance outside of electrical systems in the discussion of any driven oscillator.

Impulse

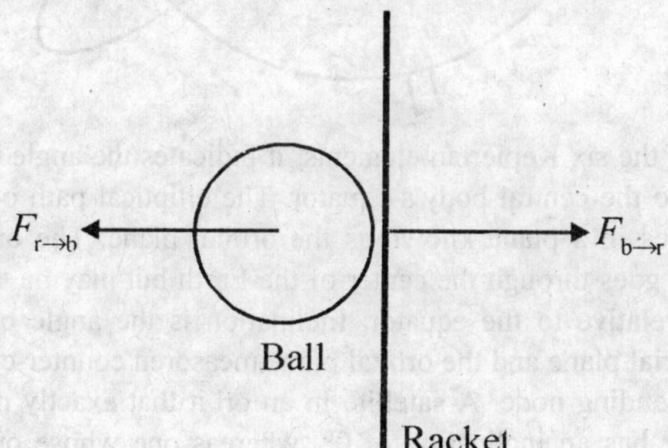

Impulse is the product of an average force and a time interval, it is not a kind of force. When a tennis ball is struck with a racket, the force on the ball changes magnitude rapidly and the impact is all over in a moment. How can we deal with such a situation? At any instant during the collision the force the racket applies to the ball is matched by the reaction force from the ball on the racket. Represent these two forces with symbols $F_{r \to b}$ for the force of the racket on the ball and $F_{b®r}$ for the reaction force of the ball on the racket.

In Situ

Latin for "in original place." Refers to measurements made at the actual location of the object or material measured.

Inclination (Aka I)

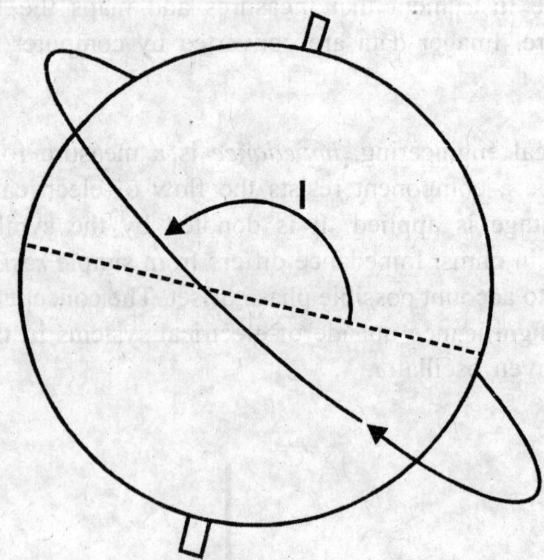

One of the six Keplerian elements, it indicates the angle of the orbit plane to the central body's equator. The elliptical path of a satellite orbit lies in a plane known as the orbital plane. The orbital plane always goes through the center of the Earth but may be tilted at any angle relative to the equator. Inclination is the angle between the equatorial plane and the orbital plane measured counter-clockwise at the ascending node. A satellite in an orbit that exactly matches the equator has an inclination of 0°, whereas one whose orbit crosses the Earth's poles has an inclination of 90°. Because the angle is measured in a counterclockwise direction, it is quite possible for a satellite to have an inclination of more than 90°. An inclination of 180° would mean the satellite is orbiting the equator; but in the opposite direction of the Earths' rotation. Some sun-synchronous satellites that maintain the same ground track throughout the year have inclinations of as much as 98°. U.S. scientific satellites that study the sun are placed in orbits closer to the equator; frequently at 28° inclination. Most weather satellites are placed in high-inclination orbits so they can oversee weather conditions worldwide.

Independence

The lack of any relationship between two random events.

Independent Variable

When two variables are related, one of the variable is free to roam so it is called the *independent variable*. The other variable depends on the first, so it gets the name *dependent*. In a scientific experiment, the experimenter *chooses* values for the independent variable, runs the experiment, and then measures the dependent variable. Ordered pairs of choosen and measured values (independent and dependent) are often plotted on a two dimensional graph for visualization. The independent variable is traditionally plotted on the horizontal axis.

Index of Refraction

An optical property of matter; the speed of light in a vacuum divided by the speed of light in the substance in question.

Induced Radioactivity

1. Radioactivity produced in materials, especially metals, exposed to high-energy photons or neutrons.
2. Radioactivity that is created by bombarding a substance with neutrons in a reactor or with charged particles produced by particle accelerators.

Inductance

Inductance is a physical characteristic of an inductor, which is an electrical device that produces at any time a voltage proportional to the instantaneous rate of change in current flowing through it. The symbol L is used for inductance in honour of the physicist Heinrich Lenz. The SI unit of inductance is the henry (H). In a typical inductor, whose geometry and physical properties are fixed, the voltage generated is as follows:

$$v = -L\frac{di}{dt}$$

where v is the voltage generated, measured in volts L is the *inductance* of the device, measured in henry. di/dt is the rate of change of current, measured in ampere /second Strictly speaking, the quantity just defined is called *self-inductance*, because the voltage is induced in the same conductor that carries the current. If the voltage is induced in another nearby conductor, the property is called *mutual inductance*, which has the symbol M. The above equation, with either L or M as the constant, applies to both cases. The operation of an inductor can be understood using a simple loop of wire as an example. The current flowing through the loop of wire produces a magnetic field by Ampere's law.A change in current (di/dt) results in a change in this magnetic field. This changing magnetic field causes an electromotive force, that some refer to as a counter-electromotive force because it runs against the current that induces it, in the conductor under Faraday's law of induction, which results in a voltage (v) forming in such a direction as to oppose the change in current. The constant of proportionality L, which tells us for a particular device how big a voltage should be expected for a given change in current, is called the inductance.

Induction

The production of an electric field by a changing magnetic field, or vice-versa.

Inertia

A descriptive term for that property of a body which resists change in its motion. Two kinds of changes of motion are recognized: changes in translational motion, and changes in rotational motion. In modern usage, the measure of translational inertia is mass. Newton's first law of motion is sometimes called the 'Law of Inertia', a label which adds nothing to the meaning of the first law. Newton's first and second laws together are required for a full description of the consequences of a body's inertia. The measure of a body's resistance to rotation is its *Moment of Inertia.*

Inertial Frame

1. A frame of reference that is not accelerating, one in which Newton's first law is true.

2. A non-accelerating coordinate system. One in which $F = ma$ holds, where F is the sum of all real forces acting on a body of mass m whose acceleration is $\boldsymbol{a}$. In classical mechanics, the *real forces* on a body are those which are due to the influence of another body. [Or, forces on a part of a body due to other parts of that body.] Contact forces, gravitational, electric, and magnetic forces are real. *Fictitious forces* are those which arise solely from formulating a problem in a non-inertial system, in which $ma = F +$ *(fictitious force terms)*

Information System

All of the means and mechanisms for data receipt, processing, storage, retrieval, and analysis. Information systems can be designed for storage and dissemination of a variety of data products-including primary data sets and both intermediate and final analysesand for an interface providing connections to external computers, external data banks, and system users. To be effective, the design and operation of an information system must be carried out in close association with the primary producers of the data sets, as well as other groups producing integrated analyses or intermediate products.

Infrared Radiation (IR)

Infrared is electromagnetic radiation whose wavelength spans the region from about 0.7 to 1000 micrometers (longer than visible radiation, shorter than microwave radiation). Remote-sensing instruments work by sensing radiation that is naturally emitted or reflected by the Earth's surface or from the atmosphere, or by sensing signals transmitted from a satellite and reflected back to it. In the visible and near-infrared regions, surface chemical composition, vegetation cover, and biological properties of surface matter can be measured. In the mid-infrared region, geological formations can be detected due to the absorption properties related to the structure of silicates. In the far infrared, emissions from the Earth's atmosphere and surface offer information about atmospheric and surface temperatures and water vapor and other trace consituents in the atmosphere. Since IR data are based on termpatures rather than visible radiation, the data may be obtained day or night.

INSAT

Indian National Satellite.

Insolation

Solar radiation incident upon a unit horizontal surface on or above the Earth's surface.

Instantaneous Field of View (IFOV)

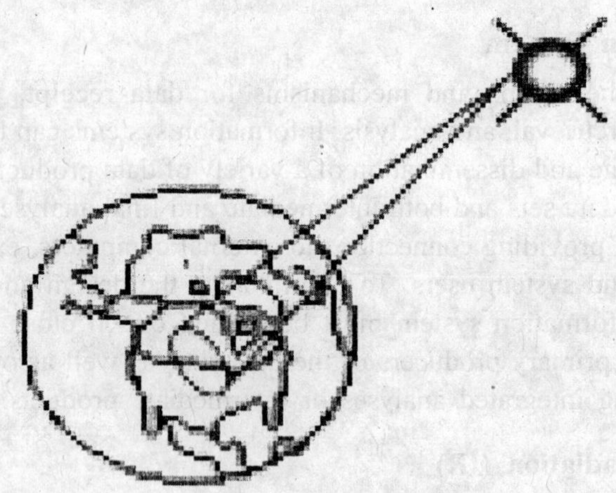

Instantaneous field of view

The field of view of a scanning detector system at a given instant.

The range of angles scanned by the system is then called the field of view, or swath width.

Integrated Circuit (IC)

A solid state electronic circuit that consists of several micro-components constructed to perform a special function.

Intensity

The amount of radiation, for example, the number of photons arriving in a given time.

Intensity Branching (%)

The intensity of a radiation emitted during radioactive decay.

Intensity Map

A color-coded map of radiation intensity as a function of position. Different colors or shades represent different intensities of observed radiation.

Intensive Variable

A measurable property of a thermodynamic system is intensive if when two identical systems are combined into one, the variable of the combined system is the same as the original value in each system. Examples: temperature, pressure.

Interaction

1. A process in which a particle decays or it responds to a force due to the presence of another particle (as in a collision). Also used to mean the underlying property of the theory that causes such effects.

2. A process in which a particle *decays* or it responds to a force due to the presence of another particle (as in a collision).

Interaction with Transparent Materials

The refractive index of a material indicates how much slower the speed of light is in that medium than in a vacuum. The slower speed of light in materials can cause refraction, as demonstrated by this prism (in the case of a prism splitting white light into a spectrum of colours, the refraction is known as dispersion). In passing through materials, light is slowed to less than *c* by the ratio called the refractive index of the material. The speed of light in air is only slightly less than *c*. Denser media, such as water and glass, can slow light much more, to fractions such as 3/4 and 2/3 of *c*. This reduction in speed is also responsible for bending of light at an interface between two materials with different indices, a phenomenon known as refraction. Since the speed of light in a material depends on the refractive index, and the refractive index depends on the frequency of the light, light at different frequencies travels at different speeds through the same material. This can cause distortion of electromagnetic waves that consist of multiple frequencies, called dispersion. On the microscopic scale, considering electromagnetic radiation to be like a particle, refraction is caused by continual absorption and re-emission of the photons that compose the light by the atoms or molecules through which it is passing. In some sense, the light itself travels only through the vacuum existing between these atoms, and is impeded by the atoms. Alternatively, considering electromagnetic radiation to be like a wave, the charges of each atom (primarily the electrons) interfere with the electric and magnetic fields of the radiation, slowing its progress.

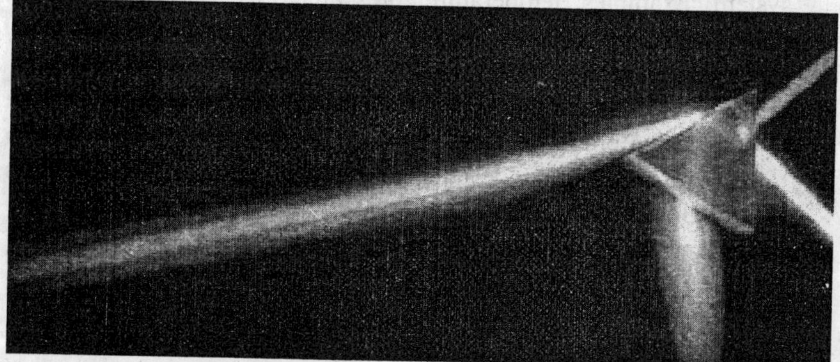

Intermediate Vector Bosons

The W and Z particles are the massive exchange particles which are involved in the nuclear weak interaction, the weak force between electrons and neutrinos. They were predicted by Weinberg, Salam, and Glashow in 1979 and measured at CERN in 1982. The prediction included a prediction of the masses of these particles as a part of the unified theory of the electromagnetic and weak forces, the electroweak unification. "If the weak and electromagnetic forces are essentially the same, then they must also have the same strength. The fact that the experimentally observed strengths seem quite different is attributed to the masses of the W and Z particles-under certain conditions a force of large strength can have the appearance of a force of small strength if the particle that carries the force is very massive. Theoretical calculations show that at a fundamental level the weak and electromagnetic forces have the same strength if the W and Z particles have masses of 80 and 90 GeV respectively." The masses measured at CERN were 82 and 93 GeV, a brilliant confirmation of the electroweak unification.

International Date Line

An imaginary line of longitude 180° east or west of the prime meridian.

International Designator

An internationally agreed-upon naming convention for satellites. The designator contains the last two digits of the launch year; the launch number of the year; and the part of the launch, i.e., "A" indicates payload, "B" the rocket booster; or second payload, etc.

International Geophysical Year (IGY)

(1957-58) The IGY was organized by the scientific community through the International Council of Scientific Unions (ICSU) . It was highlighted by international cooperation in the exploration of world-wide geophysical phenomena and by the inauguration of the space age through the launching of the first satellites (USSR's Sputnik I and US Explorer 1) to study the upper atmosphere and Earth's nearby environment.

International Space Year (ISY)

(1992) Designated the first international celebration of humanity's future in the space age. Themes included the global perspective of the space age, discovery exploration, and scientific inquiry. An important ISY scientific focus was Mission to Planet Earth. A wide range of educational programs and public events emphasized ISY's global perspective. 1992 also commemorated the 500th anniversary of Columbus' voyage to the New World and the 35th anniversary of the International Geophysical Year.

International System of Units (SI)

The International System of Units prescribes the symbols and prefixes shown in the table to form decimal multiples and submultiples of SI units. The following examples illustrate the use of these prefixes. 0.000,001 meters = 10^{-6} meters = 1 micrometer 1pm 1000 meters = 10^3 meters = 1 kilometers = 1km 1,000,000 cycles per second = 10^6 hertz = 1 megahertz = 1 MHz.

Invariant

A quantity that does not change when transformed.

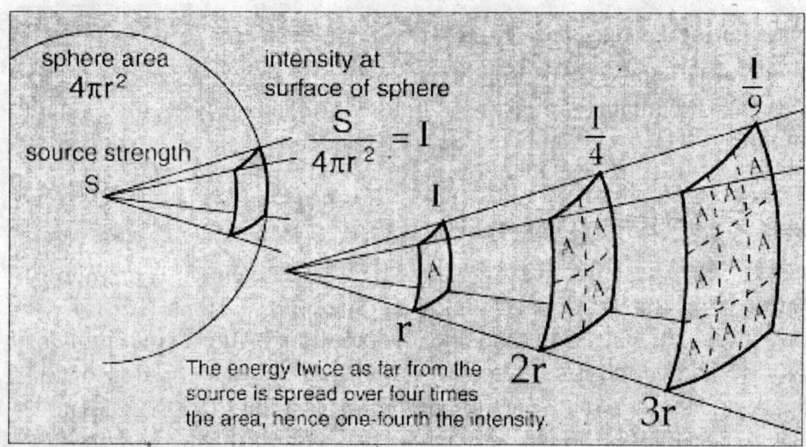

Inverse Square Law, General

Any point source which spreads its influence equally in all directions without a limit to its range will obey the inverse square law. This comes from strictly geometrical considerations. The intensity of the influence at any given radius r is the source strength divided by the area of the sphere. Being strictly geometric in its origin, the inverse square law applies to diverse phenomena. Point sources of gravitational force, electric field, light, sound or radiation obey the inverse square law. It is a subject of continuing debate with a source such as a skunk on top of a flag pole; will it's smell drop off according to the inverse square law?

Ion

1. An atomic particle that is electrically charged, either negative or positive.
2. An electrically charged atom or molecule.
3. Atom or molecule that has acquired an electric charge by the loss/gain of one or more electrons.
4. Atomic particle, atom, or chemical radical bearing an electrical *charge*, either negative or positive.
5. An atom that has lost or gained one or more *electrons* and has become electrically charged as a result.

Ionization

1. The process by which a neutral atom or molecule acquires a positive or negative *charge*.
2. The process by which *ions* are produced, typically occurring by collisions with atoms or *electrons* ("collisional ionization"), or by interaction with *electromagnetic radiation* ("photoionization").

Ionizing Radiation

1. Radiation that has enough energy to eject *electrons* from electrically neutral atoms, leaving behind *charge* atoms or ions. There are four basic types of ionizing radiation: Alpha particles (helium nuclei), beta particles (electrons), neutrons, and *gamma rays* (high frequency electromagnetic waves, x-rays, are generally identical to gamma rays except for their place of origin.) Neutrons are not themselves ionizing but their collisions with nuclei lead to the ejection of other charged particles that do cause *ionization.*

2. Radiation that is capable of producing ions either directly or indirectly.

3. Ionizing radiation is radiation with enough energy so that during an interaction with an atom, it can remove tightly bound electrons from their orbits, causing the atom to become charged or ionized. Examples are gamma rays and neutrons.

Ionosphere

The region of the Earth's upper atmosphere containing a small percentage of *free electrons* and *ions* produced by *photoionization* of the constituents of the atmosphere by solar ultraviolet radiation. The ionosphere significantly influences *radiowave* propagation of frequencies less than about 30 MHz.

IPS

Inches per second.

Irradiate

To expose to some form of radiation.

Isobars

1. Lines drawn on a weather map joining places of equal barometric pressure.

2. *Nuclides* of the same atomic mass but different *atomic number.*

Isomer

1. One of several nuclides with the same number of neutrons and protons capable of existing for a measurable time in different nuclear energy states.

2. A long-lived excited state of the *nucleus.* Arbitrarily defined in as the *Table of Isotopes* as having a half-life greater than 1 ms.

Isometric Transition

A mode of radioactive decay where a nucleus goes from a higher to a lower energy state. The mass number and the atomic number are unchanged.

Isothermal

Of or indicating equality of temperature.

Isotherms

Lines connecting points of equal temperature on a weather map.

Isotope

1. Isotopes of a given element have the same atomic number (same number of protons in their nuclei) but different atomic weights (different number of neutrons in their nuclei). Uranium-238 and uranium-235 are isotopes of uranium.
2. One of the possible varieties of atoms of a given element, having a certain number of neutrons.
3. One of two or more atoms having the same number of *protons* in its *nucleus*, but a different number of *neutrons* and, therefore, a different mass.
4. Two or more nuclides having the same *atomic number*, thus constituting the same element, but differing in the mass number. Isotopes of a given element have the same number of nuclear *protons* but differing numbers of *neutrons.* Naturally occurring chemical elements are usually mixtures of isotopes so that observed (non-integer) atomic weights are average values for the mixture.

Isthmus

Narrow strip of land located between two bodies of water; connecting two larger land areas.

ITOS (Improved TIROS Operational Satellite)

Second generation, polar-orbiting, environmental satellites utilized to augment NOAAs world-wide weather observation capabilities. ITOS were launched from 1970-1976, but eventually replaced by the third generation of polar-orbiting, environmental satellites TIROS-N (first launched in 1978).

J

Japanese National Space Development Agency (NASDA)

The agency reports to the Japanese Ministry of Science and Technology

Jet Stream

Rivers of high-speed air in the atmosphere. Jet streams form along the boundaries of global air masses where there is a significant difference in atmospheric temperature. The jet streams may be several hundred miles across and 1-2 miles deep at an altitude of 8-12 miles. They generally move west to east, and are strongest in the winter with core wind speeds as high as 250 mph. Changes in the jet stream indicate changes in the motion of the atmosphere and weather

Jet

The name physicists give to a cluster of particles emerging from a collision or *decay* all traveling in roughly the same direction and carrying a significant fraction of the energy in the event. The particles in the jet are chiefly *hadrons*.

Joint Education Initiative (JEI)

The JEI project was developed by USGS, NOAA, NASA, industry, and teachers to enable teachers and students to explore the massive quantities of Earth science data published by the U.S. Government on CD ROM. JEI encourages a research and analysis approach to science education.

Josephson Junction

Two superconductors separated by a thin insulating layer can experience tunneling of *Cooper pairs* of electrons through the junction. The Cooper pairs on each side of the junction can be represented by a wavefunction similar to a *free particle wavefunction*. In the DC Josephson effect, a current proportional to the phase difference of the wavefunctions can flow in the junction in the absence of a voltage. In the AC Josephson effect, a Josephson junction will oscillate with a characteristic frequency which is proportional to the voltage across the junction. Since frequencies can be measured with great accuracy, a Josephson junction device has become the *standard measure of voltage*.

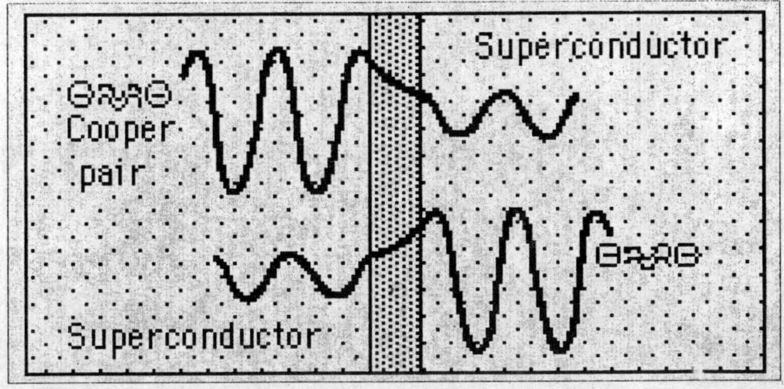

Jupiter Effect

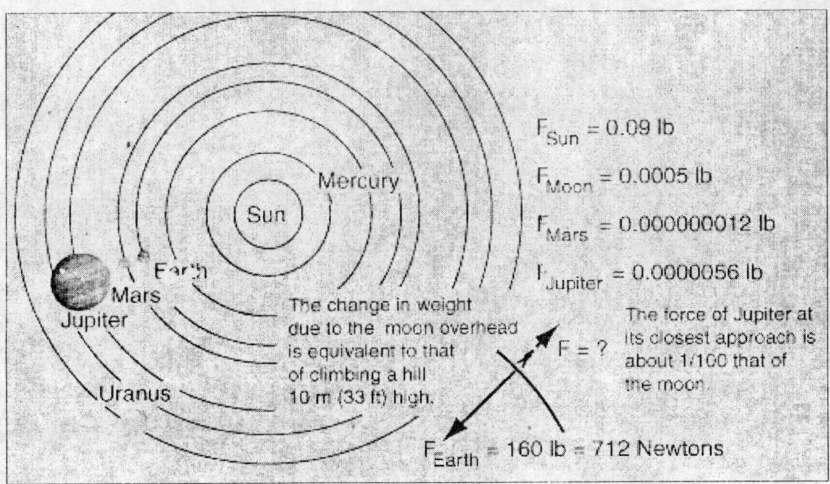

A popular book warned of disastrous gravitational effects from the special alignment of the planets Jupiter and Mars along with the *Moon*. The forces of attraction compared to that of the Earth are illustrated for a person weighing 160 lb on the earth's surface.

K

Kaon (K)

A meson containing a strange quark and an anti-up (or anti-down) quark, or an anti-strange quark and an up (or down) quark.

K-capture

The capture by an atom's nucleus of an orbital electron from the first K-shell surrounding the nucleus.

Kelvin

Abbreviated *K.* A unit of absolute temperature. Zero degrees Celsius is equal to 273.16 Kelvin.

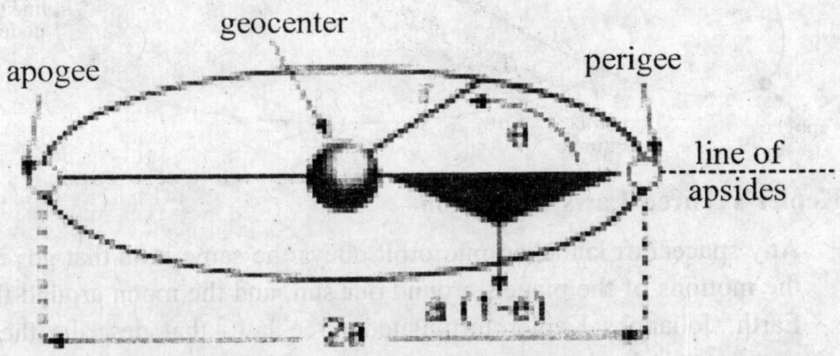

Keplerian Elements (Aka Satellite Orbital Elements)

The set of six independent constants which define an orbit-named for Johannes Kepler (1571-1630). The constants define the shape of an ellipse or hyperbola, orient it around its central body and define the position of a satellite on the orbit. The classical orbital elements are: a: semi-major axis, gives the size of the orbit, e: eccentricity, gives the shape of the orbit, *i*: inclination angle, gives the angle of the orbit plane to the central body's equator Ω: right ascension of the ascending node, which gives the rotation of the orbit plane from reference axis, ω: argument of perigee is the angle from the ascending nodes to perigee point, measured along the orbit in the direction of the satellites motion θ: true anomaly gives the location of the satellite on the orbit.

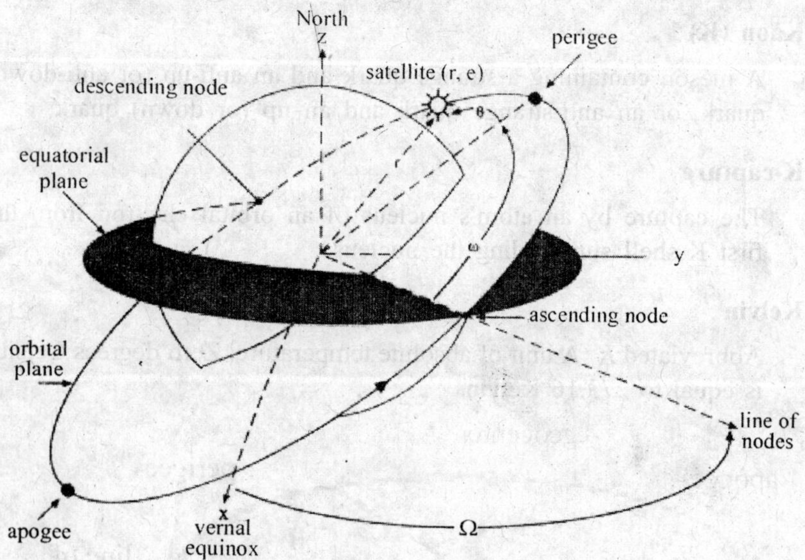

Kepler's Three Laws of Motion

Any spacecraft launched into orbit obeys the same laws that govern the motions of the planets around our sun, and the moon around the Earth. Johannes Kepler formulated three laws that describe these motions: (1) Each planet revolves around the sun in an orbit that is an ellipse with the sun as its focus or primary body Kepler postulated the lack of circular orbits-only elliptical ones-determined by gravitational perturbations and other factors. Gravitational pulls, according to Newton, extend to infinity although their forces weaken with distance and eventually become impossible to detect. Spacecraft orbiting the Earth are primarily influenced by the Earth's gravity and anomalies in its composition, but they also are influenced by the moon and sun and possibly other planets. (2) The radius vector—such as the line from the center of the sun to the center of a planet, from the center of Earth to the center of the moon, or from the center of Earth to the center of gravity of a satellite—sweeps out equal areas in equal periods of time. (3) The square of a planet's orbital period is equal to the cube of its mean distance from the sun times a constant. As extended and generalized, this means that a satellite's orbital period increases with its mean distance from the planet.

Orbital Elements

Element	conventional symbol	symbol used in GSFC computer printouts
epoch	epoch	Epoch Time, To
orbital inclination	ι	Inclination, IO
right ascension of ascending node	Ω	R.A.A.N., O0
argument of perigee	ω	(ARGP) Arg Perigee, W
eccentricity	e	(Ecce) Eccentricity, EO or e
mean motion	n	Mean Motion, NO
mean anomally	M	(MA, Phase) MeanA Anomaly, MO

KeV

One thousand electron volts.

Kilohertz (Khz)

One thousand hertz, *i.e.,* one thousand cycles per second.

Kilometer (Km)

1. Metric unit of distance equal to 3,280.8 feet or .621 statute miles.
2. Abbreviated *km.* 1 km = 1000 meters = 10^5 cm = 0.62 mile.

Kinetic Energy (K)

1. Notice that this energy of motion is proportional to the square of the speed. The unit of Joule may also be expressed as kg(m/sec) (m/sec).
2. The energy a body has by virtue of its motion. The kinetic energy is the work done by an external force to bring the body from rest to a particular state of motion.
3. The energy an object possesses because of its motion. Cf. potential energy.

Kinetic Friction

A friction force between surfaces that are slipping past each other.

Kinetic Theory

The *kinetic theory of gases* is a theory that explains the macroscopic properties of gases by consideration of their composition at a molecular level.

Klystron

An *evacuated electron* tube used as an oscillator or amplifier at microwave frequencies. In the klystron, an *electron beam* is velocity modulated (periodically bunched) to produce large amounts of power.

Knot

Unit of speed of one nautical mile (6,076.1 feet) an hour.

L

Lagrangian Mechanics

Lagrangian mechanics is a re-formulation of classical mechanics introduced by Joseph Louis Lagrange in 1788. In Lagrangian mechanics, the trajectory of an object is derived by finding the path which minimizes the action, a quantity which is the integral of the Lagrangian over time. The Lagrangian for classical mechanics is taken to be the difference between the kinetic energy and the potential energy. This considerably simplifies many physical problems. For example, consider a bead on a hoop. If one were to calculate the motion of the bead using Newtonian mechanics, one would have a complicated set of equations which would take into account the forces that the hoop exerts on the bead at each moment. The same problem using Lagrangian mechanics is much simpler. One looks at all the possible motions that the bead could take on the hoop and mathematically finds the one which minimizes the action. There are fewer equations since one is not directly calculating the influence of the hoop on the bead at a given moment.

Lake

A body of fresh or salt water entirely surrounded by land.

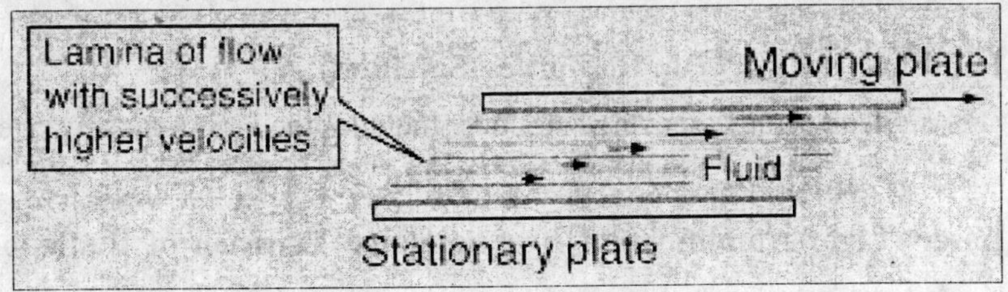

Laminar Flow

The resistance to flow in a liquid can be characterized in terms of the viscosity of the fluid if the flow is smooth. In the case of a moving plate in a liquid, it is found that there is a layer or lamina which moves with the plate, and a layer which is essentially stationary if it is next to a stationary plate.

Land Breeze

A nocturnal coastal breeze that blows from land to sea. In the evening the water may be warmer than the land, causing pressure differences. The land breeze is the flow of air from land to sea equalizing these pressure differences.

Landsat

Land Remote-Sensing Satellite, operated by the U.S. Earth Observation Satellite Company (EOSAT). Commercialized under the Land Remote-Sensing Commercialization Act of 1984, Landsat is a series of satellites (formerly called ERTS) designed to gather data on the Earth's resources in a regular and systematic manner. Objectives of the misssion are: land use inventory, geological/mineralogical exploration, crop and forestry assessment, and cartography. Landsat has a spatial resolution of 28.5 meters. Restructured Federal agency responsibilities for the Landsat program are effective for the acquisition and operation of Landsat 7. New operating policy specifies that NOAA will be responsible for satellites after they are placed in orbit, NASA will be responsible for the development and launch of Landsat 7, and that the U.S. government will provide unenhanced data to users at no cost beyond the cost of fulfilling their data request.

Landsats (Aka Earth Resources Satellites)

Any land remote-sensing satellites. Includes the U.S. Landsat system and the French SPOT

Laser (Light Amplification By Stimulated Emission of Radiation)

Active instrument that produces discretely coherent pulses of light

(light waves with no phase differences, or with predictable phases differences, are said to be coherent).

Laser Ranging

The use of lasers to measure distances.

Latitude (Aka The Geodetic Latitude)

The angle between a perpendicular at a location, and the equatorial plane of the Earth.

Law

A statement, usually mathematical, which describes some physical phenomena.

Lead Zirconate Titanate (PZT)

The piezoelectric material which has had the widest application as an ultrasonic transducer for medical diagnostic applications is lead zirconate titanate, commonly referred to as PZT, a shortened form of the chemical symbols PbZrTi

Legend

A listing that contains symbols and other information about a map.

Length

In general English usage, length (symbol: *l*) is but one particular instance of distance - an object's length is how *long* the object is– but in the physical sciences and engineering, the word *length* is in some contexts used synonymously with "distance". Height is vertical distance; width (or breadth) is a lateral distance; an object's width is less than its length. No one speaks of "the length from here to Alpha Centauri but rather of "the distance from here to Alpha Centauri but when one speaks of distance more abstractly, one says "A kilometre or a mile, is a unit of length" or "...of distance", and the two statements are synonymous. Likewise, a mountain might be a mile in height. Length is the metric of one dimension of space. The metric of space itself is volume, or $(length)^3$. Length is commonly considered to be one of the fundamental units, meaning that it cannot be defined in terms of other dimensions. However, a set of units can be constructed where units of length can be derived from fundamental physical constants.

Lens

A transparent object with two refracting surfaces. Usually the surfaces are flat or spherical (spherical lenses). Sometimes, to improve image quality. Lenses are deliberately made with surfaces which depart slightly from spherical (aspheric lenses).

LEP

The Large Electron Positron Collider at the CERN laboratory in Geneva, Switzerland.

Lepton

1. A fundamental fermion that does not participate in strong interactions. The electrically-charged leptons are the electron, the muon, the tau, and their antiparticles. Electrically-neutral leptons are called neutrinos.
2. A fundamental matter particle that does not participate in strong interactions. The *charge* leptons are the *electron* (e), the muon (μ), the tau (τ) and their *antiparticles*. Neutral leptons are called neutrinos (η).

LHC

The Large Hadron Collider at the CERN laboratory in Geneva, Switzerland. LHC will collide protons into protons at a center-of-mass energy of about 14 TeV. When completed in the year 2005, it will be the most powerful particle accelerator in the world. It is hoped that it will unlock many of the secrets of particle physics.

Light

1. (1) Form of radiant energy that acts upon the retina of the eye, optic nerve, etc., making sight possible. This energy is transmitted at a velocity of about 186,000 miles per second by wavelike or

vibrational motion. (2) A form of radiant energy similar to this, but not acting on the normal retina, such as ultraviolet and infrared radiation. Interplay between light rays and the atmosphere cause us to see the sky as blue, and can result in such phenomena as glows, halos, arcs, flashes, and streamers.

2. Anything that can travel from one place to another through empty space and can influence matter, but is not affected by gravity.

Light and Electromagnetic Waves

Electrically charged particles are constantly emitting (or absorbing) photonic fluid, which is more commonly known as light So how is light related to electromagnetic waves? E-M waves are undulatory movement patterns of light which can always be observed to be emitted by electric charges undergoing acceleration,. If a charged particle is at rest, then it does not emit electromagnetic waves. Instead, it is surrounded by an electrostatic field. If the charged particle is in inertial motion, then the electrostatic field is joined by a magnetostatic field. These pair of static fields produce a movement of electromagnetic energy (i.e. a field of non-zero Poynting vectors), which is similar to an electromagnetic wave, except that the fields are not oscillating.

Light Cone

The interval AB in the diagram to the right is 'time-like'. I.e. there is a frame of reference in which event A and event B occur at the same location in space, separated only by occurring at different times. If A precedes B in that frame, then A precedes B in all frames. It is hypothetically possible for matter (or information) to travel from A to B, so there can be a causal relationship (with A the cause and B the effect). The interval AC in the diagram is 'space-like'. *i.e.* there is a frame of reference in which event A and event C occur simultaneously, separated only in space. However there are also frames in which A precedes C (as shown) and frames in which C precedes A. Barring some way of traveling faster than light, it is not possible for any matter (or information) to travel from A to C or from C to A. Thus there is no causal connection between A and C.

Light-Slowing Experiments

Refractive phenomena, such as this rainbow, are due to the slower speed of light in a medium (water, in this case). In a sense, any light travelling through a medium other than a vacuum travels below c as a result of refraction. However, certain materials have an exceptionally high refractive index: in particular, the optical density of a Bose-Einstein condensate can be very high. In 1999, a team of scientists led by Lene Hau were able to slow the speed of a light beam to about 17 metres per second, and, in 2001, they were able to momentarily stop a beam. In 2003, Mikhail Lukin, with scientists at Harvard University and the Lebedev Institute in Moscow, succeeded in completely halting light by directing it into a mass of hot rubidium gas, the atoms of which, in Lukin's words, behaved "like tiny mirrors", due to an interference pattern in two "control" beams.

Light Year

A light year is the distance light can travel in a year. Light travels at 186,282 miles per second, so one can see that this is truly a gigantic distance. Yet in some respects it is still quite small. The nearest star to our sun is over four light years away, and the galaxy itself is about 100,000 light years across.

Lightning

A discharge of atmospheric electricity accompanied by a vivid flash of light. During thunderstorms, static electricity builds up within the clouds. A positive charge builds in the upper part of the cloud, while a large negative charge builds in the lower portion. When the difference between the positive and negative charges becomes great, the electrical charge jumps from one area to another; creating a lightning bolt. Most lightning bolts strike from one cloud to another; but they also can strike the ground. These bolts occur when positive charges build up on the ground. A negative charge called the "faintly luminous streamer" or "leader" flows from the cloud toward the ground. Then a positively charged leader; called the return stroke, leaves the ground and runs into the cloud. What is seen as a lightning bolt is actually a series of downward-striking leaders and upward-striking return strokes, all taking place in less than a second. Lightning bolts can heat the air to temperatures hotter than the surface of the sun. This burst of heat makes the air around the bolt expand explosively producing the sound we hear as thunder Since light travels a million times faster than sound, we see lightning bolts before we hear their thunderclaps. By counting the seconds between a flash of lightning and the thunderclap and dividing by five, we can determine the approximate number of miles to the lightning stroke.

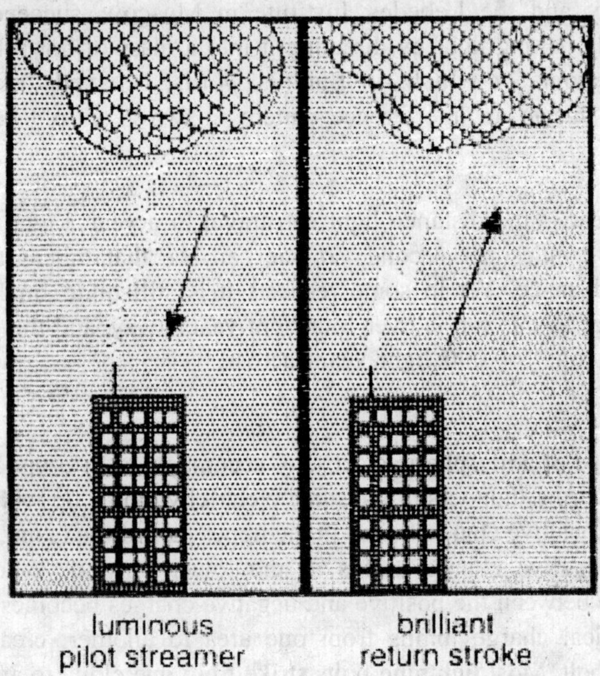

Limb Viewing (Occultation)

The process of viewing the atmosphere at a tangent to the Earth's surface. The viewing signal, from a star or another satellite, is occulted or obscured by the intervening atmosphere. The absorption of light from the sun or star provides information on the properties of the atmosphere at different heights. Limb viewing instruments can also sense infrared or microwave-emitted radiation from the atmosphere.

Linacs

An abbreviation for linear accelerator, that is, an accelerator that has no bends in it.

Linear Accelerator

A type of particle accelerator in which *charged* particles are accelerated in a straight line, either by a steady electrical field or by means of radiofrequency *electric fields*. In the latter variety, the passage of the particle is synchronized with the phase of the accelerating field. The SLAC Linear Accelerator (linac) is a two-mile long accelerator, consisting of a cylindrical, disc-loaded, copper waveguide placed on concrete girders in a tunnel about 25 feet underground.

Linear Magnification

is the ratio of the size of the object to the size of the image.

Linearization Around Operating Point

When faced with a new circuit, the software first tries to find a steady state solution. This is a solution where all nodes conform to Kirchhoff's Current Law *and* the voltages across and through each element of the circuit conform to the voltage/current equations governing that element. Once the steady state solution is found, the operating points of each element in the circuit are known. For a small signal analysis, every non-linear element can be linearized around its operation point to obtain the small-signal estimate of the voltages and currents. This is an application of Ohm's Law. The resulting linear circuit matrix can be solved with Gauss-Jordan elimination.

Line-of-Apsides (Aka Major-Axis of the Ellipse)

The straight line drawn from the perigee (point of orbit closest to Earth) to the apogee (point of orbit farthest from Earth) is the line-of-apsides.

Line-of-Nodes

The line created by the intersection of the equatorial plane and the orbital plane.

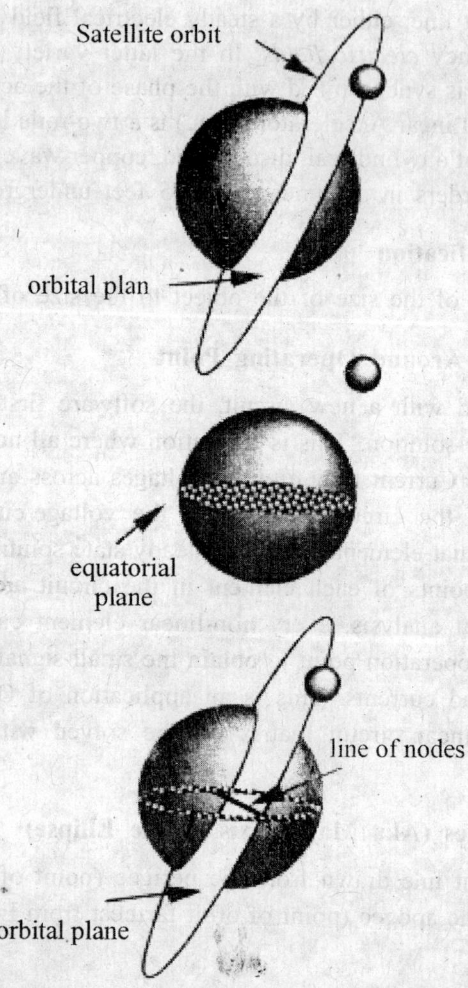

Line-of-Sight

Area within which visible contact can be made. For example, NOAA polar-orbiting satellites continuously transmit the APT signal. Radio reception of the APTsignal is possible only when the satellite is above the horizon of a particular location (not obstructed by the Earth's surface), with a line-of-sight contact with the satellite.

Logarithm

Exponent of the power to which it is necessary to raise a fixed number (the base) to produce the given number. For example, the logarithm of 100 (base 10) is 2 because $10^2 = 100$.

Longitude

The angular distance from the Greenwich meridian (00), along the equator This can be measured either east or west to the 180th meridian (180°) or 0° to 360° W

Lorentz Transformation

The transformation between frames in relative motion.

Loss of Signal (LOS)

The inability to receive a satellite signal because the satellite's orbital path has taken it below the antenna's horizon. This term is relevant to all satellites except geostationary.

Loudspeaker Sound Contours

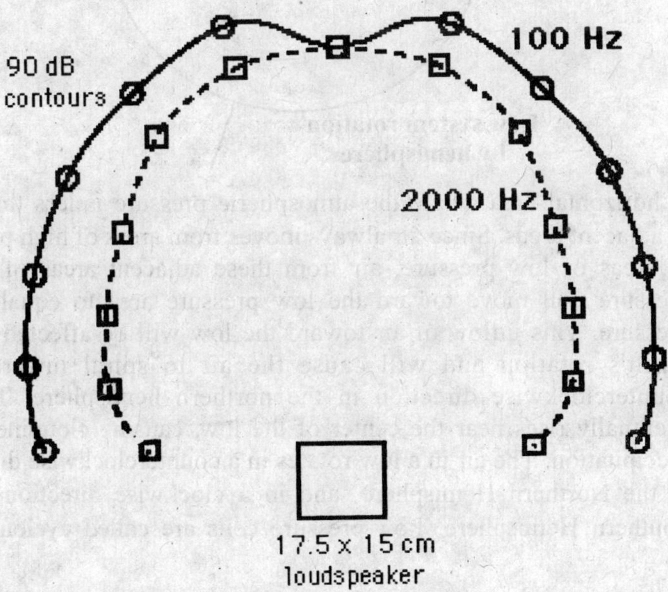

One consequence of *diffraction* is that sound from a *loudspeaker* will spread out rather than just going straight ahead. Since the bass frequencies have longer wavelengths compared to the size of the loudspeaker, they will spread out more than the high frequencies. The curves at left represent equal *intensity* contours at 90 *decibels* for sound produced by a small enclosed loudspeaker. It is evident that the high frequency sound spreads out less than the low frequency sound. These equal intensity curves were measured in an undergraduate sound laboratory experiment.

Low or Low-Pressure System

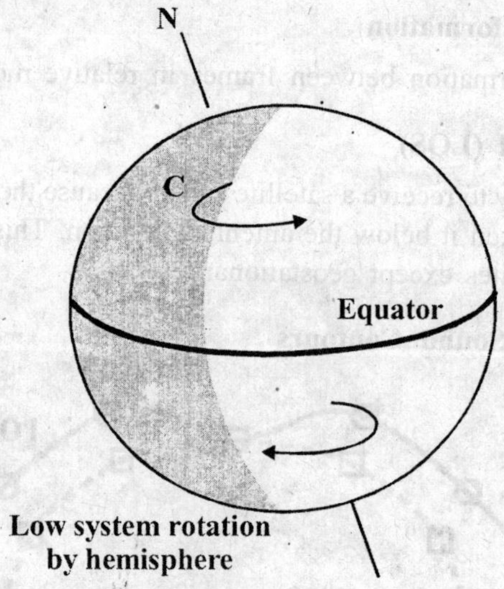

Low system rotation by hemisphere

A horizontal area where the atmospheric pressure is less than it is in adjacent areas. Since air always moves from areas of high pressure to areas of low pressure, air from these adjacent areas of higher pressure will move toward the low pressure area to equalize the pressure. This inflow of air toward the low will be affected by the Earth's rotation and will cause the air to spiral inward in a counterclockwise direction in the northern hemisphere. The air eventually rises near the center of the low, causing cloudiness and precipitation. The air in a low rotates in a counterclockwise direction in the Northern Hemisphere, and in a clockwise direction in the Southern Hemisphere. Low-pressure cells are called cyclones.

Low

A logic state corresponding to a binary "0". Satellite imagery is displayed on a computer monitor by a combination of highs and lows.

M

Macro

A prefix meaning 'large'.

Macroscopic

A physical entity or process of large scale, the scale of ordinary human experience. Specifically, any phenomena in which the individual molecules and atoms are neither measured, nor explicitly considered in the description of the phenomena.

Magnetic Dipole

An object, such as a current loop, an atom, or a bar magnet, that experiences torques due to magnetic forces; the strength of magnetic dipoles is measured by comparison with a standard dipole consisting of a square loop of wire of a given size and carrying a given amount of current.

Magnetic Field

1. A field of force, defined in terms of the torque exerted on a test dipole.
2. A field of force that is generated by electric currents. The Sun's average large-scale magnetic field, like that of the Earth, exhibits a north and a south pole linked by lines of magnetic force.
3. In physics, a magnetic field is an entity produced by moving electric charges (electric currents) that exerts a force on other moving charges. (The quantum-mechanical spin of a particle produces magnetic fields and is acted on by them as though it

were a current; this accounts for the fields produced by "permanent" ferromagnets.) A magnetic field is a vector field : it associates with every point in space a vector that may vary in time. The direction of the field is the equilibrium direction of a compass needle placed in the field. Magnetic field is usually denoted by the symbol B Historically, B was called the *magnetic flux density* or the magnetic induction, and H = B/μ was called the *magnetic field* (or *magnetic field strength*), and this terminology is still often used to distinguish the two in the context of magnetic materials (non-trivial permeability i). Otherwise, however, this distinction is often ignored, and *both* symbols are frequently referred to as the magnetic field. (Some authors call H the *auxiliary field*, instead.) In SI units, *B* and *H* are measured in teslas (T) and amperes /meter (A/m), respectively; or, in cgs units, in gauss (G) and oersteds (Oe), respectively.

Magnetic Field Lines

Imaginary lines that indicate the strength and direction of a *magnetic field*. The orientation of the line and an arrow show the direction of the field. The lines are drawn closer together where the field is stronger. Charged particles move freely along magnetic field lines, but are inhibited by the magnetic force from moving across field lines.

Magnetic Monopole

In physics, *magnetic monopole* is a term describing a hypothetical particle that could be quickly clarified to a person familiar with magnets but not electromagnetic theory as "a magnet with only one pole. In more accurate terms, it would have net "magnetic charge". Interest in the concept stems from particle theories like Grand Unified Theories and superstring theories that predict either the existence or the possibility of magnetic monopoles.

Magnetism

In physics, magnetism is a phenomenon by which materials exert an attractive or repulsive force on other materials. Some well known materials that exhibit easily detectable magnetic properties are iron, some steels, and the mineral lodestone; however, all materials are influenced to one degree or another by the presence of a magnetic field, although in most cases the influence is too small to detect without special equipment. Magnetic forces are fundamental forces that arise due to the movement of electrical charge. Maxwell's equations describe the origin and behavior of the fields that govern these forces. Thus, magnetism is seen whenever electrically charged particles are in motion. This can arise either from movement of electrons in an electric current, resulting in 'electromagnetism', or from the quantum-mechanical orbital motion (there is no orbital motion of electrons around the nucleus) like planets around the sun, but there is an "effective electron velocity") and spin of electrons, resulting in what are known as 'permanent magnets'.

Magnetohydrodynamics

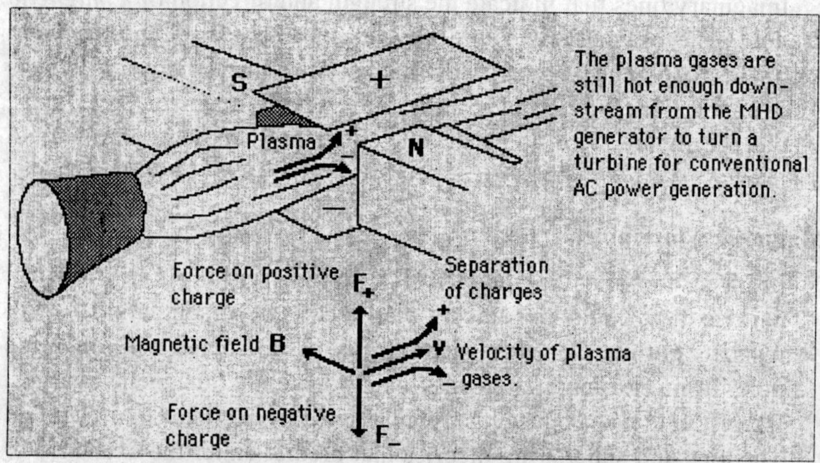

A magnetohydrodynamic generator has been described as a magnet on the tail of a jet engine. A super-hot plasma is created, ionizing the atoms of the fuel mixture. The magnetic field deflects positive and negative charges in different directions. Collecting plates for the charges provide a DC voltage. Magnetohydrodynamics as an electricity generation process holds the possibility of very efficient fuel utilization because the extremely high temperatures at which it operates correlate to a high Carnot efficiency. Its practical application has been slow in coming because of a number of problems.

Magnetosphere

Region surrounding a celestial body where its magnetic field controls the motions of charged particles. The Earth's magnetic field is dipolar in nature. That is, it behaves as if produced by a giant bar magnet located near the center of the planet with its north pole tilted several degrees from Earth's geographic north pole. The Earth's magnetic field presents an obstacle to the solar wind, as a rock in a running stream of water This obstacle is called a bow shock. The bow shock slows down, heats, and compresses the solar wind, which then flows around the rest of Earth's magnetic field.

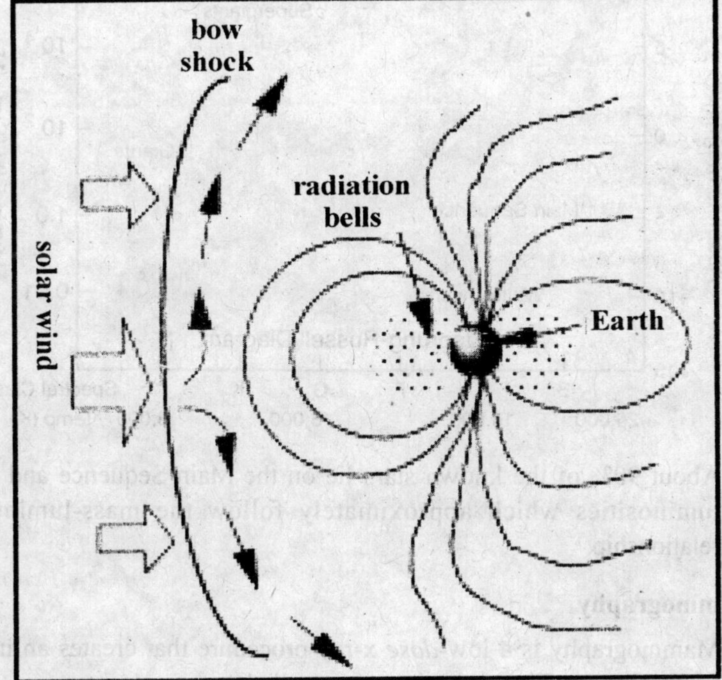

Magnification

1. The factor by which an image's linear size is increased (or decreased).

2. Two kinds of magnification are useful to describe optical systems and they must not be confused, since they aren't synonymous. Any optical system which produces a real image from a real object is described by its *linear magnification*. Any system which one looks through to view a virtual image is described by its *angular magnification*. These have different definitions, and are based on fundamentally different concepts.

Magnitude

The "amount" associated with a vector; the vector stripped of any information about its direction.

Main Sequence

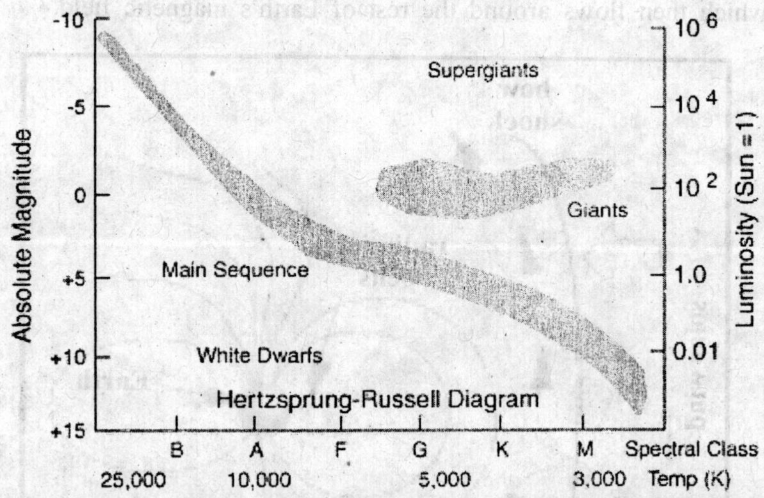

About 90% of the known stars lie on the Main Sequence and have luminosities which approximately follow the mass-luminosity relationship.

Mammography

Mammography is a low-*dose* x-ray procedure that creates an image of the breast. The x-ray image is called a mammogram.

MAPS-NET

Maryland Pilot Earth Science and Technology Education NETwork. NASA-sponsored education project designed to complement NASAs Mission to Planet Earth. MAPS-NET has been developed to enrich math and science curricula and enhance teacher preparation in Earth system science. Middle and high school teachers learn about Earth sciences and satellite direct readout at graduate-level summer workshops; academia, federal agencies, and the private sector form the support network.

Mass

1. A numerical measure of how difficult it is to change an object's motion. (In the context of relativity, some books use the word "mass" to mean what we refer to as mass multiplied by gamma.)

2. The rest mass of a particle is the mass defined by the energy of the isolated (free) particle at rest, divided by the speed of light squared. When particle physicists use the word "mass" they always mean the "rest mass" of the object in question.

3. One of the undefined qualities of physics, mass is the measure of inertia.

4. *For other uses, see Mass (disambiguation)* Mass is a property of physical objects that, roughly speaking, measures the amount of matter they contain. It is a central concept of classical mechanics and related subjects. Strictly speaking, there are three different quantities called *mass*: *Inertial mass* is a measure of an object's inertia : its resistance to changing its state of motion when a force is applied. An object with small inertial mass changes its motion more readily, and an object with large inertial mass does so less readily. *Passive gravitational mass* is a measure of the strength of an object's interaction with the gravitational field. Within the same gravitational field, an object with a smaller passive gravitational mass experiences a smaller force than an object with a larger passive gravitational mass. (This force is called the weight of the object. In informal usage, the word "weight" is often used synonymously with "mass", because the strength of the gravitational field is roughly constant everywhere on the surface of the Earth. In physics, the two terms are distinct: an object will have a larger weight if it is placed in a stronger gravitational field, but its passive gravitational mass remains unchanged.) *Active gravitational mass* is a measure of the strength of the gravitational field due to a particular object. For example, the gravitational field that one experiences on the Moon is weaker than that of the Earth because the Moon has less active gravitational mass. Although inertial mass, passive gravitational mass and active gravitational mass are conceptually distinct, no experiment has ever unambiguously demonstrated any difference between them. One of the consequences of the equivalence of inertial mass and passive gravitational mass is the fact, famously demonstrated by Galileo Galilei, that objects with different masses fall at the same rate, assuming factors like air resistance are negligible. The theory of general relativity, the most accurate theory of gravitation known to physicists to date, rests on the assumption that inertial and passive gravitational mass are *completely* equivalent. This is known as the weak equivalence principle. Classically, active and passive gravitational mass were equivalent as a consequence of Newton's third law,

but a new axiom is required in the context of relativity's reformulation of gravity and mechanics. Thus, standard general relativity also assumes the equivalence of inertial mass and active gravitational mass; this equivalence is sometimes called the strong equivalence principle.

Mass-Luminosity Relationship

For main sequence stars, the luminosity increases with the mass with the approximate power law:

$$\frac{L}{L_s} = \left[\frac{M}{M_s}\right]^{3.5}$$

The expression uses luminosities and masses compared to those of the Sun.

Mass, Momentum and Energy

In addition to modifying notions of space and time, special relativity forces one to reconsider the concepts of mass, momentum, and energy, all of which are important constructs in Newtonian mechanics. Special relativity shows, in fact, that these concepts are all different aspects of the same physical quantity in much the same way that it shows space and time to be interrelated. There are a couple of (equivalent) ways to define momentum and energy in SR. One method uses conservation laws. If these laws are to remain valid in SR they must be true in every possible reference frame. However, if one does some simple thought experiments using the Newtonian definitions of momentum and energy one sees that these quantities are not conserved in SR. One can rescue the idea of conservation by making some small modifications to the definitions to account for relativistic velocities. It is these new definitions which are taken as the correct ones for momentum and energy in SR.

Mass Number

1. The number of protons plus the number of neutrons in a nucleus; approximately proportional to its atomic mass.
2. The sum of the number of *neutrons* and *protons* in a *nucleus*.

Mathematical Dimensions

In mathematics, no definition of dimension adequately captures the concept in all situations where we would like to make use of it. Consequently, mathematicians have devised numerous definitions of dimension for different types of spaces. All, however, are ultimately based on the concept of the dimension of Euclidean n-space E^n. The point E^0 is 0-dimensional. The line E^1 is 1-dimensional. The plane E^2 is 2-dimensional. And in general E^n is n-dimensional. A tesseract is an example of a four-dimensional object.

Matter

1. Anything that is affected by gravity.
2. We call the commonly observed particles such as protons, neutrons and *electrons* matter particles, and their *antiparticles* are then *antimatter*.

Maxwell's Equations

Maxwell's equations are the set of four equations, attributed to James Clerk Maxwell, that describe the behavior of both the electric and magnetic fields, as well as their interactions with matter. Maxwell's four equations express, respectively, how electric charges produce electric fields (Gauss' law), the experimental absence of magnetic charges, how currents produce magnetic fields (Ampere's law), and how changing magnetic fields produce electric fields (Faraday's law of induction). Maxwell, in 1864, was the first to put all four equations together and to notice that a correction was required to Ampere's law: changing electric fields act like currents, likewise producing magnetic fields. Furthermore, Maxwell showed that the four equations, with his correction, predict waves of oscillating electric and magnetic fields that travel through empty space at a speed that could be predicted from simple electrical experiments-using the data available at the time, Maxwell obtained a velocity of 310,740,000 m/s. Maxwell (1865) wrote: *This velocity is so nearly that of light, that it seems we have strong reason to conclude that light itself (including radiant heat, and other radiations if any) is an electromagnetic disturbance in the form of waves propagated through the electromagnetic field according to electromagnetic laws.*

Mean Anomaly (Aka MO or MA or Phase)

Specifies the mean location (true anomaly specifies the exact location) of a satellite on an orbit ellipse at a particular time, assuming a constant mean motion throughout the orbit. Epoch specifies the particular time at which the satellites position is defined, while mean anomaly specifies the location of the satellite at epoch. Mean anomaly is measured from 0° to 360o during one revolution. It is defined as 0° at perigee, and hence is 180° at apogee.

Mean Motion (Aka NO)

Averaged speed of a satellite in a non-circular orbit (i.e., *eccentricity* 0). Satellites in circular orbits travel at a constant speed. Satellites in non-circular orbits move faster when closer to the Earth, and slower when farther away Common practice is to compute the mean motion (average the speed), which is measured in revolutions per day.

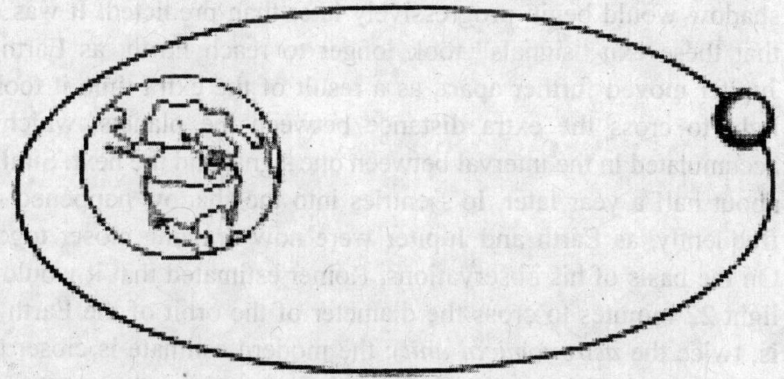

Mean motion, averaged speed in elliptical orbit

Mean-life

The *average* time during which a system, such as an atom, nucleus, or elementary particle, exists in a specified form. Also known as the average life.

Measurement of the Speed of Light

Isaac Beeckman, a friend of Descartes, proposed an experiment (1629) in which one would observe the flash of a *cannon* reflecting off a *mirror* about one mile away. *Galileo* proposed an experiment (1638), with an apparent claim to have performed it some years earlier, to measure the speed of light by observing the delay between uncovering a *lantern* and its perception some distance away. Descartes criticised this experiment as superfluous, in that the observation of eclipses, which had more power to detect a finite speed, gave a negative result. This experiment was carried out by the Accademia del Cimento of Florence in 1667, with the lanterns separated by about one mile. No delay was observed. *Robert Hooke* explained the negative results as Galileo had: by pointing out that such observations did not establish the infinite speed of light, but only that the speed must be very great. The first quantitative estimate of the speed of light was made in *1676* by *Ole Rømer* who was studying the *motions* of *Jupiter*'s satellite *Io* with a *telescope.* It is possible to time the *revolution* of Io because it is entering/exiting Jupiter's shadow at regular *intervals.* Rømer observed that Io revolved around Jupiter once every 42.5 *hours* when *Earth* was closest to Jupiter. He also observed that, as Earth and Jupiter moved apart, Io's exit from the shadow would begin progressively later than predicted. It was clear that these exit "signals" took longer to reach Earth, as Earth and Jupiter moved further apart, as a result of the extra time it took for light to cross the extra distance between the planets, which had accumulated in the interval between one signal and the next. Similarly, about half a year later, Io's entries into the shadow happened more frequently, as Earth and Jupiter were now drawing closer together. On the basis of his observations, Rømer estimated that it would take light 22 minutes to cross the diameter of the orbit of the Earth (that is, twice the *astronomical unit*); the modern estimate is closer to 16 minutes and 40 seconds. Around the same time, the astronomical unit was estimated to be about 140 million kilometres. The astronomical unit and Rømer's time estimate were combined by *Christiaan Huygens,* who estimated the speed of light to be 1000 Earth diameters per minute. This is about 220,000 kilometres per second (136,000 miles per second), well below the currently accepted value, but still very much faster than any physical phenomenon then known. *Isaac Newton* also accepted the finite speed. In his book "*Opticks*" he, in fact, reports the more accurate value of 16 minutes per diameter, which it seems he inferred for himself (whether from Rømer's data, or otherwise, is not known). The same effect was subsequently observed by Rømer for a "spot" rotating with the surface of Jupiter. And later observations also showed the effect with the three other Galilean moons, where it was more difficult to observe, thus laying to rest some further objections that had been raised.

Measurement System Integrity

The tracking and documentation over the long term of all causes of error or uncertainty in a final data-analysis product. These include instrument calibration, adequacy of measurement validation, data coverage and sampling density, availability and quality of ancillary data, procedures for data analysis and reduction, the results of checks against independent measurement, and quantitative error analysis.

Measurement Validation

The establishment of confidence in the numerical relationship between the calibrated sensor output and the actual variable being measured.

Medical Physics

This is a very diverse field that applies the knowledge gained in other areas of physics (such as *high-energy Physics*) to heal people. *Radiation therapy* is one example. *CAT scans*, mammography, and other *x-ray* imaging techniques are diagnostic techniques that have also been developed by physicists working in medicine. Another important example is the technology that went into building the particle accelerator at SLAC, which has been adapted for use in hospitals to treat cancer patients with beams of *electrons* and *x-rays*.

Megaton

An explosive force equal to one million metric tons of TNT. The energy released in the explosion of one megaton of TNT is equal to 4.2×10^{22} *ergs*.

Membrane Transport

The transport of water and other types of molecules across membranes is the key to many processes in living organisms. Many of these transport processes proceed by *diffusion* through membranes which are selectively permeable, allowing small molecules to pass but blocking larger ones. These processes, including *osmosis* and dialysis, are sometimes called passive transport since they do not require any active role for the membrane. Other types of transport, called active transport, involve properties of the membrane to selectively "pump" certain types of molecules across the membrane.

Mercator Projection

A method of making maps in which the Earth's surface is shown as a rectangle with the meridians as parallel straight lines spaced at equal intervals and the parallels of latitude as parallel straight lines intersecting the meridians at right angles. Areas away from the equator appear larger than they are, with the greatest distortion near the poles.

Meson

1. A hadron made from an even number of quark constituents The basic structure of most mesons is one quark and one antiquark.
2. A *hadron* with the basic structure of one quark and one antiquark.

Mesopause

The upper boundary of the mesosphere where the temperature of the atmosphere reaches its lowest point.

Mesosphere

The atmospheric layer above the stratosphere, extending from about 50 to 85 kilometers altitude. The temperature generally decreases with altitude.

Metadata

Information describing the content or utility of a data set. For example, the dates on which data were procured are metadata.

Meteor

The former Soviet Union's series of polar-orbitmg weather satellites. The Meteor satellites transmit images in a system compatible with the NOAA polar-orbiting satellites.

Meteorology

Study of the atmosphere and its phenomena.

METEOSAT

METEorological SATellite. Europe's geostationary weather satellite, launched by the European Space Agency and now operated by an organization called Eumetsat. METEOSAT transmits at 1691 and 1694.5 MHz.

Metsat

Generic term for meteorological (weather) satellites.

MeV (Mega Electron Volt)

1. Energy equal to that acquired by a particle with one electronic *charge* in passing through a potential difference of one million volts.
2. One million electron volts.

MHz (megahertz)

10^6 hertz.

Micro

A prefix meaning 'small', as in 'microscope', 'micrometer', 'micrograph'. Also, a metric prefix meaning 10^{-6}.

Microcurie (μCi)

One millionth of a curie (3.7 x 10^4 disintegrations per second).

Micrometer (pm, aka micron)

One millionth of a meter; used to measure wavelengths in the electromagnetic spectrum.

Micron

One millionth of a meter; also known as a micrometer.

Microprocessor

Controlling unit of a microcomputer; laid out on a tiny silicon chip and containing the logical elements for handling data, performing calculations, carrying out stored instructions, etc.

Microscopic

A physical entity or process of small scale, too small to directly experience with our senses. Specifically, any phenomena on the molecular and atomic scale, or smaller.

Microwave

1. Electromagnetic radiation with wavelengths between about 1000 micrometers and one meter.
2. Waves of electromagnetic radiation that oscillate from approximately 10^9 to 3 x 10^{11}Hz. (cycles per second). SLAC's *klystrons* produce microwaves of 2,856 MHz. Møller Scattering.

   Scattering of electrons by electrons.

Middle Infrared

Electromagnetic radiation between the near infrared and the thermal infrared, about 2-5 micrometers.

Milk

To elute a cow.

Milky Way Galaxy

Several hundred billion stars make up our galaxy, stretching over some 100,000 light years or about 30 kiloparsecs in a flattened disk which is about 10,000 light years (3 kpc) thick at the center. The sun is some 8 kiloparsecs out from the center, about two-thirds of the way out. About 6000 stars are visible with the naked eye.

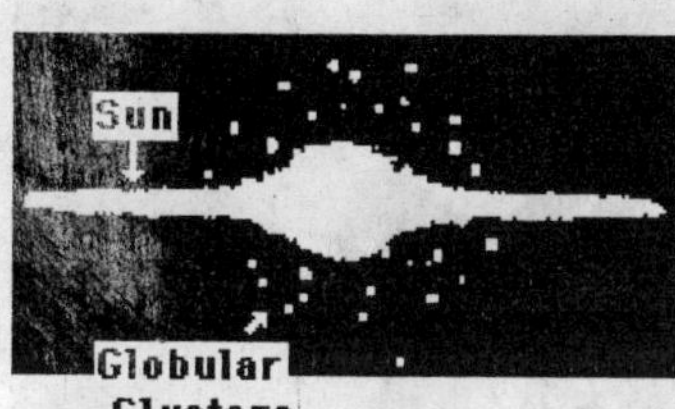

Two small irregular galaxies called the "Magellanic Clouds" are relatively near the Milky Way. The Large Magellanic Cloud is at about 160,000 light years and the Small Magellanic Cloud is at about 200,000 light years from us. The Andromeda galaxy is the nearest large galaxy at about 900 kiloparsecs or 2.9 million light years.

Millibar (Mb)

One thousandth of a bar; a unit of atmospheric pressure. The average atmospheric pressure at sea level is 1.01325 bars or 1013.25mb.

Millirem

A unit for measuring a person's exposure to radioactivity; cf rem.

Minigenerator

A trademark of Union Carbide Corporation that is used to identify radioisotope generator systems for educational use.

MIR

For 14 years the ambitious Soviet MIR space station stayed aloft, reaching a number of milestones in the habitation of space and providing a large amount of experience in the continuous operation of a space station. It's last few years were troubled by many maintenance problems and the plan as of the end of 2000 is to ditch it in the Pacific Ocean during 2001.

Mission to Planet Earth (MTPE)

International research program to understand our planet's environment as a system. A major challenge of MTPE is to observe, understand, model, assess, and eventually predict global change. Meeting this challenge will help to evaluate the impact that human activity (e.g., clearing forests and burning fossil fuels) has on our environment, and to distinguish human-induced changes from the effects of natural events (e.g. volcanic eruptions, erosion). NASA's MTPE uses space-, aircraft-, and ground~based measurements to provide the scientific basis for ~nderstanding global change. The program will produce longterm global maps of clouds, land and ocean vegetation, atmospheric ozone, sea-surface temperature, and other global processes necessary to understand the state of the Earth and to detect any patterns of change. This information will be available to scientist5 and policy makers through the Earth Observing System Data and Information System (EOSDIS). The centerpiece of NASAs MTPE will be the Earth Observing System (EOS), a series of satellites planned for launch beginning in 1998. Measurements from EOS will be complemented by the Earth Probes, a series of discipline-specific satellites and instruments designed to observe Earth processes where smaller platforms and/or different orbits from EOS are required. Planned Earth Probes will measure tropical rainfall, ocean productivity, ozone, and ocean surface winds. In addition, MTPE includes current NASA Earth science missions collecting important data on the global environment, such as the Upper Atmosphere Research Satellite (UARS) and the Ocean Topography Experiment (TOPEX/POSEIDON), Space Shuttle experiments such as ATLAS, and aircraft campaigns.

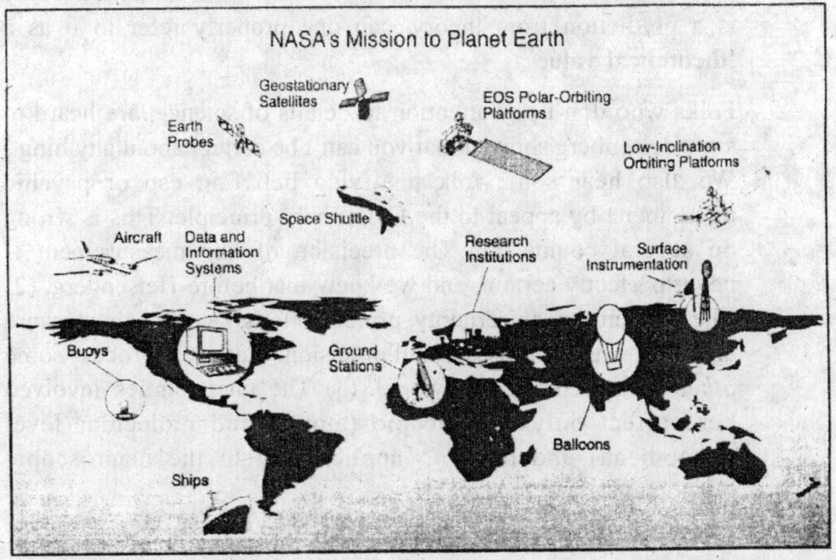

Misuse Aert

1. Be careful that the reader does not confuse this with the colloquial usage: 'One factor in the success of this experiment was...'

2. In elementary lab manuals one often sees: experimental error = |your value - book value| /book value. This should be called the *experimental discrepancy.*

3. A very common mistake found in textbooks is to speak of 'flow of current'. Current itself is a flow of charge; what, then, could 'flow of current' mean? It is either redundant, misleading, or

wrong. This expression should be purged from our vocabulary. Compare a similar mistake: 'The velocity moves West.'

4. Do not call an authoritative or 'book' value of a physical quantity a *theoretical* value, as in: 'We compared our experimentally determined value of index of refraction with the theoretical value and found they differed by 0.07.' The value obtained from index of refraction tables comes *not* from theory, but from experiment, and therefore should not be called *theoretical.* The word *theoretically* suffers the same abuse. Only when a numeric value is a prediction from theory, can one properly refer to it as a 'theoretical value'.

5. Folks who don't pay attention to details of science, are heard to say 'Heisenberg showed that you can't be certain about anything.' We also hear some folk justifying belief in esp or psychic phenomena by appeal to the Heisenberg principle. This is wrong on several counts. (1) The precision of *any* measurement is never perfectly certain, and we knew that before Heisenberg. (2) The Heisenberg uncertainty principle tells us we can measure anything with arbitrarily small precision, but in the process some *other* measurement gets worse. (3) The uncertainties involved here affect only microscopic (atomic and molecular level phenomena) and have no applicability to the macroscopic phenomena of everyday life.

6. Many books emphasize that the mole is 'just a number,' a measure of the number of particles in a collection. They say that one can have a mole of *any* kind of particles, baseballs, atoms, stars, grains of sand, etc. It doesn't have to be molecules. This is misleading. To say that the mole is 'just a number' is simply wrong, from physical, pedagogical, philosophical and historical points of view. There's no physical significance to a mole of stars or a mole of grains of sand, or a mole of people. The physical significance of the mole as a measure of quantity arises *only* when dealing with physical laws about matter on the molecular scale. The only physical and chemical *laws* which use the mole are those dealing with gases, or systems behaving like gases.

7. One hears some folks with superficial minds say 'Einstein showed that everything is relative.' In fact, special relativity shows that only certain measurable things are relative, but in a precisely and mathematically specific way, and other things are, *not* relative, for all observers agree on them.

8. Radioactivity is a *process*, not a *thing*, and not a *substance*. It is just as incorrect to say 'U-235 emits radioactivity' as it is to say 'current flows.' A malfunctioning nuclear reactor does not *release radioactivity*, though it may *release radioactive materials* into the surrounding environment. A patient being treated by radiation therapy does not *absorb radioactivity*, but does absorb some of the *radiation* (alpha, beta, gamma) given off by the radioactive materials being used. This misuse of the word *radioactivity* causes many people to incorrectly think of radioactivity as something one can *get* by being near radioactive materials. There is only one process which behaves anything like that, and it is called *artificially induced radioactivity*, a process mainly carried out in research laboratories. When some materials are bombarded with protons, neutrons, or other nuclear particles of appropriate energy, their nuclei may be transmuted, creating unstable isotopes which are radioactive.

9. Sometimes the word *electricity* is colloquially misused as if it named a physical quantity, such as 'The capacitor stores electricity,' or 'Electricity in a resistor produces heat.' Such usage should be *avoided!* In all such cases there's available a more specific or precise word, such as 'The capacitor stores *electrical energy*,' 'The resistor is heated by the electric *current*,' and 'The utility company charges me for the *electric energy* I use.' (I am not being charged based on the *power*, so these companies shouldn't call themselves *Power* companies. Some already have changed their names to something like '... Energy')

Misuse Example

'The earth's auroras-the northern and southern lights-illustrate how energy from the sun travels to our planet.' This sentence blurs understanding of the process by which energetic charged particles from the sun interact with the earth's magnetic field and our atmosphere to result in the aurorae. Whenever one hears people speaking of 'energy fields', 'psychic energy', and other expressions treating energy as a 'thing' or 'substance', you know they aren't talking physics, they are talking moonshine. In certain quack theories of oriental medicine, such as *qi gong* (pronounced *chee gung*) something called *qi* is believed to circulate through the body on specific, mappable pathways called meridians. This idea pervades the contrived explanations/rationalizations of acupuncture, and the *qi* is generally translated into English as *energy*. No one has ever found this so-called 'energy', nor confirmed the uniqueness of its meridian pathways, nor verified, through proper double-blind tests, that any therapy or treatment based on the theory actually works. The proponents of *qi* can't say whether it is a fluid, gas, charge, current, or something else, and their theory requires that it doesn't obey any of the physics of known carriers of energy. But, as soon as we hear someone talking about it as if it were a *thing* we know they are not talking science, but quackery. The statement 'Energy is a property of a body' needs clarification. As with many things in physics, the size of the energy depends on the coordinate system. A body moving with speed V in one coordinate system has kinetic energy *½m*V^2. The same body has zero kinetic energy in a coordinate system moving along with it at speed V. Since no inertial coordinate system can be considered 'special' or 'absolute', we shouldn't say 'The kinetic energy of the body is ...' but should say 'The kinetic energy of the body moving in this reference frame is...'

Mks System

The use of metric units based on the meter, kilogram, and second. Example: meters per second is the mks unit of speed, not cm/s or km/hr.

MKSA

The system of physical units based on the fundamental metric units: meter kilogram, second and ampere.

Model (noun)

A mathematical representation of a process, system, or object developed to understand its behavior or to make predictions. The representation always involves certain simplifications and assumptions.

Modem (modulator/demodulator)

Device that allows two computers (which use binary data in the form of bits) to communicate using a telephone line (which uses tones). When the computer is transmitting data, the modem is needed to modulate binary data into tones. When receiving data, the device is needed to demodulate the tones to obtain the binary data required by the computer. Since the computer must be both a transmitter and receiver of data, the modem must be able to modulate and demodulate data.

Modern Physics

Most of classical physics is concerned with matter and energy on the normal scale of observation; by contrast, much of modern physics is concerned with the behavior of matter and energy under extreme conditions or on the very large or very small scale. For example, atomic and nuclear physics studies matter on the smallest scale at which chemical elements can be identified. The physics of *elementary particles* is on an even smaller scale, being concerned with the most basic units of matter; this branch of physics is also known as high-energy physics because of the extremely high energies necessary to produce many types of particles in large *particle accelerators* On this scale, ordinary, commonsense notions of space, time, matter, and energy are no longer valid. The two chief theories of modern physics present a different picture of the concepts of space, time, and matter from that presented by classical physics. The *quantum theory* is concerned with the discrete, rather than continuous, nature of many phenomena at the atomic and subatomic level, and with the complementary aspects of particles and waves in the description of such phenomena. The theory of *relativity* is concerned with the description of phenomena that take place in a frame of reference that is in motion with respect to an observer; the special theory of relativity is concerned with relative uniform motion in a straight line and the general theory of relativity with accelerated motion and its connection with gravitation. Both the quantum theory and the theory of relativity find applications in all areas of modern physics.

Modern Physics

The physics developed since about 1900, which includes relativity and quantum mechanics.

Modulation

Variation in the frequency of a radio wave in accordance with some other impulse. Modulation is essential to communication systems in which a number of different signals must all share the same medium. One way this sharing can be accomplished is to place each signal in its own band of frequencies in the medium. Amplitude modulation and frequency modulation are two ways in which signals can be moved within the frequency domain to accomplish placement and sharing. The combining of a number of signals to share a communication medium by dividing it into different frequency bands for each signal is called frequency-division multiplexing. *Amplitude modulation (AM)* is technologically quite simple, and the bandwidth of the amplitude-modulated carrier is at most twice the bandwidth of the modulating signal. However; an amplitude-modulated carrier is very prone to the effects of additive noise. *Frequency modulation (FM)* is more complicated than amplitude modulation, and the bandwidth of the frequency-modulated carrier can be many times that of the modulating signal. However; the process of demodulating a frequency-modulated carrier eliminates much of the deleterious effects of additive noise. This trade-off between bandwidth and noise reduction characterizes most communication situations.

Mole

The term *mole* is short for the name *gram-molar-weight*; it is *not* a shortened form of the word *molecule*. (However, the word *molecule* does also derive from the word *molar*.)

Molecular Mass

The *molecular mass* of something is the mass of one mole of it (in cgs units), or *one kilomole* of it (in MKS units). The units of molecular mass are gram and kilogram, respectively. The cgs and MKS values of molecular mass are numerically equal. The molecular mass is *not* the mass of one molecule. Some books still call this the *molecular weight*. One dictionary definition of *molar* is 'Pertaining to a body of matter as a whole: contrasted with *molecular* and *atomic*.' The mole is a measure appropriate for a *macro*scopic amount of material, as contrasted with a *micro*scopic amount (a few atoms or molecules).

Molecule

A group of atoms stuck together.

Momentum

1. A measure of motion, equal to mv for material objects.
2. Momentum is a property of any moving object. For a slow moving object it is given by the mass times the velocity of the object. For an object moving at close to the speed of light this definition gets modified. The total momentum is a conserved quantity in any process. Physicists use the letter p to represent momentum, presumably because m was already used for mass, n for number, and o is too much like zero.

Monochromatic Laser Light

The light from a laser typically comes from one atomic transition with a single precise wavelength. So the laser light has a single spectral color and is almost the purest monochromatic light available

Monsoon

Heavy winds characterized by a pronounced seasonal change in direction. Winds usually blow from land to sea in the winter; while in the summer; the flow reverses and precipitation is more common. Monsoons are most typical in India and southern Asia.

Monte Carlo Calculations

There is a gaming aspect to Monte Carlo calculations. Every simulation is based up events that happen randomly, and so the outcome of a calculation is not always absolutely predictable. This element of chance reminds one of gambling and so the originators of the Monte Carlo technique, Ulam and von Neumann, both respectable scientists, called the technique *Monte Carlo* to emphasize its gaming aspect.

Montreal Protocol

An international agreement to drastically reduce CFC production, the Protocol was adopted in Montreal in 1987. It was significantly strengthened at a subsequent meeting in London in 1990 that called for a complete elimination of CFCs by the year 2000. The agreement was again amended by a Meeting of the Parties in Copenhagen in November 1992. Consumption of controlled substances—such as CFCs and halons—was greatly reduced or eliminated, and many accountability dates were moved forward, often from January 2000 to 1 January 1996.

Motion

This article is about motion in physics, motion (democracy) and Apple Motion. In physics, motion means a change in the position of a body with respect to time, as measured by a particular observer in a particular frame of reference. Until the end of the 19th century, Newton's laws of motion, which he posited as axioms or postulates in his famous *Principia,* were the basis of what has since become known as classical physics. Calculations of trajectories and forces of bodies in motion based on Newtonian or classical physics were very successful until physicists began to be able to measure and observe very fast physical phenomena. At very high speeds, the equations of classical physics were not able to calculate accurate values. To address these problems, the ideas of Albert Einstein concerning the fundamental phenomena of motion were adopted in lieu of Newton's. Whereas Newton's laws of motion assumed absolute values of space and time in the equations of motion, Einstein's theories assumed relative values for these concepts. Because

Einstein's equations yielded accurate results at high speeds and Newton's did not, Einstein's theory of relativity is now accepted as explaining bodies in motion. However, as a practical matter, Newton's equations are much easier to work with than Einstein's and therefore are more often used in applied physics and engineering. Because motion is defined as the proportion of space to time, these concepts are prior to motion, just as the concept of motion itself is prior to force. In other words, the properties of space and time determine the nature of motion and the properties of motion, in turn, determine the nature of force. Therefore, relative space and relative time result in relative motion, which means that the unit values of space and time can change for observers moving at high speeds relative to each other. These concepts have led physicists in general to conclude that only relative motion can be measured and that absolute motion is meaningless.

Mosaic

A composite picture built up from a number of image segments. An example of a mosaic is the WEFAX transmission, which includes both polar and mercator mosaics derived from TIROS-N/NOAA polar orbit image data.

Mountain and Valley Breezes

System of winds that blow downhill during the night (mountain breeze) and uphill during the day (valley breeze).

Multiplexer

A device that combines several separate communications signals into one and outputs them on a single line.

Multispectral Scanner (MSS)

A line-scanning instrument flown on Landsat satellites that continually scans the Earth in a 185 km (100 nautical miles) swath. On Landsats 1, 2, 4, and 5, the MSS had four spectral bands in the visible and near-infrared with an IFOV of 80 meters. Landsat-3 had a fifth band in the thermal infrared with an IFOV of 240 meters.

Muon

1. The second flavour of charged leptons (in order of increasing mass), with electric charge -1.
2. The second *lepton* (in order of increasing mass), with electric *charge* -1.

Muon Chamber

1. The outer layers of a particle detector capable of registering tracks of charged particles. Except for the chargeless neutrinos, only muons reach this layer from the collision point.
2. The outer layers of a particle *detector* capable of registering tracks of *charged* particles. The detector is designed so that the only charged particles that can get out to this layer are muons.

N

Nadir

Point on Earth directly beneath a satellite, the opposite of zenith.

Negative Temperatures

At low temperatures, particles tend to move to their lowest energy states. As you increase the temperature, particles move into higher and higher energy states. As the temperature becomes infinite, the number of particles in the lower energy states and the higher energy states becomes equal. In some situations, it is possible to create a system in which there are more particles in the higher energy states than in the lower ones. This situation can be described with a negative temperature. A substance with a negative temperature is not colder than absolute zero, but rather it is hotter than infinite temperature. The previous section described how heat is stored in the various translational, vibrational, rotational, electronic, and nuclear modes of a system. The macroscopic temperature of a system is related to the total heat stored in all of these modes and in a normal system thermal energy is constantly being exchanged between the various modes. However, for some cases it is possible to isolate one or more of the modes. In practice the isolated modes still exchange energy with the other modes, but the time scale of this exchange is much slower than for the exchanges within the isolated mode. One example is the case of nuclear spins in a strong external magnetic field. In this case energy flows fairly rapidly among the spin states of interacting atoms, but energy transfer between the nuclear spins and other modes is relatively slow. Since the energy flow is predominantly within the spin system, it makes sense to think of a spin temperature that is distinct from the temperature due to other modes. Based on Equation 7, we can say a positive temperature corresponds to the condition where entropy increases as thermal energy is added to the system. This is the normal condition in the macroscopic world and is always the case for the translational, vibrational, rotational, and non-spin related electronic and nuclear modes. The reason for this is that there are an infinite number of these types of modes and adding more heat to the system increases the number of modes that are energetically accessible, and thus the entropy. However, for the case of electronic and nuclear spin systems there are only a finite number of modes available (often just two, corresponding to spin up and spin down). In the absence of a magnetic field, these spin states are degenerate, meaning that they correspond to the same energy. When an external magnetic field is applied, the energy levels are split, since those spin states that are aligned with the magnetic field will have a different energy than those that are anti-parallel to it.

Naked Singularity

A singularity from which the universe is "unshielded" because there is no *event horizon*. The physical consequences of "naked" singularities are hotly debated among physicists. To quote the renowned mathematician and physicist **Roger Penrose**: It is sometimes said that if naked singularities do occur, then this would be disastrous for physics. I do not share this view. We already have the example of the *big bang* singularity in the remote past, which seems not to be avoidable. The "disaster" to physics occured right at the beginning. Surely the presence of naked singularities arising occasionally in collapse under much more "controlled" circumstances would be the very reverse of a disaster. The effects of such singular occurences could then be accessible *now*. Theories of singularities would be open to observational test. The initial mystery of creation, therefore, would no longer be able to hide in the obscurity afforded by its supposed uniqueness.

Nanometer (nm)

One billionth of a meter Nanometers are used to measure wavelengths in the electromagnetic spectrum.

NASA Centers

The ten major NASA Centers are: Ames Research Center (ARC) Located at Moffett Field, California. ARC is active in aeronautical research, life sciences, space science, and technology research. The Center houses the world's largest wind tunnel and the world's most powerful supercomputer system. The Dryden Flight Research Center, Edwards Air Force Base, California, formerly part of ARC, became a separate entity March 1994. Since the 1940s, this Mojave desert site has been a testing ground for high-performance aircraft and is one of two prime landing sites for the Space Shuttle. *Goddard Space Flight Center (GSFC)* Goddard was NASAs first major scientific laboratory devoted entirely to the exploration of space. Located in Greenbelt, Maryland, GSFC's responsibilities include design and construction of new scientific and applications satellites, as well as tracking and communication with existmg satellites in orbit. GSFC is the lead center for the Earth Observing System, a key element of Mission to Planet Earth. GSFC also directs operations at the Wallops Flight Facility on Wallops Island, Virginia, which each year launches some 50 scientific missions to suborbital altitudes on small sounding rockets. *Jet Propulsion Laboratory (JPL)* Located in Pasadena, California, JPL is operated under contract to NASA by the California Institute of Technology. Its primary focus is the scientific study of the solar system, including exploration of the planets with automated probes. Most of the lunar and planetary spacecraft of the 1960s and 1970s were developed at JPL. JPL also is the control center for the worldwide Deep Space Network, which tracks all planetary spacecraft.

Johnson Space Center (JSC) Johnson Space Center, located between Houston and Galveston, Texas, is the lead center for NASAs manned space flight program. JSC has been Mission Control for all piloted space flights since 1965, and now manages the Space Shuttle program. JSC's responsibilities include selecting and training astronauts; designing and testing vehicles and other systems for piloted spa·e flight; and planning and executing space flight missions. The center has a major role in developing the Space Station. In addition, JSC directs operations at the White Sands Test Facility in

New Mexico, which conducts Shuttle-related tests. The nearby White Sands Missile Range also serves as a backup landing site for the Space Shuttle. *Kennedy Space Center (KSC)* Located near Cape Canaveral, Florida, KSC is NASAs primary launch site. The Center handles the preparation, integration, checkout, and launch of space vehicles and their payloads. All piloted space missions since the Mercury program have been launched from here, including Gemini, Apollo, Skylab, and Space Shuttle flights. KSC is the Shuttle's home port, where orbiters are serviced and outfitted between missions, and then assembled into a complete Shuttle "stack" before launch. The Center also manages the testing and launch of unpiloted space vehicles from an array of launch complexes, and conducts research programs in areas of life sciences related to human spaceflight. *Langley Research Center (LaRC)* Oldest of NASA's field centers, LaRC is located in Hampton, Virginia, and focuses primarily on aeronautical research. Established in 1917 by the National Advisory Committee for Aeronautics, the Center currently devotes two-thirds of its programs to aeronautics, and the rest to space. LaRC researchers use more than 40 wind tunnels to study improved aircraft and spacecraft safety, performance, and efficiency. *Lewis Research Center (LeRC)* Lewis Research Center; located outside Cleveland, Ohio, conducts a varied program of research in aeronautics and space technology. Aeronautical research includes work on advanced materials and structures for aircraft. Space-related research focuses primarily on power and propulsion. Another significant area of research is in energy and power sources for spacecraft, including the Space Station, for which LeRC is developing the largest space power system ever designed.

Marshall Space Flight Center (MSFC) The MSFC, located in Huntsville, Alabama, is responsible for developing spacecraft hardware and systems, and is perhaps best known for its role in building the Saturn rockets that sent astronauts to the Moon during the Apollo program. It is NASAs primary center for space propulsion systems and plays a key role in the development of payloads to be flown on the shuttle (such as Spacelab). MSFC also manages two other NASA sites: the Michoud Assembly Facility in New Orleans where the Shuttle's external tanks are manufactured, and the Slidell Computer Complex in Slidell, Louisiana, which provides computer support to Michoud and to NASAs John C. Stennis Space Center.

Stennis Space Center (SSC) This Center, located on Mississippi's Gulf Coast, is NASAs prime test facility for large liquid propellant rocket engines and propulsion systems. The main mission of the Center is to support testing, on a regular basis, of the Space Shuttle's main propulsion system. SSC is responsible for a variety of research programs in the environmental sciences and the remotesensing of Earth resources, weather; and oceans, and is the lead NASA Center for the commercialization of space remote sensing.

NASA Prediction Bulletins

Reports published by NASA's Goddard Space Flight Center providing the latest orbit information on satellites. The report gives information in three parts: (1) two line orbital elements, (2) longitude of the south to north equatorial crossings, and (3) longitude and heights of the satellite crossings for other latitudes.

National Aeronautics and Space Administration (NASA)

U.S. Civilian Space Agency created by Congress. Founded in 1958, NASA belongs to the executive branch of the Federal Government. NASAs mission to plan, direct, and conduct aeronautical and space activities is implemented by NASA Headquarters in Washington, D.C., and by ten major centers spread throughout the United States. Dozens of smaller facilities, from tracking antennas to Space Shuttle landing strips to telescopes are located around the world. The agency administers and maintains these facilities, builds and operates launch pads, trains astronauts, designs aircraft and spacecraft, and sends satellites into Earth orbit and beyond, and processes, analyzes, and distributes the resulting data and information. NASA shares responsibility for aviation and space activities with other federal agencies, including the Departments of Commerce, Transportation, and Defense. Much of the work on major projects such as the Space Shuttle and the Space Station is done in the private sector by aerospace companies under government contract. From its inception,

NASA has been directed to pursue the expansion of human knowledge of phenomena in the atmosphere and space. NASA's programs of basic and applied research extend from microscopic sub-atomic particles to galactic astronomy. In addition to enhancing scientific knowledge, thousands of the technologies developed for aerospace have resulted in commercial applications. Science offices at NASA Headquarters carry out a wide range of research activities to fulfill NASAs science goals. Science offices within NASA are: Office of Mission to Planet Earth (MTPE) focuses on the "home planet" as a dynamic system of land, ocean, atmosphere, and life that can be investigated on a global scale from space using remote-sensing tools. *Office of Life and Microgravity Sciences and Applications* explores the basic physics of how solids, liquids, and gases behave in space; seeks an understanding of the basic mechanisms that underlie space adaptation-developing more effective countermeasures to mitigate the physiological effects of space flight; and studies the role of gravity on life. *Office of Space Science* includes the Space Physics and Astrophysics Division which studies the entire universe of stars and galaxies, including the sun. The Solar System Exploration division has launched spacecraft to all the known planets except Pluto in its quest to study the solar system.

National Center for Atmospheric Research (NCAR)

Non-profit organization dedicated to furthermg understanding of the Earth's atmosphere. Located in Boulder; Co., NCAR is operated by the University Corporation for Atmospheric Research (UCAR) and sponsored by the National Science Foundation (NSF).

National Oceanic and Atmospheric Administration (NOAA)

NOAA was established in 1970 within the U.S. Department of Commerce to ensure the safety of the general public from atmospheric phenomena and to provide the public with an understanding of the Earth's environment and resources. NOAA includes: the National Ocean Service which charts the oceans and waters of the U.S. and manages 265,000 acres of estuarine reserves; the National Marine Fisheries Service which maintains the world's largest and most complex marine fisheries management system; the NOAA Corps which operates 18 NOAA research and survey ships and flies 15 NOAA aircraft; and the Office of Oceanic and Atmospheric Research which supports experiments, laboratories, and the National Sea Grant College Program, among other efforts. NOAA has two main components: the National Weather Service (NWS), and the National Environmental Satellite, Data, and Information Service (NESDIS). The National Weather Service provides weather watch and warning services to the public through 57 Weather Service Forecast Offices (WSFO) and over 10O smaller local Weather Service Offices (WSOs) nationwide. Three national forecasting centers provide general and specialized guidance to WSFOs using computer forecast models, satellite data, and conventional surface and upper air observations from around the world. The centers are: National Meteorological Center; Camp Springs, Maryland; National Severe Storms Forecast Center; Kansas City, Missouri; National Hurricane Center; Coral Gables, Florida. NWS River Forecast Centers (RFCs) provide river stage and flood forecasts. NESDIS provides support to the Weather Service forecast mission by operating a series of environmental satellites and disseminating satellite imagery and derived products to the National Centers and WSFOs. NESDIS operates three national data and information centers; the National Geophysical Data Center; the National Climatic Data Center (NCDC), and the National Oceanographic Data Center (NODC). NOAA organizations perform numerous services in addition to monitoring weather conditions. They assess crop growth and other agricultural conditions, sense shifting ocean currents, and measure surface temperatures of oceans and land. They relay data from surface instruments that sense tide conditions, Earth tremors, river levels, and precipitation.

National Space Science Data Center (NSSDC)

The NSSDC provides on-line and off-line access to a wide variety of astrophysics, space plasma and solar physics, lunar and planetary and Earth science data from NASA space flight missions, in addition to selected other data, models, and software. Located at Goddard Space Flight Center (GSFC) in Greenbelt, Maryland, the NSSDC is sponsored by the Information Systems Office of NASAs Office of Space Sciences. NSSDC on-line data and services are currently free of charge, offline support (e.g., replications and mailing of magnetic tapes) are available for the cost of fulfilling the request. The NSSDC Master Catalog (NMC) provides an on-line listing of available data sets and the forms that the data are available in (such as CD-ROM), and provides information about the spacecraft and experiments (including past, present, and future NASA and non-NASA) from which these data were obtained. The on-line NASA Master Directory (NMD) identifies and briefly describes data of potential interest to the NASA research community, and where possible, provides electronic links to publicly-accessible data at sites world-wide. On-line information services are made available through the menu-based NSSDC Online Data Information Service (NODIS).

Natural Abundance

Percentage of an element occurring on earth in a particular stable isotopic form.

Nautical Mile

A unit of distance (U.S.) equal to exactly 1.852 kilometers or about 6076.1 feet. A nautical mile is approximately equal to 1/60 of a degree or 1 minute of arc of a great circle of the Earth (i.e., 1 minute of arc of latitude or of longitude at the equator).

NCDC

National Climatic Data Center; located in Asheville, North Carolina.

Near Infrared

Electromagnetic radiation with wavelengths from just longer than the visible (about 0.7 micrometers) to about two micrometers.

Nephanalysis

A type of analysis using satellite cloud pictures to study the relationship between cloud forms and storm systems. In classical mythology Nephele was a woman Zeus formed from a cloud.

Nepheloccygia

Clouds that resemble recognizable shapes.

NESDIS

National Environmental Satellite Data and Information Service.

Neutral

Having a net charge equal to zero. Unless otherwise specified, it usually refers to electric charge.

Neutrino

1. A lepton with no electric charge. Neutrinos participate only in weak and gravitational interactions and are therefore very difficult to detect. There are three known types of neutrinos, all of which are very light and could possibly have zero mass.

2. A *lepton* with no electric *charge* Neutrinos participate only in weak (and gravitational) *interactions* and therefore are very difficult to detect. There are three known types of neutrino, all of which have very low or possibly even zero mass.

3. An electrically neutral particle with negligible mass. It is produced in many nuclear reactions such as in beta decay.

4. A baryon with electric charge zero; it is a fermion with a basic structure of twc down quarks and one up quark (held together by gluons). The neutral component of an atomic nucleus is made from neutrons. Different isotopes of the same element are distin-guished by having different numbers of neutrons in their nucleus.

5. A *baryon* with electric *charge* zero. Its basic structure is two *down quarks a*d one up quark.

6. An uncharged particle, the other types that nuclei are made of.

7. One of the basic particles which make up an atom. A neutron and a proton have about the same weight, but the neutron has no electrical charge.

Newton's first and second laws of motion

F = *d(mv)/dt. F* is the *net* (total) force acting on the body of mass m. The individual forces acting on m must be summed vectorially. In the special case where the mass is constant, this becomes *F = ma.*

8. An electrically neutral particle with negligible mass. It is produced in many nuclear reactions such as in *beta decay*. In our diagrams, it is represented by this:

Neutron Decay

Nuclear decay by emission of a *neutron.*

Neutron Degeneracy

Neutron degeneracy is a stellar application of the Pauli Exclusion Principle, as is electron degeneracy. No two neutrons can occupy identical states, even under the pressure of a collapsing star of several solar masses. For stellar masses less than about 1.44 solar masses (the Chandrasekhar limit), the energy from the gravitational collapse is not sufficient to produce the neutrons of a neutron star, so the collapse is halted by electron degeneracy to form white dwarfs. Above 1.44 solar masses, enough energy is available from the gravitational collapse to force the combination of electrons and protons to form neutrons. As the star contracts further, all the lowest neutron energy levels are filled and the neutrons are forced into higher and higher energy levels, filling the lowest unoccupied energy levels. This creates an effective pressure which prevents further gravitational collapse, forming a neutron star. However, for masses greater than 2 to 3 solar masses, even neutron degeneracy can't prevent further collapse and it continues toward the black hole state.

Neutron Star

1. If a star's mass is too great, its nuclear matter will be compressed beyond the limits given by a *white dwarf.* The electrons and protons of the star's matter will combine to form neutrons, and

the star will in some cases possess regions that are more dense than an atomic nucleus. In a sense, Neutron Stars are like giant atomic nuclei - although the physics of so large an object as a neutron star will have many important differences.

2. For a sufficiently massive star, an iron core is formed and still the gravitational collapse has enough energy to heat it up to a high enough temperature to either fuse or fission iron. Either in the aftermath of a supernova or in just a collapsing massive star, the energy gets high enough to break down the iron into alpha particles and other smaller units, and still the pressure continues to build. When it reaches the threshold of energy necessary to force the combining of electrons and protons to form neutrons, the electron degeneracy limit has been passed and the collapse continues until it is stopped by neutron degeneracy. At this point it appears that the collapse will stop for stars with mass less than two or three solar masses, and the resulting collection of neutrons is called a neutron star. The periodic emitters called pulsars are thought to be neutron stars. If the mass exceeds about three solar masses, then even neutron degeneracy will not stop the collapse, and the core shrinks toward the black hole condition. This neutron degeneracy radius is about 20 km for a solar mass, compared to about earth size for a solar mass white dwarf. The density is quoted as about a billion tons per teaspoonful compared to 5 tons per teaspoonful for the white dwarf. Pasachoff suggests that neutron stars may be crystalline with crusts on the order of 100 meters thick and an atmosphere a few centimeters thick. They may have 1011x the earths gravity and a powerful magnetic field. A neutron star might have an atmosphere a few centimeters thick and mountain ranges poking up a few centimeters through the atmosphere. A neutron star is thought to be about 1/100,000 the diameter of the Sun, and a nucleus is on the order of 100,000 times smaller than an atom.

Neutron

1. An electrically neutral elementary particle. A neutron is 1839 times heavier than an *electron.*

2. One of the basic particles which make up an atom. A neutron and a *proton* have about the same weight, but the neutron has no electrical charge.
3. Neutrons are neutral particles that are normally contained in the nucleus of all atoms and may be removed by various interactions or processes like collision and fission

Newton's First Law

Law of Inertia. This law is also called the Law of Inertia or Galileo's Principle. Alternative Formulations. Every body's *center of mass* continues in its state of rest, or of uniform motion in a right [straight] *line*, unless it is compelled to change that state by forces impressed upon it. *A body's center of mass remains at rest, or moves in a straight line (at a constant velocity, v), unless acted upon by a net outside force.* In calculus notation, this may be expressed as:

$$\frac{d}{dt}v = 0$$

Despite the fact that Newton's First Law appears to be a special case of Newton's Second Law, the First Law defines the reference frames in which the other two laws are valid. These reference frames are called inertial reference frames or Galilean reference frames, and are moving at constant velocity, that is to say, without acceleration. (Note that an object may have a constant speed and yet have a non-zero acceleration, as in the case of uniform circular motion. This means that the surface of the Earth is not an inertial reference frame, since the Earth is rotating on its axis and orbits around the Sun. However, for many experiments, the Earth's surface can safely be assumed to be inertial. The error introduced by the acceleration of the Earth's surface is minute.) In less formal terms, Aristotle thought that things stood still if you left them alone, that to be at rest was natural, and that movement needed a cause. It would be natural to think thus, as any movement (except for that of celestial objects, which were deemed perfect) that one observes eventually stops because of friction. But Galileo's experiments, with a ball rolling down an inclined plane , found that "Things travel naturally at a steady speed (which may or may not be zero), if left alone". Moving from Aristotle's "A body's natural state is at rest" to Galileo's discovery (Newton's First Law) was one of the most profound and important discoveries in physics. In everyday life, the force of friction usually acts upon moving objects, slowing them down and eventually bringing them to rest. Newton described a mathematical model from which one could derive the motions of bodies from elementary causes: *forces*.

Newton's Law of Universal Gravitation

All bodies attract each other with what is called gravitational attraction. This applies to the largest stars as well as the smallest particles of matter The force of attraction between two small bodies (or between two spherical bodies of any size) is proportional to the product of their masses and inversely proportional to the square of the distance between their centers. In other words, the closer two bodies are to each other; the greater their mutual attraction. As a result, to stay in orbit, a satellite needs more speed in a low than a high orbit. Kepler's three laws of planetary motion, which had been derived empirically by Johannes Kepler; were obtained with mathematical rigor as a consequence of Newton's law of universal gravitation in conjunction with his three laws of motion.

Newton's Laws of Motion

1. Newton's three laws of motion are: Every body continues in a state of uniform motion in a straight line unless acted upon by some external force. The time rate of change of momentum (mass x velocity) is proportional to the impressed force. In the usual case where the mass does not change, this law can be expressed in the familiar form: force = mass × acceleration or F = ma. To every force or action, there is always an equal and opposite reaction. Kepler's three laws of planetary motion, which had been derived empirically by Johannes Kepler; were obtained with mathematical rigor as a consequence of Newton's law of universal gravitation in conjunction with his three laws of motion.

2. *Newton's laws of motion* are the three scientific laws which Isaac Newton discovered concerning the behaviour of moving bodies. These laws are fundamental to classical mechanics. Newton first published these laws in *Philosophiae Naturalis Principia Mathematica* (1687) and used them to prove many results concerning the motion of physical objects. In the third volume (of the text), he showed how, combined with his law of universal gravitation, the laws of motion would explain Kepler's laws of planetary motion.

Newton's Second Law

Fundamental law of dynamics. Alternative formulations: *The rate of change in momentum is proportional to the net force acting on the object and takes place in the direction of the force. The acceleration of an object of constant mass is proportional to the resultant force acting upon it.* The quantity *m*, or mass, in the above equation is is a characteristic of the object. For an object of constant mass *m* (a constant of proportionality) the more net force acts on an object, the greater the change in its acceleration will be. This equation, therefore, indirectly defines the concept of mass. In the equation, $F = ma$, a is directly measurable but F is not. The second law only has meaning if we are able to assert, in advance, the value of F. Rules for calculating force include Newton's law of universal gravitation. But $\mathbf{F} = \mathbf{ma}$ is not always valid. In general both the mass of the object and its velocity can be variable. For this case:

$$\mathbf{F} = \frac{d}{dt}(m\mathrm{v}) = m\frac{d\mathrm{v}}{dt} + \mathrm{v}\frac{dm}{dt} = m\mathrm{a} + \mathrm{v}\frac{dm}{dt}$$

This equation works in cases when the mass is variable. This equation is also valid in special relativity if we express the momentum as

$$\mathrm{p} = \gamma m\mathrm{v},$$

where γ is the well known

$$\frac{1}{\sqrt{1-\frac{v^2}{c^2}}}.$$

The physical meaning behind this equation is important as it implies that *objects interact by exchanging momentum, and they do this via a force.* Taken together with Newton's Third Law of Motion, Newton's Second Law implies the Law of Conservation of Momentum.

Newton's Third Law of Motion

When body A exerts a force on body B, then B exerts and equal and opposite force on A. The two forces related by this law act on *different bodies*. The forces need not be *net* forces.

Newton's Third Law : Law of Reciprocal Actions

Alternative formulations: Whenever one body exerts force upon a second body, the second body exerts an equal and opposite force upon the first body. Momentum is conserved. The very common formulation "for every action there is an equal and opposite reaction" should be avoided, as it is, at best, ambiguous and confusing. A better formulation would be that when there exists a force acting on a body A, due to another body B, there exists also a reciprocal force, acting on body B, due to the existence of body A. These formulations imply that if you strike an object with a force of 200 N, then the object also strikes you (with a force of 200 N). Not only do planets accelerate toward stars; but, stars accelerate toward planets. The reaction force has the opposite direction of action, and is of the same type and magnitude as the original force. However, it doesn't necessarily "line up" in space with the action. One example of this is a force on an electric dipole due to a point charge, when the dipole points in a direction perpendicular to the line connecting the point charge and the dipole. The force on the dipole due to the point charge is perpendicular to the line connecting them, so there is a reaction force on the point charge in the opposite direction, but these two force vectors are parallel and, even when extended to a line, they never cross each other in space.

NGDC

National Geophysical Data Center; located in Boulder; Colorado.

Nibble

Four bits of data.

Nimbus Satellite Program

A NASA program to develop observation systems meeting the research and development requirements of atmospheric and Earth scientists. The Nimbus satellites, first launched in 1964, carried a number of instruments: microwave radiometers, atmospheric sounders, ozone mappers, the Coastal Zone Color Scanner (CZCS), infrared radio-meters, etc. Nimbus-7, the last in the series, provided significant global data on sea-ice coverage, atmospheric temperature, atmospheric chemistry (*i.e.* ozone distribution), the Earth's radiation budget, and sea-surface temperature.

NOAA

Operational designation for the U.S. polar-orbiting meteorological satellites. Current NOAA spacecraft are variations of the TIROS-N bus.

Non-ionizing Radiation

Non-ionizing radiation is radiation without enough energy to remove tightly bound electrons from their orbits around atoms. Examples are microwaves and visible light.

Noninertial Frame

An accelerating frame of reference, in which Newton's first law is violated.

Nonlinear Process

In linear processes the output is directly proportional to its input. For example, the pressure of a gas in a fixed volume is directly proportional to its temperature. In nonlinear processes this direct proportionality is lost. Two basic types of equations, linear and non-linear, describe such processes mathematically. Linear equations, like : y=mx+b (the formula for a straight line) are generally easy to solve, whereas nonlinear equations, such as : xy^3 + x^2 +

(y-x)^2 = 6, are much harder to crack, and the solutions, if they can be found at all, can behave in unexpected ways. Linear processes can be accounted for by the sum of their parts and are easy to predict. Not so for nonlinear processes: they tend to be complex; their outcomes can be difficult to predict and often display so-called chaotic behavior, and the mathematical equations describing them can in some cases be very hard to solve. This is especially so in the case of general relativity. The Einstein Equations contain thousands of terms in many variables, not just x and y, and these terms are nonlinear. For all but the simplest spacetimes it's impossible to solve them precisely using traditional, *analytical* methods. However, weak spacetime curvature, e.g. near the earth, is close to linear: the tidal pulls of the sun and moon add up to result in the oceans' tides. But near a black hole, Einstein's Equations predict that the curvature of spacetime is highly non-linear. At the center of the black hole, distance and time become infinitely stretched! Furthermore, the nonlinearity of the equations can lead to strange effects such as the formation of black holes where none exist initially. Nonlinearity complicates life, but makes it more interesting!

Non-stochastic Effect

Non-stochastic effects are effects that can be related directly to the dose received. The effect is more severe with a higher dose, *i.e.*, the burn gets worse as dose increases. It typically has a threshold, below which the effect will not occur. A skin burn from radiation is a non-stochastic effect.

Nonthermal Particle

A particle that is not part of a *thermal gas*. These particles cannot be described by a conventional temperature.

Nonthermal Radiation

Radiation emitted by *nonthermal, electrons*.

Nonuniform Circular Motion

Circular motion in which the magnitude of the velocity vector changes.

Normal Force

The force that keeps two objects from occupying the same space.

Normalization

The property of probabilities that the sum of the probabilities of all possible outcomes must equal one.

Notation

When a body releases heat into its surroundings, $Q<0$. When a body absorbs heat from its surroundings, $Q>0$. Total heat, heat transfer rate, and heat flux are all notated with different permutations of the letter Q. They are often confusingly switched in different contexts. *Total heat* is notated as Q, and is measured in joules in SI units.

NRA

NASA Research Announcement.

NREN

National Research and Education Network.

NSFNET

National Science Foundation NETwork.

NSR

The Nuclear Science Reference file is a compilation of about 160,000 references relevant to nuclear structure and decay.

N-Type Semiconductor

The addition of pentavalent *impurities* such as antimony, arsenic or phosphorous contributes free electrons, greatly increasing the conductivity of the *intrinsic semiconductor*. Phosphorous may be added by diffusion of phosphine gas (PH_3).

Fig on next page.

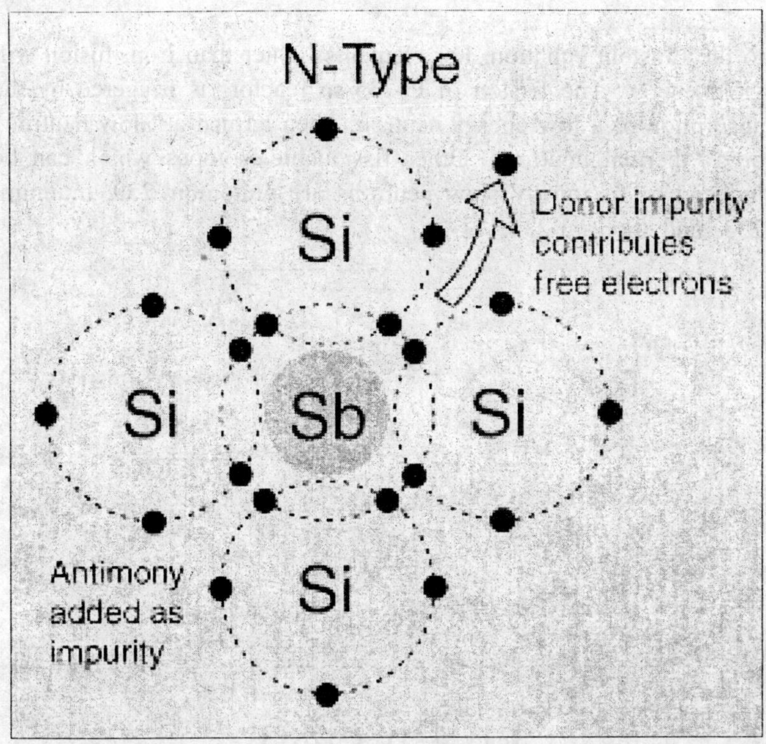

Nuclear Binding Energy

Nuclei are made up of protons and neutron, but the mass of a nucleus is always less than the sum of the individual masses of the protons and neutrons which constitute it. The difference is a measure of the nuclear binding energy which holds the nucleus together.

Nuclear Fission

If a massive nucleus like uranium-235 breaks apart (fissions), then there will be a net yield of energy because the sum of the masses of the fragments will be less than the mass of the uranium nucleus. If the mass of the fragments is equal to or greater than that of iron at the peak of the binding energy curve, then the nuclear particles will be more tightly bound than they were in the uranium nucleus, and that decrease in mass comes off in the form of energy according to the Einstein equation. For elements lighter than iron, fusion will yield energy. The fission of U-235 in reactors is triggered by the absorption of a low energy neutron, often termed a "slow neutron" or a "thermal neutron". Other fissionable isotopes which can be induced to fission by slow neutrons are plutonium-239, uranium-233, and thorium-232.

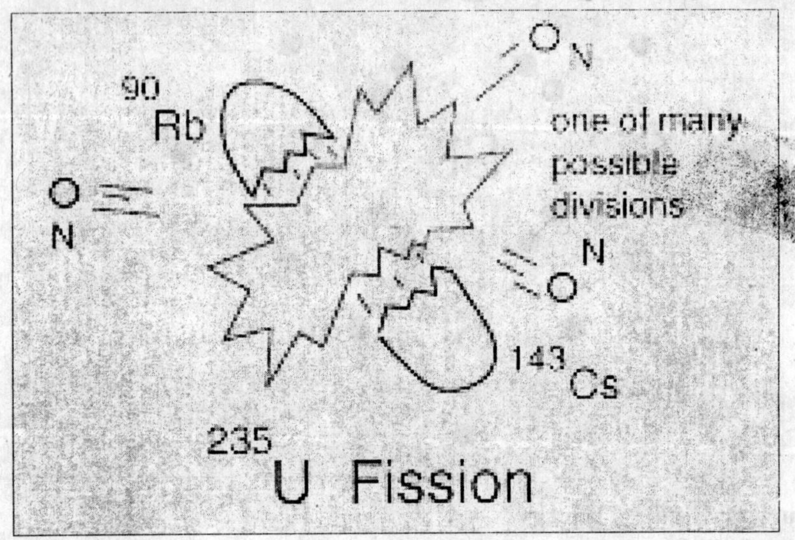

Nuclear Reactor

A device in which a fission chain reaction can be initiated, maintained, and controlled. Its essential components are fissionable fuel, moderator, shielding, control rods, and coolant.

Nucleon

A constituent of the nucleus; that is, a proton or a neutron.

Nucleonics

The science, technology, and application of nuclear energy.

Nucleus

1. A collection of neutrons and protons that forms the core of an atom (plural: nuclei).

2. The core of the atom, where most of its mass and all of its positive charge is concentrated. Except for hydrogen, it consists of protons and neutrons.

3. The core of the atom, where most of its mass and all of its positive charge is concentrated. Except for hydrogen, it consists of *proton* and *neutrons.*

4. The positively charged core of an atom, consisting of *protons* and *neutrons* (except for hydrogen), around which *electrons* orbit.

Nuclide

Any species of atom that exists for a measurable length of time. A nuclide can be distinguished by its atomic weight, atomic number, and energy state.

O

Oasis

spot in a desert made fertile by water; which normally originates as groundwater.

Oblique

Describes a force that acts at some other angle, one that is not a direct repulsion or attraction.

Occluded Front (Occlusion)

A composite of two fronts formed as a cold front overtakes a warm front. A cold occlusion results when the coldest air is behind the cold front. The cold front undercuts the warm front and, at the Earth's surface, coldest air replaces less-cold air. A warm occlusion occurs when the coldest air lies ahead of the warm front. Because the cold front can not lift the colder air mass, it rides piggyback up on the warm front over the coldest air

Ocean

The salt water surrounding the great land masses. The land masses divide the ocean into several distinct portions, each of which also is called an ocean. The oceans include the Pacific Ocean, the Atlantic Ocean, the Indian Ocean, and the Arctic Ocean.

Ohm

1. The metric unit of electrical resistance, one volt per ampere.
2. The unit of electrical resistance, equal to the resistance of a circuit in which an electromotive force of one volt maintains a

current of one ampere. Named for German physicist Georg S. Ohm (1787-1854).

Ohmic

Describes a substance in which the flow of current between two points is proportional to the voltage difference between them.

Ohm's Law

$V = IR$, where V is the potential across a circuit element, I is the current through it, and R is its resistance. This is not a generally applicable definition of resistance. It is only applicable to *ohmic* resistors, those whose resistance R is constant over the range of interest and V obeys a strictly linear relation to I. Materials are said to be *ohmic* when V depends linearly on R. Metals are ohmic so long as one holds their temperature constant. But changing the temperature of a metal changes R slightly. Therefore such a device as an electric light bulb increases its temperature as it warms up, which is why it glows slightly brighter for a very brief time just after it is turned on. For non-ohmic resistors, R is a function of current and the definition $R = dV/dI$ is far more useful. This is sometimes called the *dynamic resistance*. Solid state devices such as thermistors are non-ohmic, and non-linear. A thermistor's resistance decreases as it warms up, so its dynamic resistance is negative. Tunnel diodes and some electrochemical processes have a complicated I-V curve with a negative resistance region of operation. The dependence of resistance on current is partly due to the change in the device's temperature with increasing current, but other subtle processes also contribute to change in resistance in solid state devices.

Open Circuit

A circuit that does not function because it has a gap in it.

Operational Definition

1. A definition that states what operations should be carried out to measure the thing being defined.

2. A definition which describes an *experimental procedure* by which a numeric value of the quantity may be determined. *Example:* Length is operationally defined by specifying a *procedure* for subdividing a standard of length into smaller units to make a measuring stick, then laying that stick on the object to be measured, etc.. *Very* few quantities in physics need to be operationally defined. They are the *fundamental* quantities, which include length, mass and time. Other quantities are defined from these through mathematical relations.

Optical Radiation

Electromagnetic radiation (light) that is visible to the human eye.

Optical Sign Conventions

In introductory (freshman) courses in physics a sign convention is used for objects and images in which the lens equation must be written $1/p + 1/q = 1/f$. Often the rules for this sign convention are presented in a convoluted manner. A simple and easy to remember rule is this: p is the *object-to-lens* distance. q is the *lens to image* distance. The coordinate axis along the optic axis is in the direction of passage of light through the lens, this defining the *positive* direction. Example: If the axis and the light direction is left-to-right (as is usually done) and the object is to the left of the lens, the object-to-lens distance is positive. if the object is to the right of the lens (virtual object), the object-to-lens distance is negative. It works the same for images. For refractive surfaces, define the surface radius to be the directed distance from a surface to its center of curvature. Thus a surface convex to the incident light is positive, one concave to the incident light is negative. The surface equation is then $n/s + n'/s' = (n'-n)/R$ where s and s' are the object and image distances, and n and n' the refractive index of the incident and emergent media, respectively. For mirrors, the equation is usually written $1/s + 1/s' = 2/R = 1/f$. A diverging mirror is convex to the incoming light, with negative f. From this fact we conclude that R is also negative. This form of the equation is consistent with that of the lens equation, and the interpretation of sign of focal length is the same also. But violence is done to the definition of R we used above, for refraction. One can say that the mirror *folds* the length axis at the mirror, so that emergent rays to a real image at the left represent a positive value of s'. We are forced also to declare that the mirror also flips the sign of the surface radius. For reflective surfaces, the radius of curvature is defined to be the directed distance from a surface to its center of curvature, *measured with respect to the axis used for the emergent light.* With this qualification the convention for the signs of s' and R is the same for mirrors as for refractive surfaces. In advanced optics courses, a cartesian sign convention is used in which all things to the left of the lens are negative, all those to the right are positive. When this is used, the lens equation must be written $1/p + 1/f = 1/q$. (The sign of the $1/p$ term is opposite that in the other sign convention). This is a particularly meaningful version, for $1/p$ is the measure of vergence (convergence or divergence) of the rays as they enter the lens, $1/f$ is the amount the lens changes the vergence, and $1/q$ is the vergence of the emergent rays.

Orbit

The path described by a heavenly body in its periodic revolution. Earth satellite orbits with inclinations near 0° are called equatorial orbits because the satellite stays nearly over the equator. Orbits with inclinations near 90° are called polar orbits because the satellite crosses over (or nearly over) the north and south poles.

Orbital Plane

An imaginary gigantic flat plate containing an Earth satellite's orbit. The orbital plane passes through the center of the Earth.

Orbital Period

The amount of time it takes a spacecraft or other object to travel once around it's orbit.

Osmosis

If two solutions of different concentration are separated by a semi-

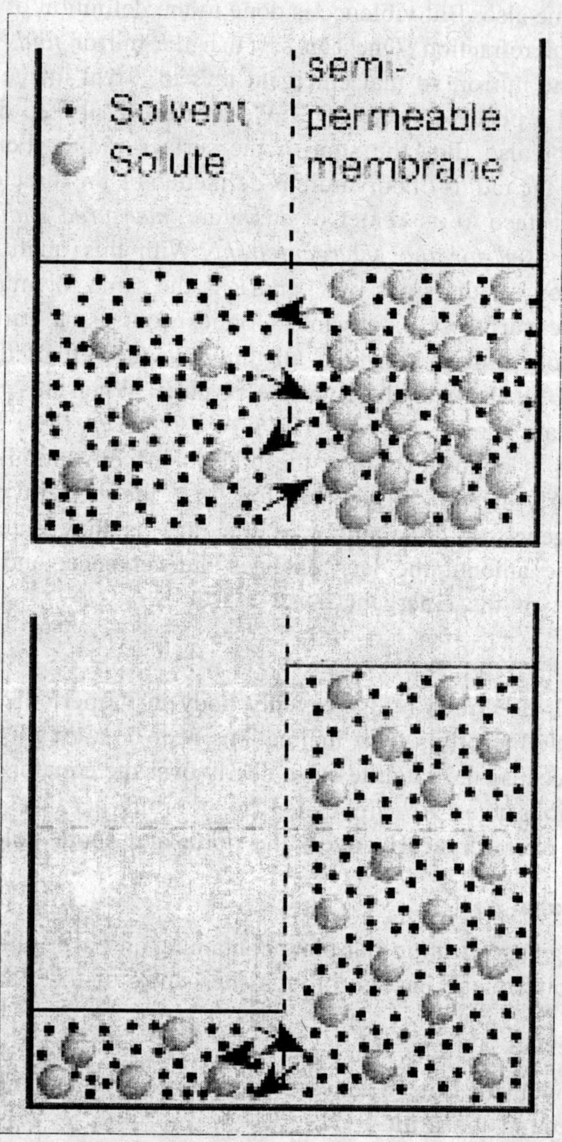

permeable membrane which is permeable to to the smaller solvent molecules but not to the larger solute molecules, then the solvent will tend to *diffuse* across the membrane from the less concentrated to the more concentrated solution. This process is called osmosis.

Osmosis is of great importance in biological processes where the solvent is water. The transport of water and other molecules across biological *membranes* is essential to many processes in living organisms. The energy which drives the process is usually discussed in terms of *osmotic pressure*.

Ozone Hole

A large area of intense stratospheric ozone depletion over the Antarctic continent that typically occurs annually between late August and early October; and generally ends in mid November. This severe ozone thinning has increased conspicuously since the late seventies and early eighties. This phenomenon is the result of chemical mechanisms initiated by man-made chlorofluorocarbons. Continued buildup of CFCs is expected to lead to additional ozone loss worldwide. The thinning is focused in the Antarctic because of particular meteorological conditions there. During Austral spring (September and October in the Southern Hemisphere) a belt of stratospheric winds encircles Antarctica essentially isolating the cold stratospheric air there from the warmer air of the middle latitudes. The frigid air permits the formation of ice clouds that facilitate chemical interactions among nitrogen, hydrogen, and chlorine (elevated from CFCs) atoms, the end product of which is the destruction of ozone.

Ozone Layer

The layer of ozone that begins approximately 15 km above Earth and thins to an almost negligible amount at about 50 km, shields the Earth from harmful ultraviolet radiation from the sun. The highest natural concentration of ozone (approximately 10 parts per million by volume) occurs in the stratosphere at approximately 25 km above Earth. The stratospheric ozone concentration changes throughout the year as stratospheric circulation changes with the seasons. Natural events such as volcanoes and solar flares can produce changes in ozone concentration, but man-made changes are of the greatest concern.

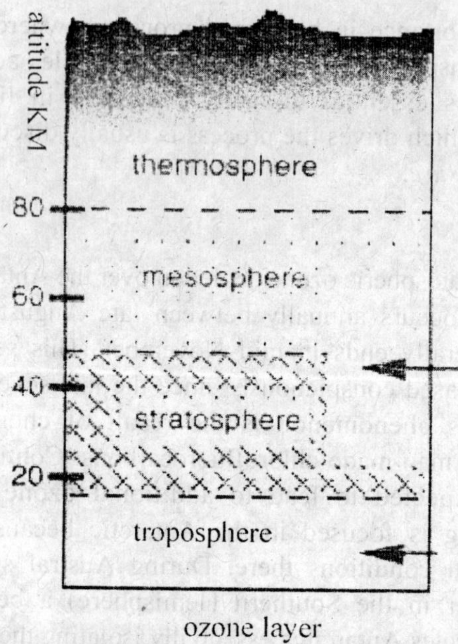

Ozone Mini-Hole(S)

Rapid, transient, polar-ozone depletion. These depletions. which take place over a 50-kilometer squared area, are caused by weather patterns in the upper troposphere. The decrease in ozone during a mini-hole event is caused by transport, with no chemical depletion of ozone. However; the cold stratospheric temperatures associated with weather systems can cause clouds to form that can lead to the conversion of chlorine compound from inert to reactive forms. These chlorine compounds can then produce longer-term ozone reductions after the mini-hole has passed.

Ozone

An almost colorless, gaseous form of oxygen with an odor similar to weak chlorine. A relatively unstable compound of three atmos of oxygen, ozone constitutes—on the average—less than one part per million (ppm) of the gases in the atmosphere (peak ozone concentration in the stratosphere can get as high as 10 ppm). yet ozone in

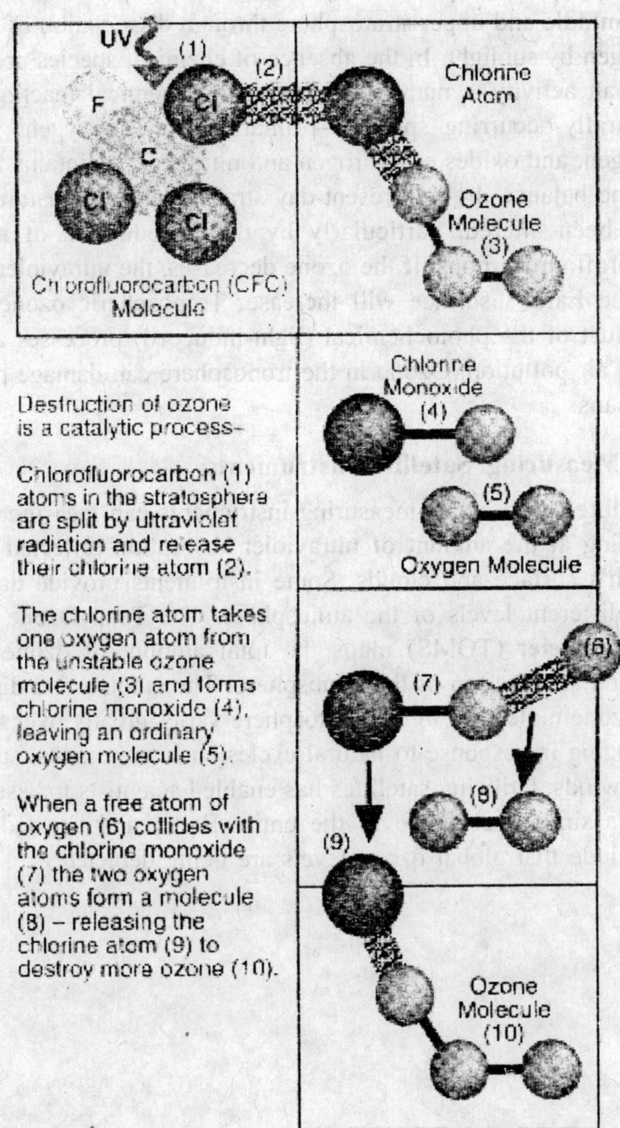

the stratosphere absorbs nearly all of the biologically damaging solar ultraviolet radiation before it reaches the Earth's surface where it can cause skin cancer, cataracts, and immune deficiencies, and can harm crops and aquatic ecosystems. Ozone is produced naturally in the middle and upper stratosphere through dissociation of molecular oxygen by sunlight. In the absence of chemical species produced by human activity, a number of competing chemical reactions among naturally-occurring species—primarily atomic oxygen, molecular oxygen, and oxides of hydrogen and nitrogen—maintains the proper ozone balance. In the present-day stratosphere, this natural balance has been altered, particularly by the introduction of man-made chloroflourocarbons. If the ozone decreases, the ultraviolet radiation at the Earth's surface will increase. Tropospheric ozone is a by-product of the photochemical (light-indueced) processes associated with air pollution. Ozone in the troposphere can damage plants and humans.

Ozone-Measuring Satellite Instruments

Satellite-based ozone-measuring instruments can measure ozone by looking at the amount of ultraviolet absorption reflected from the Earth's surface and clouds. Some instruments provide data within the different levels of the atmosphere. The Total Ozone Mapping Spectrometer (TOMS) maps the total amount of ozone between ground and the top of the atmosphere. The amount and distribution of ozone molecules in the stratosphere varies greatly over the globe, changing in response to natural cycles such as seasons, sun cycles. and winds. Utilizing satellites has enabled scientists to assess ozone levels simultaneously over the entire Earth, and has led them to conclude that global ozone levels are being depleted.

P

Pair Production and Annihilation

Whenever sufficient energy is available to provide the mass-energy, a particle and its matching antiparticle can be produced (pair production). When a particle collides with its matching antiparticle they may annihilate — which means they both disappear and their energy appears as some other particles — with balanced number of particles and *antiparticles* for each type. All conservation laws are obeyed in these processes.

Paleoclimate

Climate as it existed in the distant past, particularly before historical records.

Paleogeography

The study of ancient or prehistoric geography.

Panchromatic

Sensitive to all or most of the visible spectrum.

Parabola

The mathematical curve whose graph has y proportional to x^2.

Parallax

A nearby star's apparent movement against the background of more distant stars as the Earth revolves around the Sun is referred to as parallax. The parallax can be used to measure the distance to the few stars which are close enough to the Sun to show a measurable parallax. The distance to the star is inversely proportional to the parallax. The distance to the star in *parsecs* is given by

$$\mathbf{d} = \frac{1}{\mathbf{p}} \qquad \begin{array}{l} \text{d in parsecs} \\ \text{p in seconds of arc.} \end{array}$$

The nearest star is *proxima centauri*, which exhibits a parallax of 0.762 arcsec, and therefore is 1.31 parsecs away. Some well-known examples of distance measurement by parallax are 61 Cygni at 1/3 of an arcsec, distance 3 parsecs, and Barnard's Star at 5.9 parsecs.

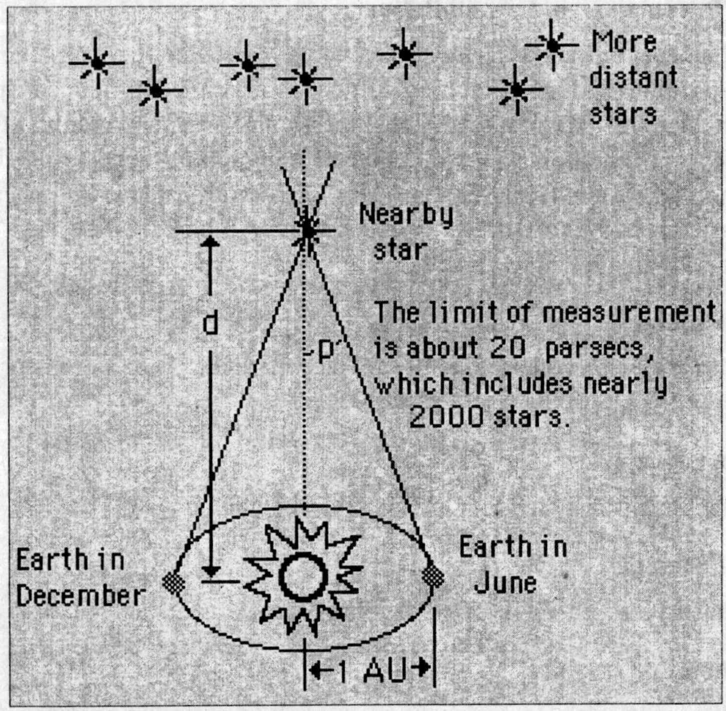

Parallel Light from a Laser

The light from a typical laser emerges in an extremely thin beam with very little divergence. Another way of saying this is that the beam is highly "collimated". An ordinary laboratory *helium-neon laser* can be swept around the room and the red spot on the back wall seems about the same size at that on a nearby wall. The high degree of collimation arises from the fact that the cavity of the laser has very nearly parallel front and back mirrors which constrain the final laser beam to a path which is perpendicular to those mirrors. The back mirror is made almost perfectly reflecting while the front mirror is about 99% reflecting, letting out about 1% of the beam. This 1% is the output beam which you see. But the light has passed back and forth between the mirrors many times in order to gain intensity by the *stimulated emission* of more photons at the same wavelength. If the light is the slightest bit off axis, it will be lost from the beam.

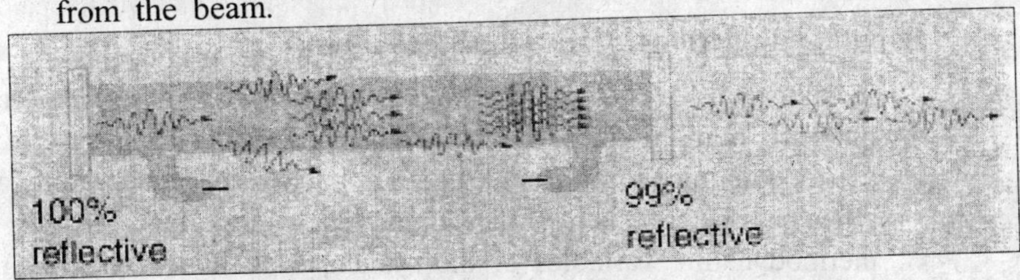

Parallelogram Law

The law applying to the addition of vectors. Each of the vectors is represented in magnitude and in direction by the sides of a parallelogram. The included diagonal of the parallelogram represents, in magnitude and in direction, the resultant of the two vectors. The other diagonal equals the difference between the two vectors.

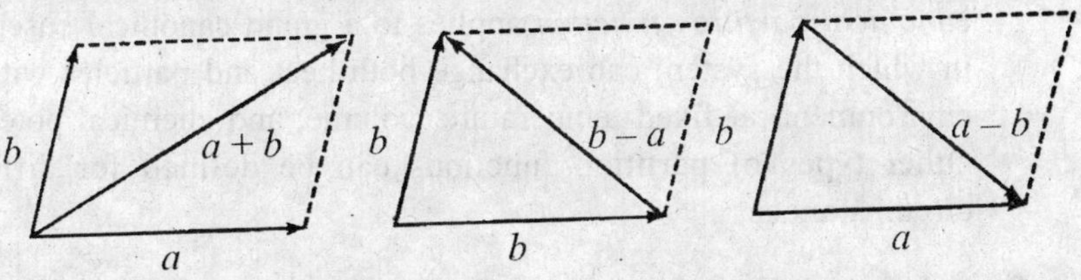

Parent

A radionuclide that decays to another nuclide which may be either radioactive or stable.

Parity

1. The addition of one or more redundant bits to information to verify its accuracy.

2. A *nucleus* or particle has odd (-) or even (+) parity according to whether or not its wave function changes sign when all of the space coordinates are changed.

Particle

1. A subatomic object with a definite mass and charge.
2. In "particle physics", a subatomic object with definite mass and *charge.*

Partition Function (Statistical Mechanics)

In statistical mechanics, the *partition function Z* is an important quantity that encodes the statistical properties of a system in thermodynamic equilibrium. It is a function of temperature and other parameters, such as the volume enclosing a gas. Most of the thermodynamic variables of the system, such as the total energy, free energy, entropy, and pressure, can be expressed in terms of the partition function or its derivatives. There are actually several different types of partition function, each corresponding to different types of statistical ensemble (or, equivalently, different types of free energy.) The *canonical partition function* applies to a canonical ensemble, in which the system is allowed to exchange heat with the environment at fixed temperature, volume, and number of particles. The *grand canonical partition function* applies to a grand canonical ensemble, in which the system can exchange both heat and particles with the environment, at fixed temperature, volume, and chemical potential. Other types of partition functions can be defined for different circumstances.

Path Integral Formulation

This article is about a formulation of quantum mechanics. For integrals along a path, also known as line or contour integrals, see Path integral.

These are just three of the paths that contribute to the quantum amplitude for a particle moving from point A at some time t0 to point B at some other time t1. Richard Feynman developed the *path integral formulation* of quantum mechanics in 1948 (some preliminaries were worked out earlier, in the course of his doctoral thesis work with John Archibald Wheeler) as a description of quantum theory corresponding to the action principle of classical mechanics. It replaces the classical notion of a single, unique history for a system with a sum, or functional integral, over an infinity of possible histories to compute a quantum amplitude.

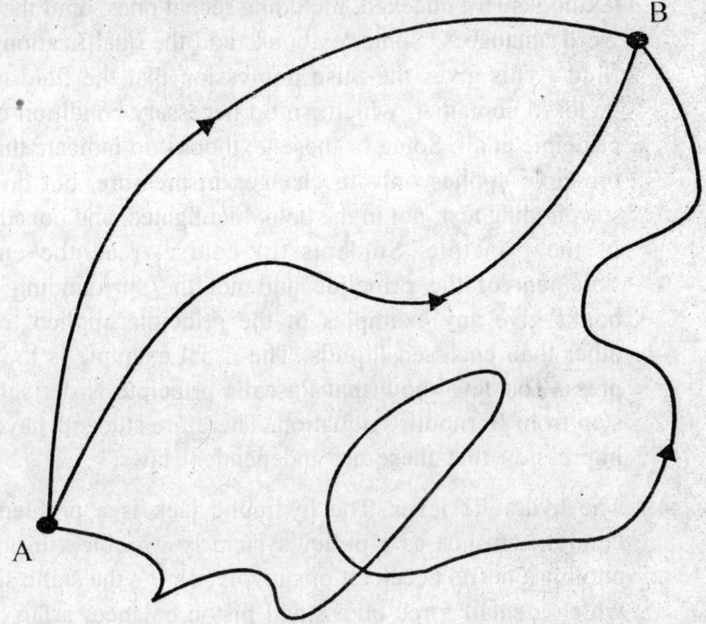

Pascal (Pa)

1. Unit of atmospheric pressure named in honor of Blaise Pascal (I 632-1 662), whose experiments greatly increased knowledge of the atmosphere. A pascal is the force of one newton acting on a surface area of one square meter. It is the unit of pressure designated by the International System. 100,OOO Pa= 1000mb 1 bar.
2. The pressure at any point in a liquid exerts force equally in all directions. This means that an infinitessimal surface area placed at that point will experience the same force due to pressure no matter what its orientation.

3. When pressure is changed (increased or decreased) at any point in a homogenous, incompressible fluid, all other points experience the same change of pressure. Except for minor edits and insertion of the words 'homogenous' and 'incompressible', this is the statement of the principle given in John A. Eldridge's textbook *College Physics* (McGraw-Hill, 1937). Yet over half of the textbooks I've checked, including recent ones, omit the important word 'changed'. Some textbooks add the qualification 'enclosed fluid'. This gives the false impression that the fluid must be in a closed container, which isn't a necessary condition of Pascal's principle at all. Some of these textbooks do indicate that Pascal's principle applies only to changes in pressure, but do so in the surrounding text, not in the bold, highlighted, and boxed statement of the principle. Students, of course, read the emphasized statement of the principle and not the surrounding text. Few books give any examples of the principle applied to anything other than enclosed liquids. The usual example is the hydraulic press. Too few show that Pascal's principle is derivable in one step from Bernoulli's equation. Therefore students have the false impression that these are independent laws.

4. The hydraulic lever. The hydraulic jack is a problem in fluid equilibrium, just as a pulley system is a problem in mechanical equilibrium (no accelerations involved). It's the static situation in which a small force on a small piston balances a large force on a large piston. No change of pressure need be involved here. A constant force on one piston slowly lifts a different piston with a constant force on it. At all times during this process the fluid is in near-equilibrium. This 'principle' is no more than an application of the definition of pressure as *F/A*, the quotient of net force to the area over which the force acts. However, it also uses the principle that pressure in a fluid is uniform throughout the fluid at all points of the same height. This hydraulic jack lifitng process is done at constant speed. If the two pistons are at different levels, as they usually are in real jacks used for lifting, there's a pressure difference between the two pistons due to height difference *(rho)gh*. In textbook examples this is generally

considered small enough to neglect and may not even be mentioned. Pascal's own discussion of the principle is not concisely stated and can be misleading if hastily read. See his *On the Equilibrium of Liquids*, 1663. He inroduces the principle with the example of a piston as part of an enclosed vessel and considers what happens if a force is applied to that piston. He concludes that each portion of the vessel is pressed in proportion to its area. He does mention parenthetically that he is 'excluding the weight of the water..., for I am speaking only of the piston's effect.'

Pascal's Principle of Hydrostatics

Pascal actually has three separate principles of hydrostatics. When a textbook refers to *Pascal's Principle* it should specify which is meant.

Passive System

A system sensing only radiation emitted by the object being viewed or reflected by the object from a source other than the system.

Pauli Exclusion Principle

1. Fermions obey a rule called the Pauli Exclusion Principle, which states that no two fermions can exist in the same state at the same time.
2. No two *fermions* of the same type can exist in the same state at the same place and time.

Payload

The instruments that are accommodated on a spacecraft.

Per Unit

In my opinion this expression is a barbarism best avoided. When a student is told that electric field is *force per unit charge* and in the MKS system one unit of charge is a coulomb (a *huge* amount) must we obtain that much charge to measure the field? Certainly not. In fact, one must take the limit of F/q as q goes to zero. Simply say:

'Force divided by charge' or 'F over q' or even 'force per charge'. Unfortunately there is no graceful way to say these things, other than simply writing the equation. *Per* is one of those frustrating words in English. The *American Heritage Dictionary* definition is: 'To, for, or by each; for every.' Example: '40 cents per gallon.' We must put the blame for *per unit* squarely on the scientists and engineers.

Percentage

Older dictionaries suggested that *percentage* be used when a non-quantitative statement is being made: 'The percentage growth of the economy was encouraging.' But use *percent* when specifying a numerical value: 'The gross national product increased by 2 percent last year.' Though newer dictionaries are more permissive, I find the indiscriminate and unnecessary use of the ugly word *percentage* to be overdone and annoying, as in 'The experimental percentage uncertainty was 9%.' Much more graceful is: 'The experimental uncertainty was 9%.' Related note: Students have the strange idea that results are *better* when expressed as percents. Some experimental uncertainties must *not* be expressed as percents. Examples: (1) temperature in Celsius or Fahrenheit measure, (2) index of refraction, (3) dielectric constants. These measurables have arbitrarily chosen 'fixed points'. Consider a 1 degree uncertainty in a temperature of 99 degrees C. Is the uncertainty 1%? Consider the same error in a measurement of 5 degrees. Is the uncertainty now 20%? Consider how much smaller the percent would be if the temperature were expressed in degrees Kelvin. This shows that percent uncertainty of Celsius and Fahrenheit temperature measurements is meaningless. However, the absolute (Kelvin) temperature scale has a physically meaningful fixed point (absolute zero), rather than an arbitrarily chosen one, and in some situations a percent uncertainty of an absolute temperature *is* meaningful.

Perigee (Aka Periapsis or Perifocus)

On an elliptical orbit path, the point where a satellite is closest to the Earth.

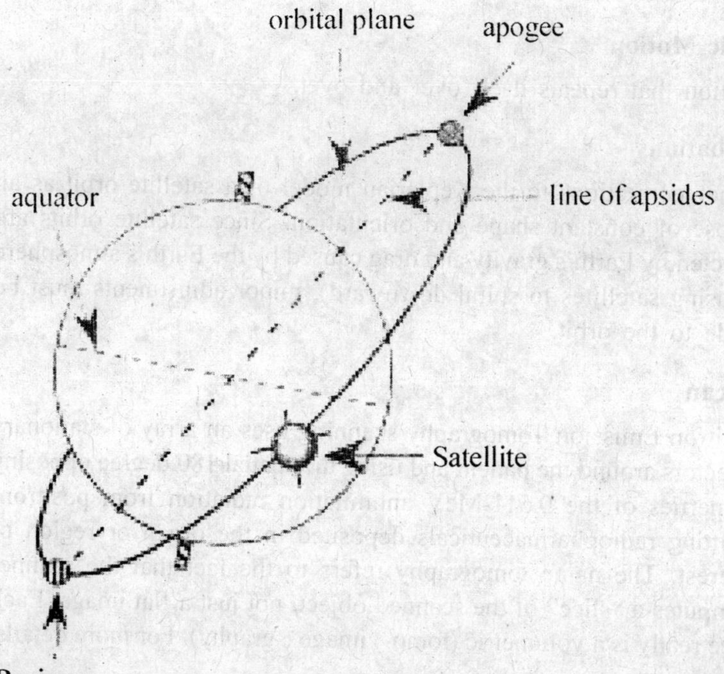

Perihelion

The point in the orbit of a planet or comet which is nearest the Sun (as opposed to the aphelion, which is the point in the orbit farthest from the Sun).

Period Decay (Aka Decay)

The tendency of a satellite to lose orbital velocity due to the influence of atmospheric drag and gravitational forces. A decaying object eventually impacts the surface of the Earth or burns up in the atmosphere. This parameter directly affects the satellite's mean motion.

Period

1. The time required for one cycle of a periodic motion (q.v.).
2. Time required for a satellite to make one complete orbit.

Periodic Motion

Motion that repeats itself over and over.

Perturbations

Minor corrections to the Keplerian model of a satellite orbit as an ellipse of constant shape and orientation. Since satellite orbits are affected by Earth's gravity and drag caused by the Earth's atmosphere (causing satellites to spiral downward), minor adjustments must be made to the orbit.

PET Scan

Positron Emission Tomography scanning uses an array of stationary detectors around the patient and using the spatial 180 degree opposing properties of the 0.511-MeV annihilation radiation from positron-emitting radiopharmaceuticals deposited in the organ or region of interest. The name tomography refers to the fact that the scanner computes a "slice" of the scanned object, not just a flat image. Each slice really is a volumetric (tomo-) image (-graphy). For more details,

pH

A symbol for the degree of acidity or alkalinity of a solution. Expressed as a negative logarithm of the hydrogen ion concentration in a solution, pH = $-\log_{10}[H+]$. If the hydrogen ion concentration of a solution increases, the pH will decrease, and vice versa. The value for pure distilled water is regarded as neutral, pH values from 0 to 7 indicate acidity, and from 7 to 14 indicate alkalinity.

Phase Difference

Phase Difference is an angle in radians, between 0 and 2p, which indicates the extent to which two wave motions of the same frequency are out of step. It is the fraction of a complete cycle, in radians, by which one signal would have to be retarded or advanced to be in phase with another signal of the same frequency. The definition and its meaning are illustrated in the diagram above. The lower signal leads the upper signal by about 3p/8 radians. 'Leading' seems the wrong way round until we remember that time later, is on the right. The upper signal peaks later; it lags the lower signal. The concept is particularly relevant to alternating currents and voltages.

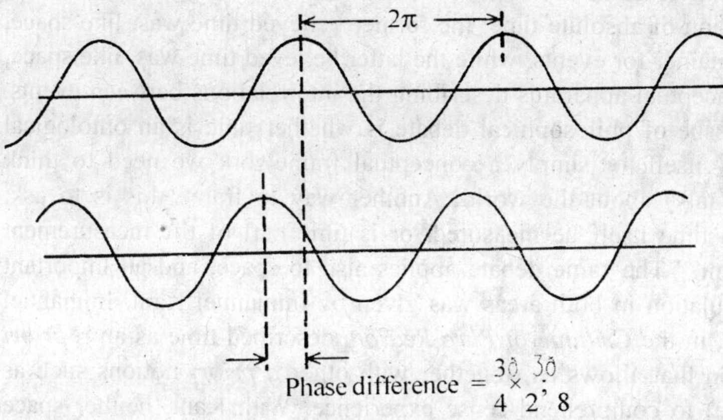

Phase Interval

In direct readout, the time between the end of a satellite image start tone and the start of the actual frame data. The phase interval repre sents white level video, interrupted by a black level pulse marking the start of each line and is used to set up phasing prior to image display

Phenology

Subdiscipline of agriculture, a science that treats relations between climate and periodic biological phenomena that are related to or caused by climatic conditions, such as the budding of trees and the migration of birds.

Philosophy of Time

Important questions in the philosophy of time include: Is time absolute or merely relational? Is time without change conceptually impossible or is there more to the idea? Does time “pass” or are the ideas of past, present and future entirely subjective, descriptions only of our deception by the senses? Zeno’s paradoxes fundamentally challenged the ancient conception of time, and thereby helped motivate the development of calculus. McTaggart believed, rather eccentrically, that time and change are illusions. Parmenides (of whom Zeno was a follower) held a similar belief based on a rather interesting argument. A point of contention between Newton and Leibniz concerned the question of absolute time: the former believed time was, like space, a container for events, while the latter believed time was, like space, a conceptual apparatus describing the interrelations between events. An issue of philosophical debate is whether time is an ontological entity itself, or simply a conceptual framework we need to think (and talk) about the world. Another way to frame this is to ask, "Can time itself be measured, or is time part of the measurement system?" The same debate applies also to space, and an important formulation in both areas was given by Immanuel Kant. Immanuel Kant, in the *Critique of Pure Reason*, described time as an *a priori* notion that allows us (together with other *a priori* notions such as space) to comprehend sense experience. With Kant, neither space nor time are conceived as substances, but rather both are elements of a systematic framework we use to structure our experience. Spatial measurements are used to quantify how far apart objects are, and temporal measurements are used to quantify how far apart events occur.

Physical Dimensions

The spacetime in which we live appears to be 4-dimensional. It is conventional (and for most practical purposes entirely sensible) to consider this as three spatial dimensions and one of time. We can move up-or-down, north-or-south, or east-or-west, and movement in any other direction can be expressed in terms of just these three. Moving down is the same as moving up a negative amount. Moving northwest is merely a combination of moving north and moving west. Time is often referred to as the 'fourth dimension'. It is somewhat different to the three spatial dimensions in that there is only one of it, and movement seems to be possible in only one direction. On the macroscopic scale that we perceive, physical processes are not symmetric with respect to time. However, at the subatomic Planck scale , almost all physical processes are time symmetric (ie. the equations used to describe these processes are the same regardless of the direction of time), although this doesn't imply that subatomic particles can move backwards in time. Theories such as string theory predict that the space we live in has in fact many more dimensions (frequently 10, 11 or 26), but that the universe measured along these additional dimensions is subatomic in size. In the physical sciences and in engineering, the *dimension* of a physical quantity is the expression of the class of physical unit that such a quantity is measured against. The dimension of speed, for example, is length divided by time. In the SI system, the dimension is given by the seven exponents of the fundamental quantities.

Phosphor

A substance that emits light when excited by radiation.

Photochemical Smog

A type of smog that forms in large cities when chemical reactions take place in the presence of sunlight, its principal component is ozone. Ozone and other oxidants are not emitted into the air directly but form from reactions involving nitrogen oxides and hydrocarbons. Because of its smog-making ability, ozone in the lower atmosphere (troposphere) is often referred to as "bad" ozone.

Photoelectric Effect

The ejection, by a photon, of an electron from the surface of an object.

Photon

1. A particle of light.
2. A quantity of electromagnetic energy. Photons have momentum but no mass or electrical charge.
3. A quantum (smallest unit in which waves may be emitted or absorbed) of light.
4. Photon is the name given to a quantum of light or other *electromagnetic radiation*. The photon energy is given in the Planck relationship. The photon is the exchange particle responsible for

the electromagnetic force The force between two electrons can be visualized in terms of a Feynman diagram as shown below. The infinite range of the electromagnetic force is owed to the zero rest mass of the photon. While the photon has zero rest mass, it has finite momentum, exhibits deflection by a gravity field, and can exert a force. The photon has an intrinsic angular momentum or "spin" of 1, so that the electron transitions which emit a photon must result in a net change of 1 in the angular momentum of the system. This is one of the "selection rules " for electron transitions.

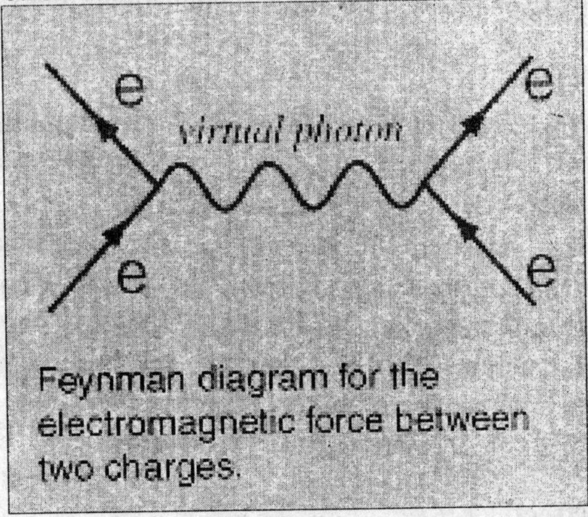

Feynman diagram for the electromagnetic force between two charges.

5. The carrier particle of electromagnetic interactions.

6. The *carrier particle* of the *electromagnetic interaction*. Depending on its frequency (and therefore its energy) photons can have different names such as visible light, X rays and *gamma rays*. We describe light in several ways. When we talk about "photons" we generally think of uncharged particles with out mass that carry energy (but be careful, there are other particles like this!). Photons of light are known by other names too, such as *gamma rays* and *x-rays*. Low-energy forms are called *ultraviolet rays*, *infrared rays*, even *radio waves*! A photon is one of the *fundamental particle* in nature and it plays an important role involving electron interactions. Photons are the most familiar

particles in everyday existence. The light we see, the radiant heat we feel, microwaves we cook with, are make use of photons of different energies. An x-ray is simply a name given to the most energetic of these particles.

7. A discrete quantity of electromagnetic energy. Short *wavelength* (high frequency) photons carry more energy than long wavelength (low *frequency*) photons.

Photosynthetically Active Radiation

Electromagnetic radiation in the part of the spectrum used by plants for photosynthesis.

Photosphere

The visible surface of the Sun. It consists of a zone in which the gaseous layers change from being completely opaque to radiation to being transparent. It is the layer from which the light we actually see (with the human eye) is emitted.

Physical Climate System

The system of processes that regulate climate, including atmospheric and ocean circulation, evaporation, and precipitation.

Physics

Branch of *science* traditionally defined as the study of *matter, energy,* and the relation between them; it was called natural philosophy until the late 19th cent. and is still known by this name at a few universities. Physics is in some senses the oldest and most basic pure science; its discoveries find applications throughout the natural sciences, since matter and energy are the basic constituents of the natural world. The other sciences are generally more limited in their scope and may be considered branches that have split off from physics to become sciences in their own right. Physics today may be divided loosely into classical physics and modern physics.

Piece-Wise Linear Approximation

This type of simulator uses piece-wise linear approximations of the equations governing the elements of a circuit. This approximation comes down to splitting the circuit into two parts: a completely linear network with a number of terminals that connect to ideal diodes. Every time a diode switches from on to off or vice versa, the linear network is configured differently. Increasing the accuracy of the simulation can be achieved by adding more detail to the approximation of the equations, this will increase the running time of the simulation. This flexibility allows an engineer to make a trade-off between simulation time and the precision of the results, something that is not easily done with the previous simulation technique.

Piezoelectric Effect

Crystals which acquire a charge when compressed, twisted or distorted are said to be piezoelectric. This provides a convenient transducer effect between electrical and mechanical oscillations. Quartz demonstrates this property and is extremely stable. Quartz crystals are used for watch crystals and for precise frequency reference crystals for radio transmitters. Rochelle salt produces a comparatively large voltage upon compression and was used in early *crystal microphones*. Barium titanate, lead zirconate, and lead titanate are ceramic materials which exhibit piezoelectricity and are used in *ultrasonic transducers* as well as microphones. If and electrical oscillation is applied to such ceramic wafers, they will respond with mechanical vibrations which provide the ultrasonic sound source. The standard piezoelectric material for medical imaging processes has been *lead zirconate titanate (PZT)*. Piezoelectric ceramic materials have found use in producing motions on the order of nanometers in the control of *scanning tunneling microscopes*.

Pink Noise

For processes of testing and equalizing rooms and auditoriums, it is convenient to have broad-band noise signals. Typically, *white noise* or pink noise is used. Whereas white noise is defined as sound with equal power per Hz in frequency, pink noise is filtered to give equal power per octave or equal power per 1/3 octave. Since the number of Hz in each successive octave increases by two, this means the power of pink noise per Hz of bandwidth decreases by a factor of two or 3 dB per octave.

Pion

1. The least massive type of meson, pions can have electric charges of +1, -1, or 0.
2. The lightest type of *mesons*. They are copiously produced in high energy particle collisions.

Pixel

Smallest part (addressable element) of an electronically-coded image, such as a computer display. Pixel is a contraction of "picture element."

Place Theory

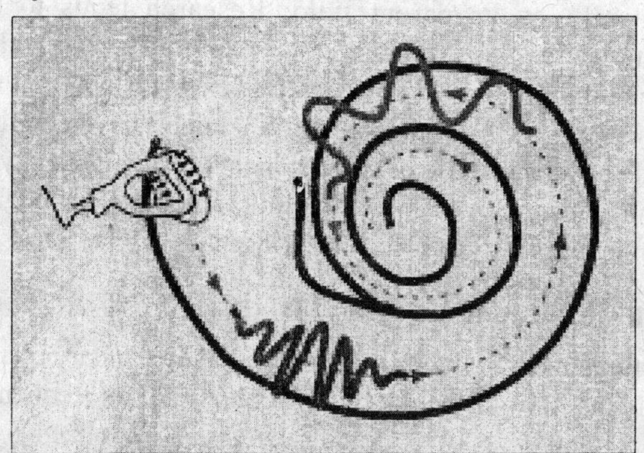

High frequency sounds selectively vibrate the *basilar membrane* of the *inner ear* near the entrance port (the oval window). Lower frequencies travel further along the membrane before causing appreciable excitation of the membrane. The basic *pitch* determining mechanism is based on the location along the membrane where the *hair cells* are stimulated. A schematic view of the place theory unrolls the *cochlea* and represents the distribution of sensitive hair cells on the *organ of Corti.* Pressure waves are sent through the fluid of the inner ear by force from the *stirrup.*

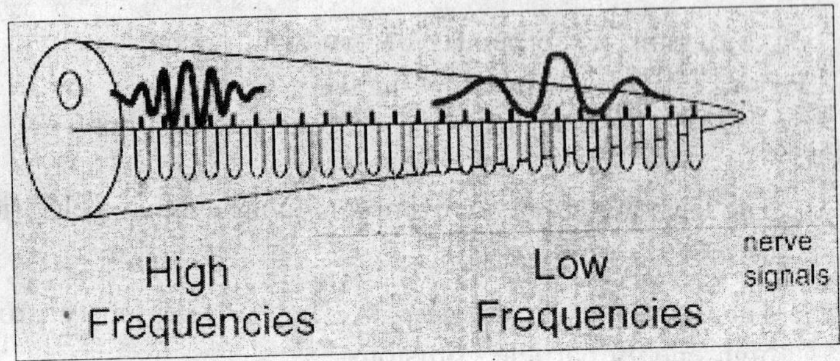

Planetary Science

Planetary science, also known as *planetology or planetary astronomy*, is the science of planets and the solar system, and incorporates an interdisciplinary approach drawing from diverse sciences. Planetary science may be considered a part of the Earth sciences, or more logically, as its parent field. Research tends to be done by a combination of astronomy, space exploration (particularly unmanned space missions), and comparative, experimental and meteorite work based on Earth. There is also an important theoretical component and considerable use of computer simulation. Planetary science studies objects ranging in size from micrometeoroids to gas giants, their composition, dynamics and history.

Planck's Constant

A fundamental physical constant, the elementary quantum of action, It is the ratio of the energy of a photon to its frequency and is equal to 6.626069×10^{-34} Joule seconds. Symbolized by *h*.

Planetary Albedo

The fraction of incident solar radiation that is reflected by a planet and returned to space. The planetary albedo of the Earth-atmosphere system is approximately 30 percent, most of which is due to backscatter from clouds in the atmosphere.

Plasma

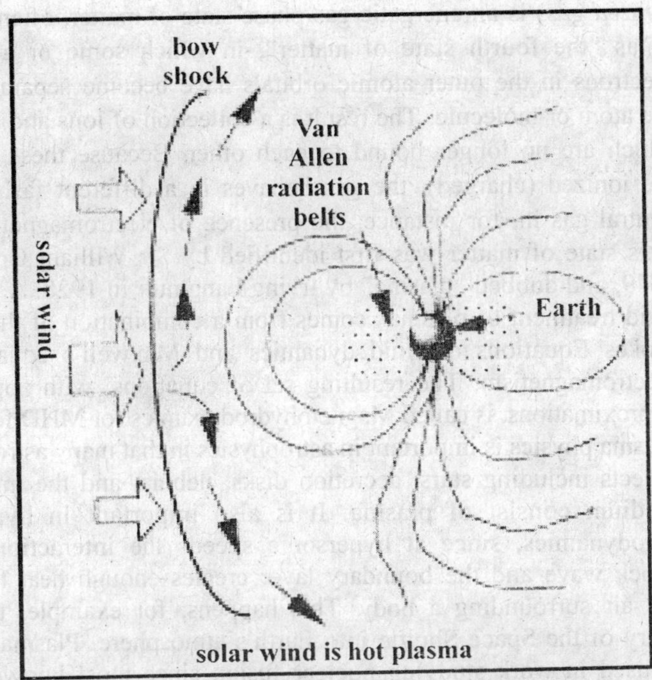

1. Plasma consists of a gas heated to sufficiently high temperatures that the atoms *ionize*. The properties of the gas are controlled by electromagnetic forces among constituent *ions* and *electrons*, which results in a different type of behavior. Plasma is often considered the fourth state of matter (besides solid, liquid, and gas). Most of the matter in the Universe is in the plasma state.

2. A fourth state of matter (in addition to solid, liquid, and gas) that exists in space. In this state, atoms are positively charged and share space with free negatively-charged electrons. Plasma can conduct electricity and interact strongly with electric and magnetic fields. The solar wind is actually hot plasma blowing from the sun.

Plasma Physics

Plasma physics is the field of physics which studies ionized gases known as plasmas. In physics and chemistry, *plasma* (also called an *ionized gas*) is an energetic gas-phase state of matter, often referred to as "the fourth state of matter", in which some or all of the electrons in the outer atomic orbitals have become separated from the atom or molecule. The result is a collection of ions and electrons which are no longer bound to each other. Because these particles are ionized (charged), the gas behaves in a different fashion than neutral gas in, for instance, the presence of electromagnetic fields. This state of matter was first identified by Sir William Crookes in 1879, and dubbed "plasma" by Irving Langmuir in 1928. A common fluid treatment of plasmas comes from a combination of the Navier Stokes Equations of fluid dynamics and Maxwell's equations of electromagnetism. The resulting set of equations, with appropriate approximations, is called Magnetohydrodynamics (or MHD for short). Plasma physics is important in astrophysics in that many astronomical objects including stars, accretion disks, nebula, and the interstellar medium consist of plasma. It is also important in hypersonic aerodynamics, since at hypersonic speeds the interaction of the shock wave and the boundary layer creates enough heat to ionize the air surrounding a body. This happens, for example, upon re-entry of the Space Shuttle into Earth's atmosphere. Plasma physics is used in work studying nuclear fusion since most known fusion reactions take place in plasma.

View of a "shot" on the Large Helical Device though a side port showing hot plasma affected by magnetic fields.

Plate Tectonics

Concept that the Earth's crust is composed of rigid plates that move over a less rigid interior.

Platforms

A satellite that can carry instruments. The same term is applied to automatic weather data transmitters installed on buoys, balloons, ships, and planes, and mounted in remote areas.

POES (Polar-orbiting Operational Environmental Satellite)

Operated by the National Oceanic and Atmospheric Administration, they are designated "NOAA satellites." Included in this group are the current series of TIROS-N satellites, the third-generation polar-orbiting environmental spacecraft operated by NOAA.

Polar Orbit

An orbit with an orbital inclination of near 90° where the satellite ground track will cross both polar regions once during each orbit. The term is used to desribe the near-polar orbits of spacecraft such as the USA's NOAA/TIROS and Landsat satellites.

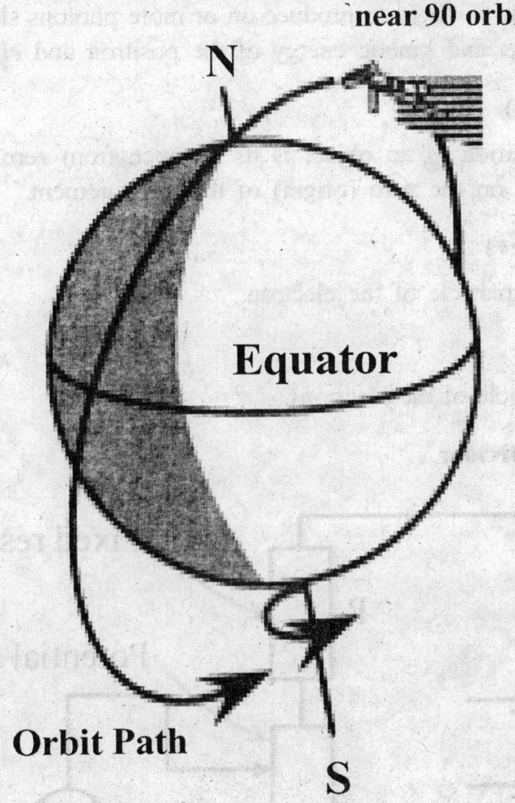

Polarization

A polarized particle *beam* is a beam of particles whose spins are aligned in a particular direction. The polarization of the beam is the fraction of the particles with the desired alignment.

Poloidal Radius

The radius of the actual loop structure. For a doughnut, it is measured from the center to the edge of the pastry (*not* from the center of the hole). Positron decay in matter by annihilation with an *electron*. Usually and "atom" of positronium (e+e-) forms which annihilates to produce two 511-keV photons. Occasionally, the positron will annihilate in flight to produce on or more photons sharing the total rest mass and kinetic energy of the positron and *electron*.

Position (x)

The position of an object is its distance from zero. The position depends on the zero (origin) of the measurement.

Positron (e+)

The antiparticle of the electron.

Positron

Antiparticle of the *electrons*.

Potential Divider

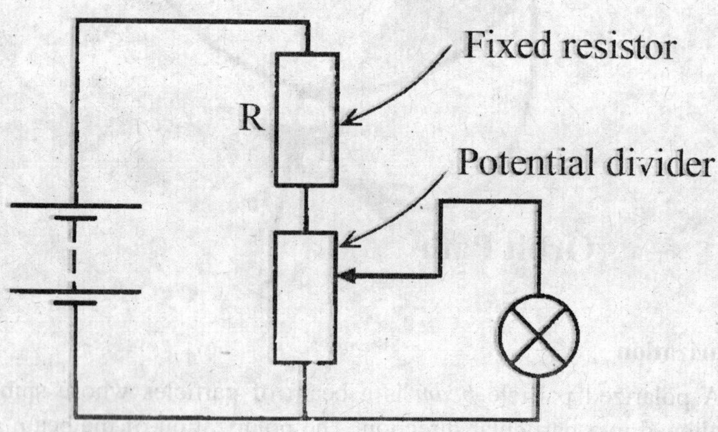

A fixed resistor which may be tapped at any point along its length to enable part of the *potential difference* across the whole resistor to be accessed. The circuit below shows how a potential divider may be used to adjust the voltage across a lamp between zero and any pre–decided value. The voltage across the lamp or any other components will not, in general, vary linearly with the distance moved by the slider. This will happen only when the resistance of the components is much larger than the end-to-end resistance of the potential divider. The series resistor R is a protection device. It may protect the component, the lamp, from the full terminal potential difference of the power supply. It may protect the power supply from the danger of delivering too much current if the component is connected straight across its output. Whatever the reason, its value must be computed beforehand.

Potential Energy

The energy having to do with the distance between to objects that interact via a noncontact force.

Power (P)

1. Power is the rate of working. When power is transmitted electrically, the amount is the product of voltage (V) with current (I).
2. The rate of transferring energy; a scalar quantity with units of watts (W).

Precession

The comparatively slow torquing of the orbital planes of all satellites with respect to the Earth's axis, due to the bulge of the Earth at the equator which distorts the Earth's gravitational field. Precession is manifest by the slow rotation of the line of nodes of the orbit (westward for inclinations less than 90° and eastward for inclinations greater than 90°).

Precipitation

Moisture that falls from clouds. Although clouds appear to float in the sky they are always falling, their water droplets slowly being pulled down by gravity. Because their water droplets are so small and light, it can take 21 days to fall 1,000 feet and wind currents can easily interrupt their descent. Liquid water falls as rain or drizzle. All raindrops form around particles of salt or dust. (Some of this dust comes from tiny meteorites and even the tails of comets.) Water or ice droplets stick to these particles, then the drops attract more water and continue getting bigger until they are large enough to fall out of the cloud. Drizzle drops are smaller than raindrops. In many clouds, raindrops actually begin as tiny ice crystals that form when part or all of a cloud is below freezing. As the ice crystals fall inside the cloud, they may collide with water droplets that freeze onto them. The ice crystals continue to grow larger; until large enough to fall from the cloud. They pass through warm air; melt, and fall as raindrops. When ice crystals move within a very cold cloud (10°F and -40°F) and enough water droplets freeze onto the ice crystals, snow will fall from the cloud. If the surface temperature is colder than 32°F the flakes will land as snow. Precipitation Weights; one raindrop .000008 lbs one snowflake .0000003 lbs one cumulus cloud 10,000,000 lbs one thunderstorm 10,000,000,000 lbs one hurricane 10,000,000,000,000 lbs.

Precise

Sharply or clearly defined. Having small experimental uncertainty. A precise measurement may still be inaccurate, if there were an unrecognized determinate error in the measurement (for example, a miscalibrated instrument).

Prevailing Westerlies

Winds in the middle latitudes (approximately 30° to 60°) that generally blow from west to east. The subtropical high pressure regions at the horse latitudes (30°) forces surface air poleward, and the rotation of the Earth causes these winds to bear to the right (east) in the Northern Hemisphere and to the left (east) in the Southern Hemisphere. This is, to some extent, an idealized picture of the atmospheric circulation. The actual circulation on individual days includes modifications and variations due to the migratory cyclones and anticyclones of middle latitudes, causing rapid and often violent weather changes, as warm semi-tropical air from the horse latitudes meets cold polar air from the high latitudes.

Preventing Heat Transfer

It is often desired to prevent heat transfer.

Prime Meridian

An imaginary line running from north to south through Greenwich, England, used as the reference point for longitude.

Printed Circuit

A fiber card. on which integrated circuits and other electronic components can be mounted. Connections between the components are etched in the correct circuit patterns.

Probability Currents

In order to describe how probability density changes with time, it is acceptable to define probability current or probability flux. The probability flux represents a flowing of probability across space. For example, consider a Gaussian probability curve centered around $x0$, imagine that $x0$ moving in a speed v toward the right. Then one may say that the probability is flowing toward right, i.e., there is a probability flux directed to the right.

Probability Distribution

A curve that specifies the probabilities of various random values of a variable; areas under the curve correspond to probabilities.

Probability

The likelihood that something will happen, expressed as a number between zero and one.

Process Study

An organized, systematic investigation of a particular process designed to identify all of the state variables involved and to establish the relationships among them. Process studies yield numerical algorithms that connect the state variables and determine their rates of change; such algorithms are essential ingredients of Earth system models.

Process

An association of phenomena governed by physical, chemical, or biological laws. An example of a process is the vertical mixing of ocean waters in the so-called surface-mixed layer; the state variables for this process include temperature, salinity in the water on a vertical scale of tens of meters, and heat flow and wind stress at the sea surface. Other examples include the volcanic deposition of dust and gases into the atmosphere, eddy formation in the atmosphere and oceans, and soil development.

Prograde Orbit

Orbits of the Earth in the same direction as the rotation of the Earth (west-to-east).

Proof

A term from logic and mathematics describing an argument from premise to conclusion using strictly logical principles. In mathematics, theorems or propositions are established by logical arguments from a set of axioms, the process of establishing a theorem being called a *proof*. The colloquial meaning of 'proof' causes lots of problems in physics discussion and is best avoided. Since mathematics is such an important part of physics, the mathematician's meaning of proof should be the only one we use. Also, we often ask students in upper level courses to do proofs of certain theorems of mathematical physics, and we are *not* asking for experimental demonstration! So, in a laboratory report, we should not say 'We proved Newton's law.' Rather say, 'Today we *demonstrated* (or *verified*) the validity of Newton's law in the particular case of...'

Proton (p)

1. The most common hadron, a baryon with electric charge +1 equal and opposite to that of the electron. Protons have a basic structure of two up quarks and one down quark (bound together

by gluons). The nucleus of a hydrogen atom is a proton. A nucleus with electric charge Z contains Z protons; therefore the number of protons is what distinguishes the different chemical elements.

2. A *baryon* with electric *charge* +1. Protons contain a basic structure of two up quarks and one *down quark*. The nucleus of a hydrogen atom is a proton. A nucleus with atomic number Z contains Z protons; therefore the number of protons is what distinguishes the different chemical elements.

3. A positively charged particle, one of the types that nuclei are made of.

4. One of the basic particles which makes up an atom. The proton is found in the nucleus and has a positive electrical charge equivalent to the negative charge of an electron and a mass similar to that of a neutron: a hydrogen nucleus.

5. A positively charged elementary particle. A proton is 1836 times heavier than an *electron*.

Proton-Proton Chain

In the Sun and other less massive stars, this chain is the primary source of heat and radiation. The proton-proton chain converts hydrogen into helium releasing energy in the form of particles and *gamma-rays*. Hydrogen is converted into helium in a chain of reactions. The first reaction takes an average of 1 billion years to occur while the others are much shorter. One step is only 1 second long. In the Sun, there are so many hydrogen nuclei that the 1 billion year waiting period does not stop it from producing tremendous radiation.

Proton Separation Energy

The energy required to remove a proton from a *nucleus*.

P-Type Semiconductor

The addition of trivalent *impurities* such as boron, aluminum or gallium to an *intrinsic semiconductor* creates deficiencies of valence electrons,called "holes". It is typical to use B_2H_6 diborane gas to diffuse boron into the silicon material.

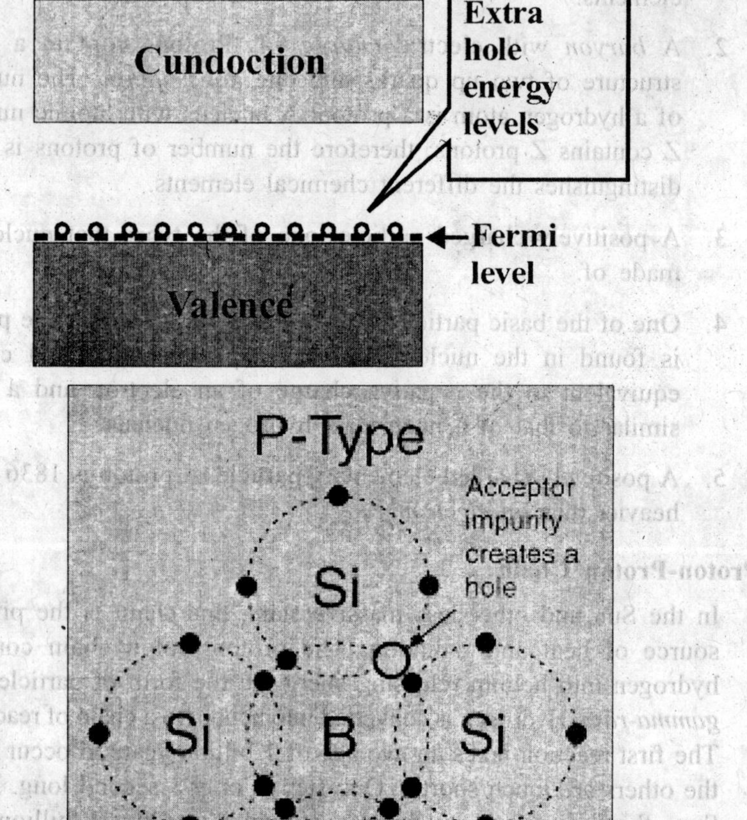

Psychrometer

An instrument designed to measure dew point and relative humidity consisting of two thermometers (one dry bulb and one wet bulb).

The dew point and humidity levels are determined by drying the wet bulb (either by fanning or whirling the instrument) and comparing the difference between the wet and dry bulbs with preexisting calculations.

Pulsar

A neutron star (burnt-out star) that emits radio waves which pulse on and off

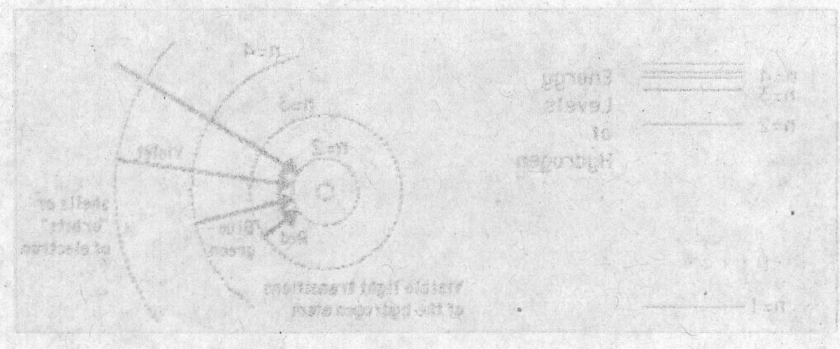

Q

Quality Factor

The number of oscillations required for a system's energy to fall off by a factor of 535 due to damping.

Quantized Energy States

The electrons in free atoms can will be found in only certain discrete energy states. These sharp energy states are associated with the orbits or shells of electrons in an atom, *e.g.*, a hydrogen atom. One of the implications of these quantized energy states is that only certain photon energies are allowed when electrons jump down from higher levels to lower levels, producing the hydrogen spectrum. The Bohr model successfully predicted the energies for the hydrogen atom, but had significant failures that were corrected by solving the Schrodinger equation for the hydrogen atom.

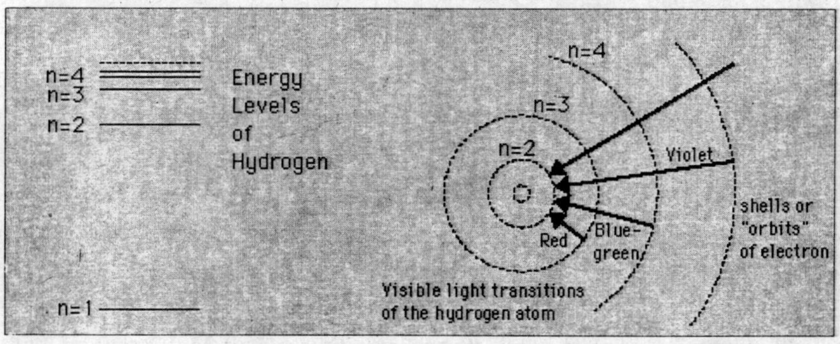

Quantized

Describes quantity such as money or electrical charge, that can only exist in certain amounts.

Quantum Chromodynamics (QCD)

The quantum theory of the strong interaction.

Quantum Electrodynamics (QED)

The quantum theory of the electromagnetic interaction.

Quantum Entanglement

Quantum Entanglement is a quantum mechanical phenomenon in which the quantum states of two or more objects have to be described with reference to each other, even though the individual objects may be spatially separated. This leads to correlations between observable physical properties of the systems. For example, it is possible to prepare two particles in a single quantum state such that when one is observed to be spin-up, the other one will always be observed to be spin-down and vice versa, this despite the fact that it is impossible to predict, according to quantum mechanics, which set of measurements will be observed. As a result, measurements performed on one system seem to be instantaneously influencing other systems entangled with it. However, classical information cannot be transmitted through entanglement faster than the speed of light. Quantum entanglement is the basis for emerging technologies such as quantum computing and quantum cryptography, and has been used for experiments in quantum teleportation. At the same time, it produces some of the more theoretically and philosophically disturbing aspects of the theory, as one can show that the correlations predicted by quantum mechanics are inconsistent with the seemingly obvious principle of local realism, which is that information about the state of a system should only be mediated by interactions in its immediate surroundings. Different views of what is actually occurring in the process of quantum entanglement give rise to different interpretations of quantum mechanics.

Quantum Field Theory

Quantum field theory (QFT) is the application of quantum mechanics to fields. It provides a theoretical framework widely used in particle physics and condensed matter physics. This framework is necessary in order to formulate consistent relativistic quantum theories, such as quantum electrodynamics, the quantum theory of electromagnetism, one of the most well-tested and successful theories in physics. However, not all quantum field theories are relativistic; the BCS theory is an example of a non-relativistic quantum field theory, which has been highly successful in describing superconductivity.

Quantum Harmonic Oscillator

The *quantum harmonic oscillator* is the quantum mechanical analogue of the classical harmonic oscillator. It is one of the most important model systems in quantum mechanics because, as in classical mechanics, a wide variety of physical situations can be reduced to it either exactly or approximately. In particular, a system near an equilibrium configuration can often be described in terms of one or more harmonic oscillators. Furthermore, it is one of the few quantum mechanical systems for which a simple exact solution is known.

Quantum Mechanical Effects

As mentioned in the introduction, there are several classes of phenomena that appear under quantum mechanics which have no analogue in classical physics. These are sometimes referred to as "quantum effects". The first type of quantum effect is the quantization of certain physical quantities. In the example we have given, of a free particle in empty space, both the position and the momentum are continuous observables. However, if we restrict the particle to a region of space (the so-called "particle in a box problem), the momentum observable will become discrete; it will only take on the values h/L, where L is the length of the box. Such observables are said to be quantized, and they play an important role in many physical systems. Examples of quantized observables include angular momentum, the total energy of a bound system, and the energy contained in an electromagnetic wave of a given frequency. Another quantum effect is the uncertainty principle, which is the phenomenon that consecutive measurements of two or more observables may possess a fundamental limitation on accuracy. In our free particle example, it turns out that it is impossible to find a wavefunction that is an eigenstate of both position and momentum. This implies that position and momentum can never be simultaneously measured with arbitrary precision, even in principle: as the precision of the position measurement improves, the maximum precision of the momentum measurement decreases, and vice versa. Those variables for which it holds (*e.g.* momentum and position, or energy and time) are canonically conjugate variables in classical physics. Another quantum effect is the wave-particle duality. It has been shown that, under certain experimental conditions, microscopic objects like atoms or electrons exhibit particle-like behavior, such as scattering. ("Particle-like" in the sense of an object that can be localized to a particular region of space.) Under other conditions, the same type of objects exhibit wave-like behavior, such as interference. We can observe only one type of property at a time.

Quantum Mechanics

1. The laws of physics that apply on very small scales. The essential feature is that electric charge, momentum, and angular momentum, as well as charges, come in discrete amounts called quanta.

2. The laws of physics that apply on very small scales. The essential feature is that energy, momentum and angular momentum as well as *charge* come in discrete amounts called quanta.

3. The wavefunctions of an electron in a hydrogen atom possessing definite energy (increasing downward: n=1,2,3,...) and angular momentum (increasing across: *s*, *p*, *d*,...). Brighter areas correspond to higher probability density for a position measurement. The angular momentum and energy are quantized, and only take on discrete values like those shown. *Quantum Mechanics* is a fundamental physical theory which extends and corrects Newtonian mechanics, especially at the atomic and subatomic levels. The term *quantum* (Latin for *how much*) refers to the discrete units that the theory assigns to certain physical quantities, such as the energy of an atom at rest. Quantum mechanics is

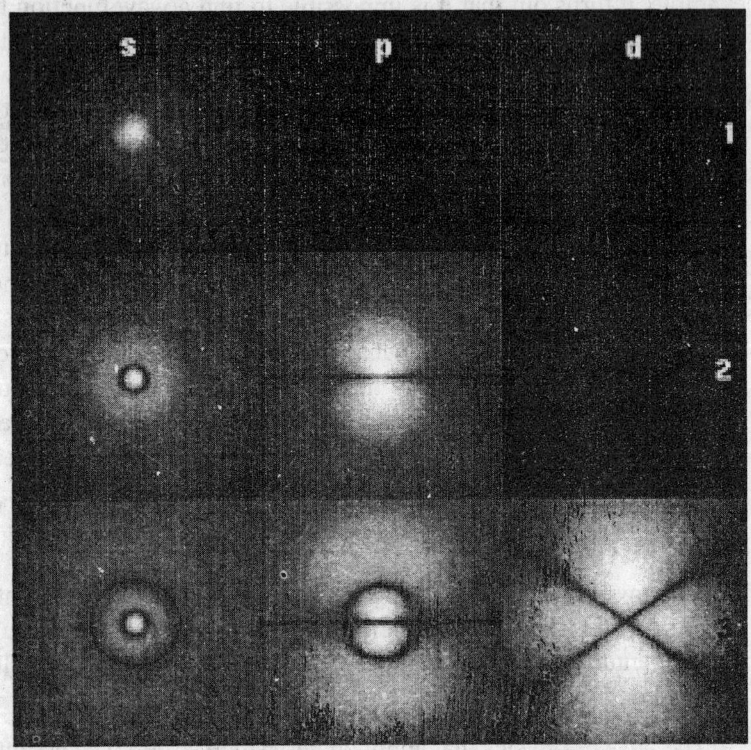

Fig. 1

a theory of mechanics, that branch of physics that deals with the motion of bodies and associated physical quantities such as energy and momentum. Mechanics can be subdivided into classical mechanics, relativistic mechanics, non-relativistic quantum mechanics, and relativistic quantum mechanics (quantum field theory). Quantum mechanics is regarded as more "fundamental" than the first two theories of mechanics, because the predictions of quantum mechanics have never been disproven after a century's worth of experiments. We will use the term "quantum mechanics" to refer to both relativistic and non-relativistic quantum mechanics; the terms *quantum physics and quantum theory* are synonymous. It should be noted, however, that certain authors refer to "quantum mechanics" in the more restricted sense of non-relativistic quantum mechanics. Quantum mechanics is the underlying framework of many fields of physics and chemistry, including condensed matter physics, quantum chemistry, and particle physics. It describes with great accuracy and precision many phenomena where classical mechanics drastically fails, including the behavior of systems at atomic length scales and below (for instance, classical mechanics is unable to account for the existence of stable atoms), as well as special macroscopic systems such as superconductors and superfluids. However, quantum mechanics also reduces to classical mechanics in some physical situations; this property of quantum mechanics was highlighted by Niels Bohr, and is known as the correspondence principle. Most physicists now believe that quantum mechanics is a correct theory of the physical world (but not near black holes or when considering the observable Universe as a whole-the domain of general relativity). The question of compatibility between quantum mechanics and general relativity remains an area of active research. Quantum mechanics predicts at least three classes of phenomena that classical mechanics and classical electrodynamics cannot account for: (i) the quantization (discretization) of certain physical quantities, (ii) wave-particle duality, and (iii) quantum entanglement.

Quantum Number

1. A number that labels a state, it denotes the number of quanta of a particular type that the state contains. Electric *charge* given as an integer multiple of the electron's charge is an example of a quantum number.

2. A numerical label used to classify a quantum state.

Quantum Processes

Quantum properties dominate the fields of atomic and molecular physics. Radiation is quantized such that for a given frequency of radiation, there can be only one value of *quantum energy* for the photons of that radiation. The energy levels of atoms and molecules can have only certain quantized values. Transitions between these quantized states occur by the photon processes *absorption, emission* and *stimulated emission.* All of these processes require that the photon energy given by the *Planck relationship* is equal to the energy separation of the participating pair of quantum energy states.

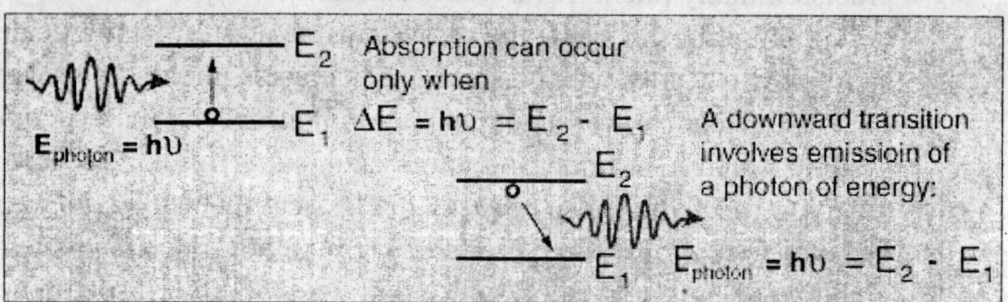

Quantum

1. When used as a noun (plural quanta): a discrete quantity of energy, momentum or angular momentum, given in units involving Planck's constant *h*. For example electromagnetic radiation of a given frequency *f* is composed of quanta (also called photons) with energy *hf*. When used as an adjective (as in quantum theory, quantum mechanics, quantum field theory): defines the theory as involving quantities which depend on Planck's constant *h*. In such theories radiation comes in discrete quanta as described above; angular momenta must be integer units of *h*, except that the intrinsic angular momenta of fundamental particles are integer multiples of 1/2*h*; and solutions for the possible states of a particle in a potential (such as the states of an electron the electrical potential due to an atomic nucleus) occur only for certain discrete energies.

2. The smallest discrete amount of any quantity.

Quark (q)

1. A fundamental fermion that has strong interactions. Quarks have electric charge of either +2/3 (up, charm, top) or -1/3 (down, strange, bottom) in units where the proton charge is 1.

2. A fundamental matter particle that has strong *interactions*. Quarks have an electric *charge* of either +2/3 (up, *charm* and top) or

-1/3 (down, strange and bottom) in units where the proton charge is 1.

Quasar

A faint blue, star-like object commonly considered to be extremely distant, probably an unusual nucleus of a galaxy. It has a tendency to *flare*.

Q-value

The energy available for decay. This energy is released by the *nucleus* mainly as *gammas, betas, neutrinos*, and/or *alpha* particles.

R

R & D

Research and Development.

Rad

1. One rad is equal to an energy *absorption* of 100 ergs in a gram of any material. An "erg" is a unit for quantifying energy (just like a "mile" is used for measuring distance.)
2. Radiation Absorbed Dose. The basic unit of an absorbed dose of ionizing radiation. One rad is equal to the absorption of 100 ergs of radiation energy per gram of matter.
3. The rad is a unit used to measure a quantity called absorbed dose. This relates to the amount of energy actually absorbed in some material, and is used for any type of radiation and any material. One rad is defined as the absorption of 100 ergs per gram of material. The unit rad can be used for any type of radiation, but it does not't describe the biological effects of the different radiations.

Radial

Parallel to the radius of a circle; the in-out direction. Cf. tangential.

Radial Electric Field

The electric field near an isolated point electric charge. The diagram below shows the *field lines* and the equipotential lines (broken lines) near an isolated positive charge. Notice how the field lines get further apart as the field weakens. Notice, also, how the spacing of the equipotentials increases if the change in potential from line to line is always the same.

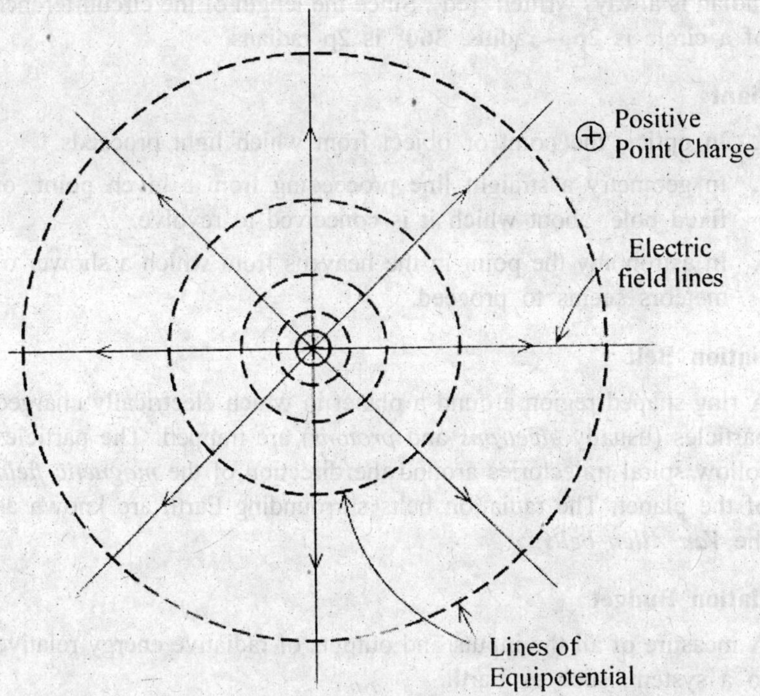

Radian

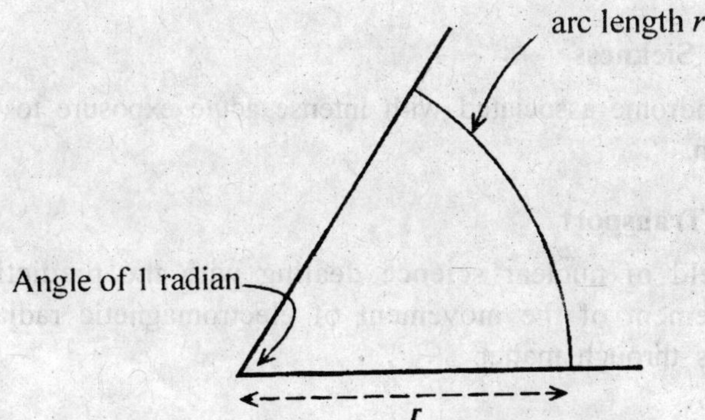

Radian is a unit of angle. One radian is the angle subtended at the centre of a circle by an arc of length one radius. The symbol for radian is always written 'red'. Since the length of the circumference of a circle is 2p – radius, 360° is 2p radians.

Radiant

1. In optics, the point or object from which light proceeds.
2. In geometry a straight line proceeding from a given point, or fixed pole, about which it is conceived to revolve.
3. In astronomy the point in the heavens from which a shower of meteors seems to proceed.

Radiation Belt

A ring-shaped region around a planet in which electrically charged particles (usually *electrons* and *protons*) are trapped. The particles follow spiral trajectories around the direction of the *magnetic field* of the planet. The radiation belts surrounding Earth are known as the *Van Allen belts*.

Radiation Budget

A measure of all the inputs and outputs of radiative energy relative to a system, such as Earth.

Radiation Oncology

Treatment of tumors with *ionization radiation.*

Radiation Sickness

The syndrome associated with intense acute exposure to ionizing radiation.

Radiation Transport

The field of nuclear science dealing with the prediction and measurement of the movement of electromagnetic radiation or particles through matter.

Radiation

1. Energy transfer in the form of electromagnetic waves or particles that release energy when absorbed by an object.
2. Radiation is energy in transit in the form of high speed particles and *electromagnetic waves*. Radiation is further defined into ionizing and non-ionizing radiation. *Lonizing Radiation* is radiation with enough energy so that during an interaction with an atom, it can remove tightly bound *electrons* from their orbits, causing the atom to become *charged* or ionized. Examples are X-rays and electrons. *Non-ionizing radiation* is radiation without enough energy to remove tightly bound electrons from their orbits around atoms. Examples are microwaves and visible light.
3. Radiation is a means of heat transfer. Radiative heat transfer is the only form of heat transfer that can occur in the absence of any form of medium and as such is the only means of heat transfer through a vacuum. Thermal radiation is a direct result of the movements of atoms and molecules in a material. Since these atoms and molecules are composed of charged particles (protons and electrons), their movements result in the emission of electromagnetic radiation, which carries energy away from the surface. At the same time, the surface is constantly bombarded by radiation from the surroundings, resulting in the transfer of energy to the surface. Since the amount of emitted radiation increases with increasing temperature, a net transfer of energy from higher temperatures to lower temperatures results. For room temperature objects (~300 K), the majority of photons emitted (and involved in radiative heat transfer) are in the infrared spectrum, but this is by no means the only frequency range involved in radiation. The frequencies emitted are partially related to black-body radiation. tter objects-a campfire is around 700 K, for instance-transfer heat in the visible spectrum or beyond. Whenever EM radiation is emitted and then absorbed, heat is transferred. This principle is used in microwave ovens, laser cutting, and RF hair removal.

Radiative Cooling

Cooling process of the Earth's surface and adjacent air; which occurs when infrared (heat) energy radiates from the surface of the Earth upward through the atmosphere into space. Air near the surface transfers its thermal energy to the nearby ground through conduction, so that radiative-cooling lowers the temperature of both the surface and the lowest part of the atmosphere.

Radiative Transfer

Theory dealing with the propagation of electromagnetic radiation through a medium.

Radio Frequency (RF)

1. A frequency that is useful for radio transmission, usually between 10 kHz and 300,000 MHz.
2. Any ac frequency at which coherent electromagnetic radiation of energy is possible. Usually considered to denote frequencies above 150 kilohertz and extending up to the infrared range.

Radio Frequency Bands

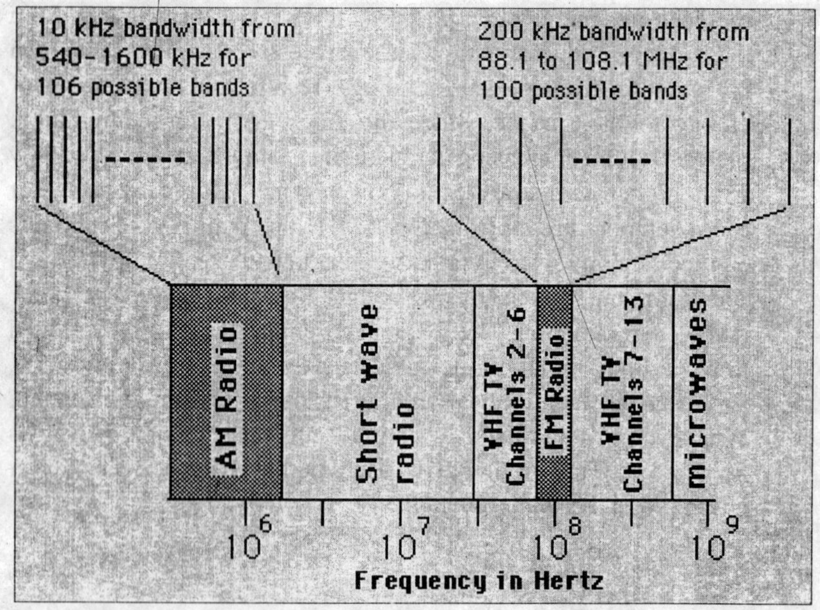

Because of the division of the FM band for the transmission of *FM stereo*, the frequency limit for music transmission is at 15 kHz. This allows high fidelity signal transmission. The operational bandwidth is limited to 150 kHz, with 25 kHz on each side of that for gaurd bands. Actually FM stereo covers 106 kHz of that. AM radio is limited to 5000 Hz maximum frequency by the width of the AM bands. Music transmission by AM radio is limited in fidelity, and adjacent stations tend to interfere with each other.

Radio Spectrum

The complete range of frequencies or wave lengths of electromagnetic waves, specifically those used in radio and television.

Radio Wave

An electrical impulse sent through the atmosphere at radio frequency.

Radioactive Contamination

Radioactive contamination is radioactive material distributed over some area, equipment or person. It tends to be unwanted in the location where it is, and has to be cleaned up or decontaminated.

Radioactive Dating

A technique for estimating the age of an object by measuring the amounts of various radioisotopes in it.

Radioactive Material

1. A material whose nuclei spontaneously give off nuclear radiation. Naturally radioactive materials (found in the earth's crust) give off alpha, beta, or gamma particles. Alpha particles are Helium nuclei, beta particles are electrons, and gamma particles are high energy photons.
2. Radioactive Material is any material that contains radioactive atoms.

Radioactive Waste

Materials which are radioactive and for which there is no further use.

Radioactive

1. A word distinguishing radioactive materials from those which aren't. Usage: 'U-235 is radioactive; He-4 is not.'

   Note: Radioactive is least misleading when used as an adjective, not as a noun. It is sometimes used in the noun form as an shortened stand-in for *radioactive material*, as in the example above.

2. Giving off or capable of giving off radiant energy in the form of particles or rays, as in alpha, beta, and gamma rays.

Radioactivity

1. The process of emitting particles from the nucleus. Usage: 'Certain materials found in nature demonstrate radioactivity.'

2. The spontaneous decay of disintegration of an unstable atomic nucleus accompanied by the emission of radiation.

3. The property of spontaneously emitting alpha, beta, and/or gamma radiation as a result of nuclear disintegration.

4. Radioactivity is the spontaneous transformation of an unstable atom and often results in the emission of radiation. This process is referred to as a transformation, a decay or a disintegrations of an atom.

Radiography

The making of shadow images on a photographic emulsion by the action of *ionization radiation.* The image is the result of the differential *attenuation* of the radiation in its passage through the object being radiographed.

Radioimmunotherapy

This type of radiotherapy is the application of monoclonal antibodies that have been tagged with high activities of suitable radionuclides. These tumor-specific antibodies are derived from the patient's own cancer and, hence, they selectively target this tumor when injected into the patient. Also known as monoclonoal antibody therapy.

Radioisotope

A radioactive isotope. A common term for a radionuclide.

Radiology

The branch of medicine that deals with the diagnostic and therapeutic applications of radiation.

Radiometer

An instrument that quantitatively measures electromagnetic radiation. Weather satellites carry radiometers to measure radiation from snow, ice, clouds, bodies of water; the Earth's surface, and the sun.

Radionuclide

1. A radioactive nuclide. An unstable isotope of an element that decays or disintegrates spontaneously, emitting radiation.
2. Radionuclides are materials that produce *ionization radiation*, such as X-rays, *gamma rays*, alpha particles, and beta particles.

Radiosonde

A balloon-borne instrument that measures meteorological parameters from the Earth's surface up to 20 miles in the atmosphere. The radiosonde measures temperature, pressure, and humidity and transmits or "radios" these data back to Earth. Upper air winds also are deter-mined through tracking of the balloon ascent. Radiosonde observations generally are taken twice a day (0000 and 1200 UTC) around the globe. NOAA's National Weather Service (NWS) operates a network of about 90 radiosonde observing sites in the U.S and its territories. When the balloons burst radiosondes return to Earth on a parachute. Approximately 25 percent are recovered and returned to NWS for reconditioning and reuse.

Radiotherapy *(also called Radiation Therapy)*

Medical therapy consisting of one or more treatments with *ionization radiation*. More information about radiotherapy is available through the *National Cancer Institute* at the *National Institutes of Health*.

Rain Forest

An evergreen woodland of the tropics distinguished by a continuous leaf canopy and an average rainfall of about 100 inches per year. Rain forests play an important role in the global environment. The Earth sustains life because of critical balances and interactions among many factors. Were there not processes at work that limit the effects of other essential processes, Earth would become uninhabitable. Destruction of tropical rain forests reduces the amount of leaf area in the tropics, and consequently the amount of carbon dioxide absorbed, causing increases in levels of carbon dioxide and other atmospheric gases. It is estimated that cutting and burning of tropical forests contributes about 20 percent of the carbon dioxide added to the atmosphere each year. The World Resources Institute and the International Institute for Environment and Development have reported that the world's tropical forests are being destroyed at the rate of fifty-four acres per minute, or twenty-eight million acres lost annually. Rain forest destruction also means the loss of wide spectrum and biological life, erosion of soil, and possible desertification.

Rain Gauge

Calibrated container that measures the amount of rainfall during a specific period of time.

RAM

Random Access Memory Computers use two types of memory RAM and ROM. RAM is the computer's working area, the primary location where the microprocessor stores the information it needs. The designation "random access" stems from the microprocessor's ability to access information in memory randomly by knowing its location, or address, rather than hunting through memory sequentially from beginning to end. Because information in RAM is stored electronically accessing data stored in RAM is much faster than getting that data from a mechanical storage device such as a disk drive. But because it is stored electronically all information in RAM is temporary (which is why you must store it on a more permanent storage capability, such as a disk).

Rate Meter

An electronic instrument that indicates, on a meter, the number of radiation induced pulses per minute from radiation detectors such as a Geiger-Muller tube.

Rate

A quantity of one thing compared to a quantity of another. [Dictionary definition] In physics the comparison is generally made by taking a quotient. Thus speed is defined to be the dx/dt, the 'time rate of change of position'.

Ratio

The quotient of two similar quantities. In physics, the two quantities must have the same units to be 'similar'. Therefore we may properly speak of the ratio of two lengths. But to say 'the ratio of charge to mass of the electron' is improper. The latter is properly called 'the specific charge of the electron.'

Reaction

Reaction forces are those equal and opposite forces of Newton's Third Law. Though they are sometimes called an *action and reaction* pair, one never sees a single force referred to as an *action force.*

Real Image

1. A place where an object appears to be, because the rays diffusely reflected from any given point on the object have been bent so that they come back together and then spread out again from the new point.

2. The point(s) to which light rays converge as they emerge from a lens or mirror.

Real Object

The point(s) from which light rays diverge as they enter a lens or mirror.

Real Time

As it happens.

Receiver Sensitivity

The ability of a receiver to detect weak signals through the noise level of the receiving system, which includes the antenna and internal thermal noise of the receiver.

Red Giant Phase of Sun

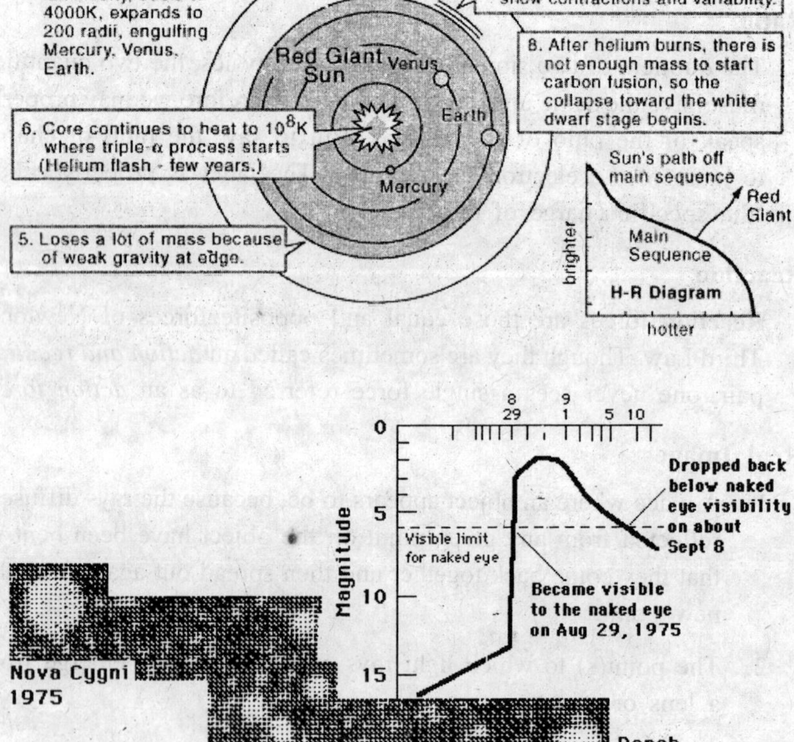

From observations of numerous other stars which appear to be similar to our Sun, it is anticipated that the Sun will eventually move upward and to the right of its current position on the main sequence and enter a red giant phase. The final stage of our Sun is anticipated to be as a white dwarf.

The brightest Nova in recent years was Nova Cygni 1975 which became visible in Cygnus in August. It reached a magnitude of 1.8 after a rise in magnitude of about 15 in two days time. This is a million-fold increase in luminosity!

Red Giants

For stars of less than 4 solar masses, hydrogen burn-up at the center triggers expansion to the red giant phase.

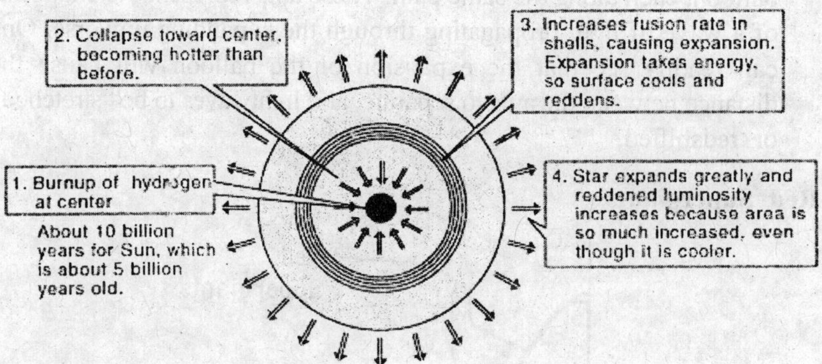

Redshift

Redshift is the lengthening of the wavelength of electromagnetic radiation (or, equivalently, the shortening of its frequency). There are three types of redshift. The (Relativistic) Doppler Effect. Named after its discoverer, Christian Doppler (1803-53), this is the change in frequency that results when the emitter is travelling away from the viewer, or when the viewer is travelling away from the emitter (in special relativity this is really the same thing). However, if the emitter and observer are moving *towards* each other (relatively, of course!), the observed radiation's frequency will be increased or *blueshifted.* Gravitational Redshift. As a result of the slowing down of time within a gravitational field, light emitted from the surface of a planet or near the surface of a black hole is reduced in frequency, or redshifted. In effect, the light's observed energy is diminished as it battles to escape the grip of gravity. Cosmological Redshift Due to the *expansion* of the universe, light may also be redshifted as viewed by an observer. This is a result of two effects, one resulting from special relativity, the other from general relativity. Imagine the source and the viewer of the light to be two dimensional creatures living on the surface of a balloon. As their universe, the balloon, begins to inflate, any two points on the surface of the balloon will acquire a relative speed and exhibit corresponding doppler shifts. The second effect is a direct result upon the wavelength. Imagine two ants crawling at the same speed across the surface of the balloon, each along the same path. These ants represent the endpoints of a wave of light propagating through the expanding universe. One can readily see that the expansion of the balloon will cause the distance between the ants to separate, and lightwaves to be "stretched" or redshifted.

Red Supergiants

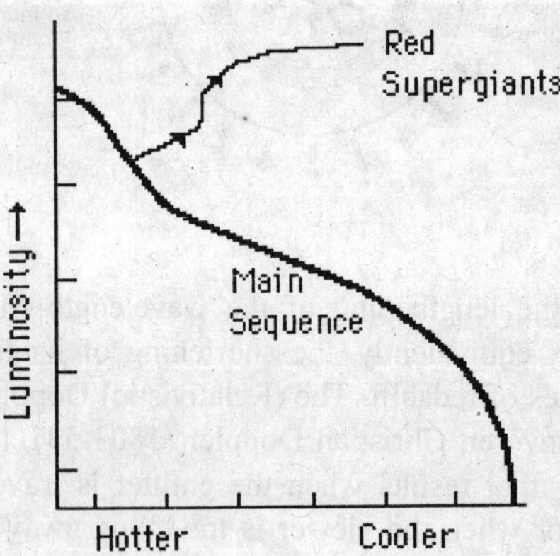

A star of 15 solar masses exhausts its hydrogen in about one-thousandth the lifetime of our sun. It proceeds through the red giant phase, but when it reaches the triple-alpha process of nuclear fusion, it continues to burn for a time and expands to an even larger volume. The much brighter, but still reddened star is called a red supergiant. Betelgeuse, at the shoulder of Orion, is the best-known example. Absolute luminosities may reach -10 magnitude compared to +5 for our sun. Some of these supergiants are unstable and form the very important Cepheid variables. In their final stages, supergiants may explode into supernovae. The collapse of these massive stars may produce a neutron star or a black hole.

Reflection

1. The return of light or sound waves from a surface. If a reflecting surface is plane, the angle of reflection of a light ray is the same as the angle of incidence.

2. What happens when light hits matter and bounces off, retaining at least some of its energy.

Reflection of Light

The change of direction of a light ray at the surface of a medium which ensures that the light ray remains in the one medium. The change of direction obeys two laws known as the laws of reflection. 1st law – The incident ray, the reflected ray and the normal to the reflecting surface at the point of incidence, all lie in the same plane. 2nd law – The angle of incidence equals the angle of reflection. The first law states that reflection is a two–dimensional phenomenon which can be represented by a two–dimensional diagram on a flat piece of paper. The second states the principle to be followed when drawing these diagrams.

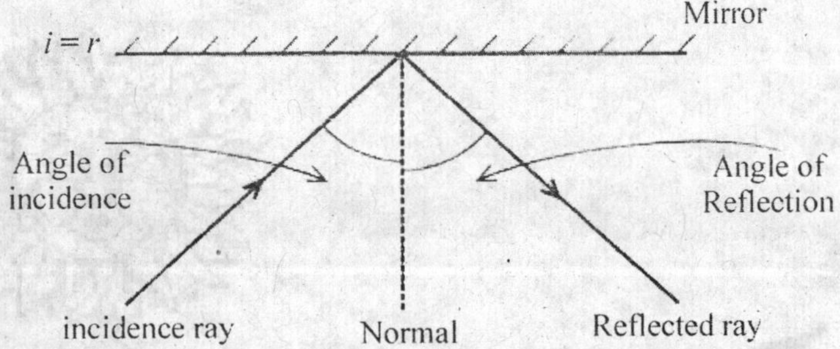

Refraction

The change in direction that occurs when a wave encounters the interface between two media.

Reference Frame

Consider a astronaut travelling in a starship, and someone sitting on the earth. According to the special theory of relativity, each perceives himself or herself to be at rest while the other is perceived to be moving. Each of these observers is said to be in his or her own "reference frame." Any two observers travelling with the same velocity (the same speed and direction) are said to be in the same reference frame. Two people travelling with a different speed and direction are said to be in different reference frames.

Relative Planet Sizes

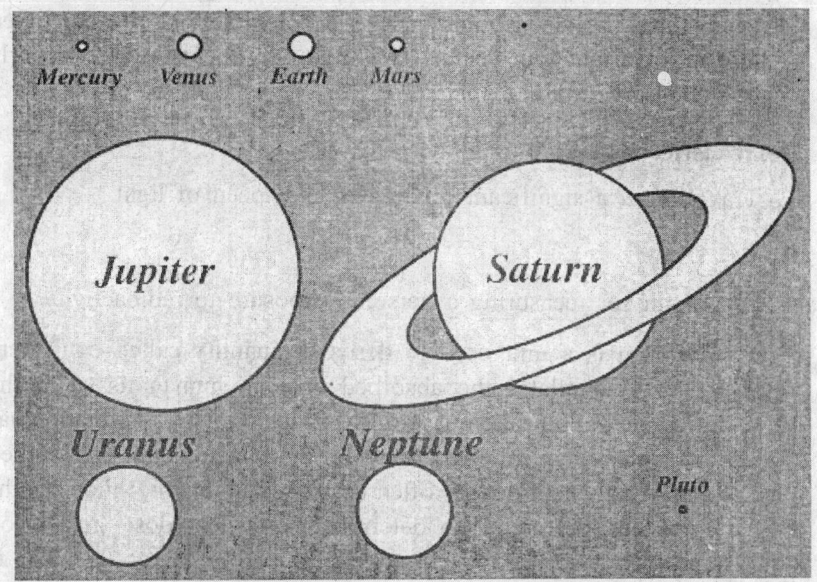

Relative Uncertainty

The uncertainty in a quantity compared to the quantity itself, expressed as a ratio of the absolute uncertainty to the size of the quantity. It may also be expressed as a percent uncertainty. The relative uncertainty is dimensionless and unitless.

Relative

Colloquially 'compared to'. In the *theory of relativity* observations of moving observers are quantitatively compared. These observers obtain different values when measuring the same quantities, and these quantities are said to be *relative*. The theory, however, shows us how the differing measured values are precisely related to the relative velocity of the two observers. Some quantities are found to be the same for all observers, and are called *invariant*. One postulate of relativity theory is that the speed of light is an invariant quantity. When the theory is expressed in four dimensional form, with the appropriate choice of quantities, new invariant quantities emerge: the world-displacement *(x + y + z +ict)*, the energy-momentum four-vector, and the electric and magnetic potentials may be combined into an invariant four-vector. Thus relativity theory might properly be called *invariance theory.*

Relativistic

Traveling at a significant fraction of the speed of light.

Rem

1. A unit for measuring a person's exposure to radioactivity.
2. The rem is a unit used to derive a quantity called equivalent dose. This relates the absorbed dose in human tissue to the effective biological damage of the radiation. Not all radiation has the same biological effect, even for the same amount of absorbed dose. Equivalent dose is often expressed in terms of thousandths of a rem, or mrem. To determine equivalent dose (rem), you multiply absorbed dose (rad) by a quality factor (Q) that is unique to the type of incident radiation.

Remapping

Flattening the Earth into a standard map projection. When the spherical Earth is photographed by satellites, areas lying near the outer edge of the picture are distorted. Remapping rectifies the distortion.

Remote Sensing

The technology of acquiring data and information about an object or phenomena by a device that is not in physical contact with it. In other words, remote sensing refers to gathering information about the Earth and its environment from a distance, a critical capability of the *Earth Observing System.* For example, spacecraft in low-Earth orbit pass through the outer thermosphere, enabling direct sampling of chemical species there. These samples have been used extensively to develop an understanding of thermospheric properties. Explorer-17, launched in 1963, was the first satellite to return quantitative measurements of gaseous stratification in the thermosphere. However, the mesosphere and lower layers cannot be probed directly in this way—global observations from space require remote sensing from a spacecraft at an altitude well above the mospause. The formidable technological challenges of atmospheric remote sensing, many of which are now being overcome, have delayed detailed study of the stratosphere and mesosphere by comparison with thermospheric research advances. Some remote-sensing systems encountered in everyday life include the human eye and brain, and photographic and video cameras.

Repulsive

Describes a force that tends to push the two participating objects apart.

Residual Interaction

Interaction between objects that do not carry a charge but do contain constituents that have that charge. Although some chemical substances involve electrically-charged ions, much of chemistry is due to residual electromagnetic interactions between electrically-neutral atoms. The residual strong interaction between protons and neutrons, due to the strong charges of their quark constituents, is responsible for the binding of the nucleus.

Residual Interaction

Interaction between objects that do not carry a *charge* but that contain constituents that do have a charge. Although some chemical substances involve electrically-charged ions, much of chemistry is due to residual *electromagnetic interactions* between electrically neutral atoms. The residual strong interaction between protons and neutrons, due to the strong charges of their quark constituents, is responsible for the binding of the nucleus.

Resistance

The ratio of the voltage difference to the current in an object made of an ohmic substance.

Resolution Cell

The smallest unit of area in an image of discrete elements. The area represented by a pixel.

Resolution

A measure of the ability to separate observable quantities. In the case of imagery, it describes the area represented by each pixel of an image. The smaller the area represented by a pixel, the more accurate and detailed the image. APT has a resolution of 4 km, i.e., each pixel represents a square, 4 km on each side. HRPT has a resolution of 1.1 km at nadir (4 km at edge of scan), and WEFAX of 8 km.

Resolution of Vector

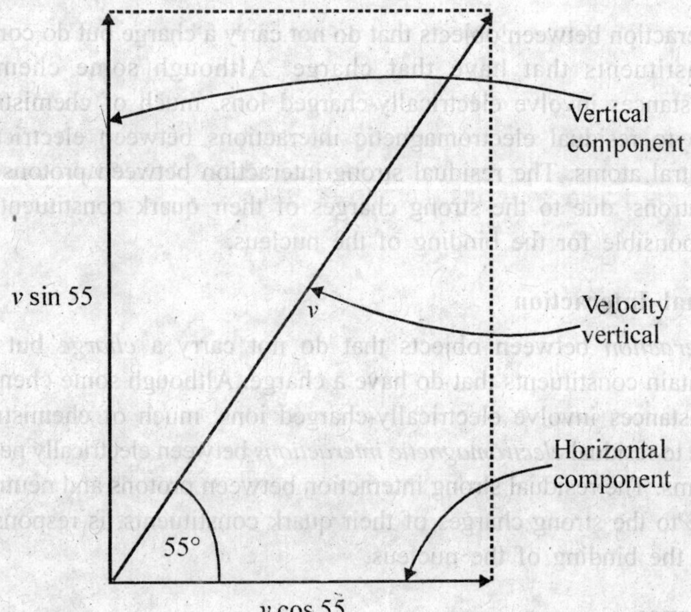

Resolution of Vector usually means finding the horizontal and vertical components of a single oblique vector. The diagram below shows a velocity vector *v* inclined at 55° to the ground. The vertical and horizontal directions are given. Dropping two perpendiculars from the far end of the vector constructs a rectangular 'parallelogram'. The horizontal component, v cos 55°, and the vertical component, *v* sin 55°, can both be read from the diagram. Resolving displacement and velocity vectors in this way simplifies a problem where the vertical motion is subject to gravitational force and acceleration while the horizontal motion continues at constant velocity.

Resolving Power

Resolving Power is a measure of the fineness of detail distinguishable in the focused image formed by an optical instrument or by something similar, such as a radio telescope. The definition has a simple mathematical form illustrated in the diagram. The resolving power of an image–forming instrument is the minimum angular distance in radians, measured at the instrument, of two small objects in the field of view which can just be distinguished as two separate objects in the image. Resolving power is limited by diffraction effects in the instrument. For an optical or radio telescope.

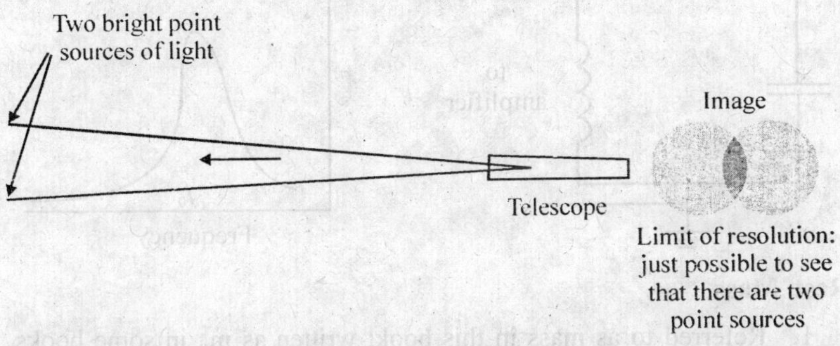

Resonance

The tendency of a vibrating system to respond most strongly to a driving force whose frequency is close to its own natural frequency of vibration.

Resonance in AC Circuits

Resonance in AC circuits is of two kinds. The impedance of a series circuit has a minimum value at the resonant frequency, the impedance of a parallel circuit has a maximum value at the resonant frequency. The parallel circuit shown in the diagram above is set up as a tuning circuit. The aerial picks up radio signals of many frequencies. The inductor has impedance w*L;* it offers a low impedance pathway to earth to the lower frequency signals. The capacitor has impedance 1/*wC;* it offers a low impedance pathway to earth to the higher frequency signals. The impedances are equal :

$$\omega L = \frac{1}{\omega C} \quad \text{and} \quad \omega = \frac{1}{\sqrt{(LC)}}$$

This will be the carrier frequency for the radio signal which sets up the maximum amplitude voltage oscillation across the parallel pair and which is passed to the next stage for amplification.

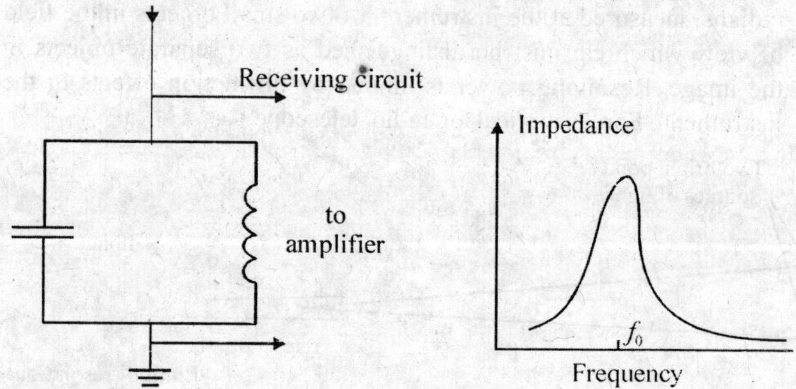

Rest Mass

1. Referred to as mass in this book; written as m_0 in some books.
2. The rest mass of a particle is the mass defined by the energy of the isolated (free) particle at rest, divided by the speed of light squared. When particle physicists use the word "mass", they always mean the "rest mass" of the object in question.

Revolution

Process of the Earth circling the sun in its orbit. Revolution determines the seasons, and the length of the year. In addition, differences in seasons occur because of Earth's inclination (tilt on its axis) of about 23.5 degrees as it revolves around the sun.

Right Ascension of Ascending Node (AkaΩ, RAAN or RA of Node). One of six Keplerian elements, it indicates the rotation of the orbit plane from some reference point. Two numbers orient an orbital plane in space; inclinations is the first, this is the second. After specifying inclination, an infinite number of orbital planes are possible. The intersection of the equatorial plane and the orbital plane (see diagram, *line of nodes* must be specified by a location on the equator that fully defines the orbital plane. The line of nodes occurs in two places. However, by convention, only the ascending node (where the satellite crosses the equator going from south to north) is specified. The descending node (where the satellite crosses the equator going from north to south) is not. Because the Earth spins, conventional latitude and longitude points are not used to separate where the lines of node occur. Instead, an astronomical coordinate system is used, known as the right-ascension/declination coordinate system, which does not spin with the Earth. Right ascension of ascending node is an angle, measured at the center of the Earth, from the vernal equinox to the ascending node. For example, draw a line from the center of the Earth to the point where the satellite crossed the equator (going from south to north). If this line points directly at the vernal equinox, then RAAN = 0°.

Roentgen (R)

The roentgen is a unit used to measure a quantity called exposure. This can only be used to describe an amount of gamma and X-rays, and only in air. One roentgen is equal to depositing in dry air enough energy to cause 2.58E-4 coulombs per kg. It is a measure of the ionizations of the molecules in a mass of air. The main advantage of this unit is that it is easy to measure directly, but it is limited because it is only for deposition in air, and only for gamma and x rays.

ROM

Read Only Memory. Refers to the computer memory chips that contain information the computer uses (along with system files) throughout the system, including the information it needs to get itself started. Information in ROM is permanent; it doesn't vanish when the power is turned off.

Rotation

Process of the Earth turning on its axis. Rotation determines day and night, and the length of the day.

S

Sailing Boats

Sailing Boats are driven by the flow of air over the sail. The diagram below shows a sailing boat seen from above. The wind flows faster over the outer, convex surface of the sail than over the inner, concave surface. According to *Bernouilli's principle,* this difference in speed sustains a difference in pressure, and the difference in pressure acting over the area of the sail gives a resultant thrust. Next, this thrust is resolved in two directions : along the keel and normal to the keel. The thrust normal keel is balanced by a normal contact force from the water. The thrust along the keel drives the boat through the water.

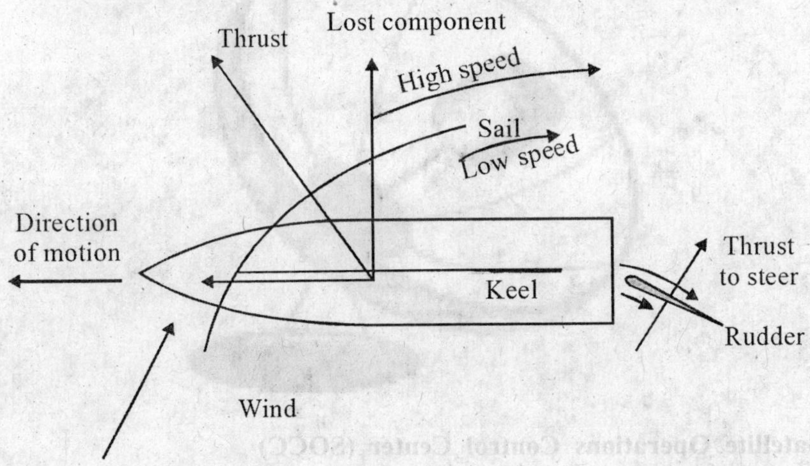

Sampling

The process of obtaining a sequence of discrete digital values from a continuous sequence of analog data.

SAR SAT

Search and Rescue Tracking System carried on NOAA polar-orbiting satellites that receives emergency signals from persons in distress. The satellites transmit these signals to ground receiving stations in the U.S. and overseas. Signals are forwarded to the nearest rescue coordination center which computes the location from which the emergency signals came and provides the coordinates of the emergency site to a rescue team.

Satellite Dish (Aka Parabolic Reflector)

Bowl shaped antennas that collect and focus the signals that a satellite beams down to Earth. The dish reflects the incoming radio frequency energy to a focal point where it can be picked up by a feedhorn antenna to transfer the RF energy to a transmission line. The bigger the dish, the greater will be the intercepted RE energy and hence, the gain. For example, a satellite dish is used to receive GOES WEFAX imagery.

Satellite Operations Control Center (SOCC)

NOAA National Environmental Satellite Data and Information Service (NESDIS) Satellite Operations Control Center located in Suitland,

Maryland. A principal operating feature of the NOAA system is the centralized remote control of the satellite through command and data acquisition (CDA) stations. The CDA stations transmit command programs to the satellite, and acquire and record meteorological and engineering data from the satellite. Data is transmitted from CDA to Suitland NESDIS Data Processing Services Subsystem (DPSS). DPSS is responsible for data processing and timely generation of meteorological products and distribution of these products.

Satellite Positioning

A procedure by which satellites are used to locate precise objects or particular points on Earth.

Satellite Revolution

The time from one perigee (the point of an elliptical orbit path where a satellite is closest to Earth) to the next.

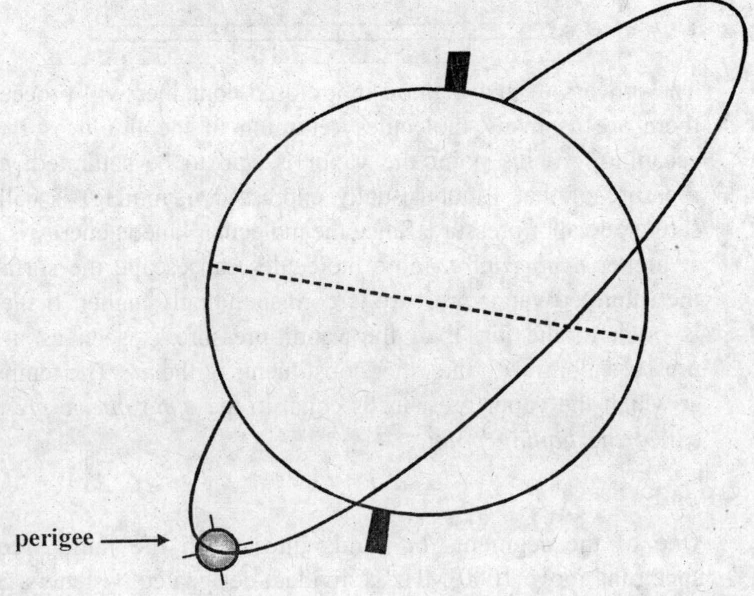

Satellite

A free-flying object that orbits the Earth, another planet, or the sun.

Saturated Vapor Pressure

The process of *evaporation* in a closed container will proceed until there are as many molecules returning to the liquid as there are escaping. At this point the vapor is said to be saturated, and the pressure of that vapor (usually expressed in mmHg) is called the saturated vapor pressure. Since the molecular kinetic energy is greater at higher *temperature*, more molecules can escape the surface and the saturated vapor pressure is correspondingly higher. If the liquid is open to the air, then the vapor pressure is seen as a partial pressure along with the other constituents of the *air*. The temperature at which the vapor pressure is equal to the *atmospheric pressure* is called the *boiling point*.

S-Band

One of the segments or bands into which the radio frequency spectrum above 1000 MHz is divided, designated by letters. Signals from GOES and other geostationary spacecraft transmitting on or near 1691 MHz are transmitting on S-Band.

Scalar

1. A quantity that has no direction in space, only an amount.
2. Any quantity that has only magnitude as opposed to both magnitude and direction. For example mass is scalar quantity. By convention in physics the word speed is a scalar quantity, having only magnitude, while the word velocity is used to denote both the speed and the direction of the motion and is thus a *vector* quantity.

Scale-limited

A measuring instrument is said to be *scale-limited* if the experimental uncertainty in that instrument is smaller than the smallest division readable on its scale. Therefore the experimental uncertainty is taken to be half the smallest readable increment on the scale.

Scaler

An electronic instrument for counting radiation induced pulses from radiation detectors such as a Geiger-Muller tube.

Scanner

A system that optically scans its detector(s) across a scene and records or stores the data in a two-dimensional format to form an image.

Scanning Radiometer

An imaging system consisting of lenses, moving mirrors, and solid-state image sensors used to obtain observations of the Earth and its atmosphere. Scanning radiometers, which are the sole imaging systems on all current operational weather satellites, have far better long4erm performance than the vidicon IV camera tubes used with eårlier spacecraft.

Scanning Tunneling Microscope

The subject of a 1986 Nobel Prize, the scånning tunneling microscope is currently the most powerful and versatile tool for imaging individual atoms. If a pointed metal probe is placed sufficiently close to a solid sample and a voltage of say 10 millivolts is applied between the probe and the surface, then electron tunneling can occur. The separation between the mini-tip of the probe and the sample must be on the order of nanometers. The exponential variation of the tunneling current with separation can show surface variations in the range 0.01 nanometers. The sharpening of the mini-tip of the probe is accomplished with electrochemical etching. This etching sharpens it to only about 1000 nm, but since the sharpening is not perfectly smooth, it leaves a surface with many mini-tips which are on the scale of nanometers.

Scattering

The process by which electromagnetic radiation interacts with and is redirected by the molecules of the atmosphere, ocean, or land surface. The term is frequently applied to the interaction of the atmosphere on sunlight, which causes the sky to appear blue (since light near the blue end of the spectrum is scattered much more than light near the red end).

Schrödinger Equation

In physics, the *Schrödinger equation*, proposed by the Austrian physicist Erwin Schrödinger in 1925, describes the time-dependence of quantum mechanical systems. It is of central importance to the theory of quantum mechanics, playing a role analogous to Newton's second law in classical mechanics. In the mathematical formulation of quantum mechanics, each system is associated with a complex Hilbert space such that each instantaneous state of the system is described by a unit vector in that space. This state vector encodes the probabilities for the outcomes of all possible measurements applied to the system. As the state of a system generally changes over time, the state vector is a function of time. The Schrödinger equation provides a quantitative description of the rate of change of the state vector.

Schrödinger Wave Equation

The state space of certain quantum systems can be spanned with a *position basis.* In this situation, the Schrödinger equation may be conveniently reformulated as a partial differential equation for a wavefunction, a complex scalar field that depends on position as well as time. This form of the Schrödinger equation is referred to as the *Schrödinger wave equation.* Elements of the position basis are called position eigenstates. We will consider only a single-particle system, for which each position eigenstate may be denoted by |**r**>, where the label **r** is a real vector. This is to be interpreted as a state in which the particle is localized at position **r**. In this case, the state space is the space of all square-integrable complex functions.

Schwarzschild Radius

The Schwarzschild radius is the radius at which *the event horizon* of a *Critical Circumference.*

Scintillation Counter

1. An instrument that detects and measures gamma radiation by counting the light flashes (scintillations) induced by the radiation.
2. A scintillation counter consists of a material that emits light when radiation passes through it. Various liquid, plastic, and crystalline materials have scintillation properties. Scintillation light is measured with photomultiplier tubes. In general the amount of scintillator light detected is proportional to the energy of the radiation.

Screaming Eagles

Cloud pattern so named because some observers maintain they can see the head of an eagle facing west in these cloud patterns. The pattern is similar to a comma, only the pattern is disorganized and not solid. Weather associated with screaming eagles consists of rain showers and gusty surface winds up to about 25 knots. The eagles can intensifv and enlarge when moving into areas east of troughs; in that case, intense thundeF storms can develop. Screaming eagles are common in the Pacific Ocean between Hawaii and the equator and are uncommon in the western Atlantic.

Sea Breeze

Local coastal wind that blows from the ocean to land. Sea breezes usually occur during the day because the heating differences of land and sea cause pressure differences. Cooler heavier air from the sea moves in to replace rising warm air on the coastline.

Sea Level

The datum against which land elevation and sea depth are measured. Mean sea level is the average of high and low tides.

Search and Rescue

International satellite-aided search and rescue project. COSPASI SARSAT satellites monitor the entire surface of the Earth, and transmit distress signals to special ground receiving stations. The receiving stations compute the location of the signal, and notifv the nearest rescue coordination center. Satellite search has cut recovery time from days to hours, and has aided downed airplanes, capsized boats, and persons in other emergencies.

Secular Equilibrium

A state of parent-daughter equilibrium which is achieved when the half-life of the parent is much longer than the half-life of the daughter. In this case, if the two are not separated, the daughter will eventually be decaying at the same rate at which it is being produced. At this point, both parent and daughter will decay at the same rate until the parent is essentially exhausted.

Semiconductor Detector

Radiation striking very pure Ge and Si semiconductor detectors can excite a large number of *electrons* into the conduction band leading to a measurable current. This current is proportional to the energy of the radiation. Semiconductor detectors can be used to accurately measure the energy and intensity of radiation.

Semi-Major Axis (Aka A)

One of the six Keplerian elements, it indicates the size of an orbit.

The semi-major axis is one half of the longest diameter of an orbital ellipse, e.g., one-half of the distance between the apogee and perigee of an Earth orbit. (The semi-major axis is related to the orbital period and mean motion by Kepler's third law.

Sensor Calibration

The relationship between input and output for a given measurement.

Sensor

Device that produces an output (usually electrical) in response to stimulus such as incident radiation Sensors aboard satellites obtain information about features and objects on Earth by detecting radiation reflected or emitted in different bands of the electromagnetic spectrum. Analyzing the transmitted data provides valuable scientific information about Earth. Weather satellites commonly carry radiometers, which measure radiation from snow ice, clouds, and bodies of water Spaceborne radars are used for Earth observations, bouncing radar waves off land and ocean surfaces to study sea-surface conditions, ice thickness, and land surface features. A wind scatterometer is a special type of radar designed to measure ocean surface winds indirectly by boun& mg signals off the water and measuring them from various angles. Infrared (IR) detectors measure heat generated by Earth features in the IR band of the spectrum. Photographic reconnaissance sensors in their simplest form are large telescope-camera systems used to view objects on Earths' surface. The bigger the lens, the smaller the object that can be detected. Camera-telescope systems now incorporate all sorts of Sophisticated electronics to produce better images, but even these systems need cloudless skies, excellent lighting, and good color contrast between objects and their surroundings to detect objects the size of a basketball. Some of the satellites produce film images that must be returned to Earth, but a more convenient method is to record the image as a series of digital code numbers, then reconstruct the image from the electronic code using a computer at a ground station.

Sharpening of Pitch Perception

The high *pitch resolution* of the ear suggests that only about a dozen *hair cells*, or about three tiers from the four banks of cells are associated with each distinguishable *pitch*. It is hard to conceive of a mechanical resonance of the basilar membrane that sharp. So we look for enhancements of the basic *place theory* of pitch perception. There must be some mechanism which sharpens the response curve of the *organ of Corti*, as suggested schematically in the diagram.

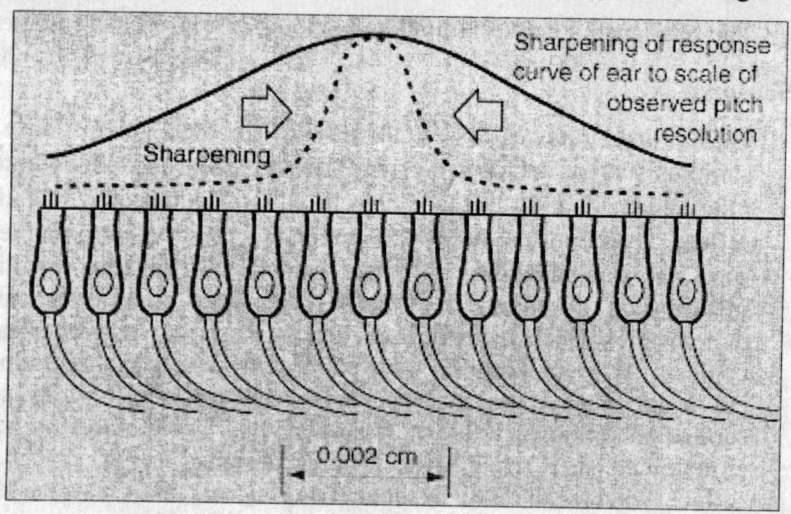

Shield

A mass of attenuating material used to prevent or reduce the passage of radiation or particles.

Shielding

A protective barrier, usually a dense material, which reduces the passage of radiation from radioactive materials to the surroundings.

Short Circuit

A circuit that does not function because charge is given a low-resistance "shortcut" path that it can follow, instead of the path that makes it do something useful.

Shower

(also called Electromagnetic Cascade Shower). Electrons can create

photons by interacting with a medium. In a similar way, photons can create electrons and their *antiparticles* , positrons, by interacting with a medium. So, imagine a very high-energy electron, of the sort used at SLAC, impinging on some material. The electron can set photons into motion and these photons can, in turn, set electrons and positrons into motion, and this process can continue to repeat. One high-energy electron can set thousands of particles into motion. Albert Einstein's famous relation governing the equivalence of matter and energy ($E = mc^2$) governs this process — namely, matter (electrons and positrons) can be creased from pure energy and vice versa. The particle creation process only stops when the energy runs out.

SI Prefixes

Many units are broken down into smaller units or expressed as multiples, using standard metric prefixes. As examples, a kilo-becquerel (kBq) in 1000 becquerels, a millirad (mrad) is 10^{-3} rad, a microrem (μrem) is 10^{-6} rem, a nanogram is 10^{-9} grams, and a picocurie is a 10^{-12} curies.

Sievert (Sv)

The sievert is a unit used to derive a quantity called equivalent dose. This relates the absorbed dose in human tissue to the effective biological damage of the radiation. Not all radiation has the same biological effect, even for the same amount of absorbed dose. Equivalent dose is often expressed in terms of millionths of a sievert, or micro-sievert. To determine equivalent dose (Sv), you multiply absorbed dose (Gy) by a quality factor (Q) that is unique to the type of incident radiation. One sievert is equivalent to 100 rem.

Signal

Electrical impulses, sound or picture elements, etc., received or transmitted. Signals can exist in many different forms and media (electrical/wires, acoustic/air light/transparent fibers, etc.), but all signals will vary with time. The signal shape plotted as a function of time is called the waveshape or waveform. Some waveforms are repetitive or periodic, that is, a small segment of the waveform repeats itself regularly. Other waveforms, such as noise, are nonperiodic or aperiodic. All waveforms can be distilled into the combination of pure waves called sine waves. The frequency of a sine wave is the rate at which the fundamental shape repeats itself. Most signals occupy a limited range of frequencies between a lower limit and an upper limit. This range or band of frequencies occupied by a signal is called the bandwidth of the signal. Communication medium or channel can pass only a specific range or band of frequencies, which is called the bandwidth of the channel. The bandwidths of the channel and the signal determine the number and types of signals that can be transmitted by a particular communication channel. Signals often are too small and need to be made larger through a process called amplification. The amount of amplification is measured in decibels. However amplification is an imperfect process, and inadvertently introduces various distortions, noise, and bandwidth limitations. Often, multiple signals must share the same medium. One way the sharing can be accomplished is to place each signal in its own band of frequencies within the total band of the medium. The combining of a number of signals to share a medium by dividing it into different frequency bands for each signal is called frequency-division multiplexing. Frequency-division multiplexing requires the ability to move signals around so that each multiplexed signal occupies its own band. This is accomplished through a process called modulation, in which a high-frequency sine wave carries the signal into the specified band. Either the amplitude or the frequency of the carrier wave can be varied, or modulated, in synchrony with the information-bearing signal. These methods are called amplitude modulation (AM) and frequency modulation (FM). FM is the more complex process of the two, and the bandwidth of the FM carrier can be many times that of the modulating signal. The process of demodulating a frequency-modulated signal eliminates much of the deleterious effects of additional noise. (The trade-off between bandwidth and noise immunity characterizes most communication systems. Both are analog modulation schemes for multiplexing signals in the frequency spectrum.) Digitizing a signal requires a number of steps and results in a binary digital signal that takes on one of two discrete values. This process results in considerable immunity to additive noise, but requires a considerable increase in bandwidth.

Signal-to-noise Ratio (SNR)

In decibels (dB), the difference between the amplitude of a desired radio frequency (RF) signal and the internal or external RF noise level in a system. A negative SNR indicates the signal is below the system noise level and unusable. The greater the positive SNR, the less effect noise will have on the final quality. SNR of at least + 12dB is necessary to produce imagery with minimal noise effects.

Significant Figures

Digits that contribute to the accuracy of a measurement.

Silicon and Germanium

Solid state electronics arises from the unique properties of silicon and germanium, each of which has four *valence electrons* and which form *crystal lattices* in which substituted atoms (*dopants*) can dramatically change the electrical properties.

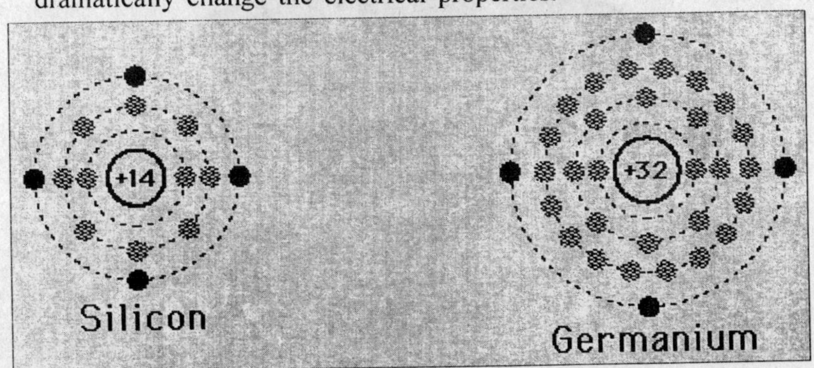

Silicon Vertex Detector

This is a silicon based detector similar to that in a digital camera. It provides precision particle tracking by connecting the dots due to a particle passing through its multiple layers. This allows one to reconstruct any vertex from which two or more tracks emerge.

Such a vertex, if outside the beam collision region, indicates the position of a particle decay.

Simple Harmonic Motion

Motion whose x-t graph is a sine wave.

Since pink noise has relatively more bass than white noise, it sounds more like the roar of a waterfall than like the higher hissing sound of white noise. Pink noise is often the choice for equalizing auditoriums. Real-time analyzers can be set up so that they display a straight horizontal line when they receive pink noise.

Simultaneity and Causality

Special relativity holds that events that are simultaneous in one frame of reference need not be simultaneous in another frame of reference.

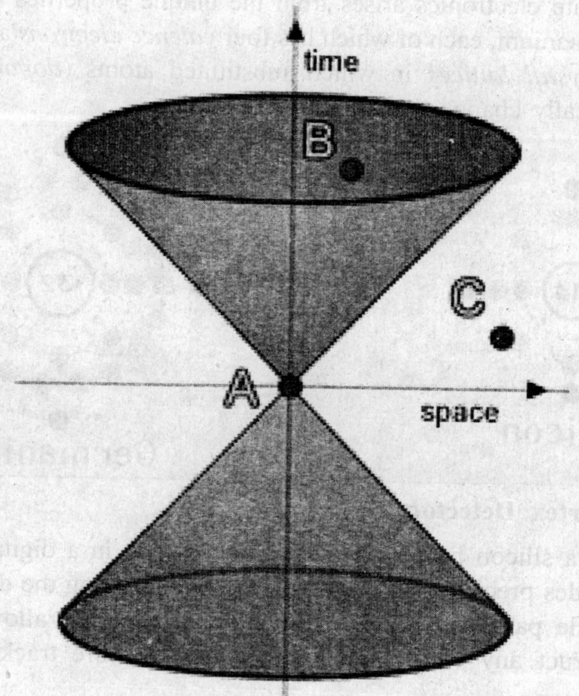

Sine Wave

A smoothly varying wave that repeats itself; its frequency is the rate at which the fundamental shape repeats itself. Any waveform can be distilled into a combination of pure sine waves of varying frequencies and amplitudes.

Singularity

In the center of the mathematical model of a black hole is a singularity which has the shape of a point (or a ring if the hole is rotating), at which the *curvature* of *spacetime* becomes infinitely large. A singularity represents a great difficulty for theoreticians because it is impossible to predict how a singularity will affect objects in its causal future. If *cosmic censorship* is true, then this needn't cause any trouble because they will only be found inside *event horizons.*

Sink

1. A point at which field vectors converge.
2. The process of providing storage for a substance. For example, plants—through photosynthesis—transform carbon dioxide in the air into organic matter which either stays in the plants or is stored in the soils. The plants are a sink for carbon dioxide.

Sirius-A

1. The star Sirius, referred to as Sirius-A, is perhaps most notable for the study of the "companion of Sirius" or *Sirius-B* which was the first example of a *white dwarf* star to be studied. Sirius itself is one of the brightest stars in the sky, being only 8.6 light-years away from us. It is also notable for being the subject of one of the first serious studies of the *carbon cycle* of *nuclear fusion*. It is much hotter than our Sun and it was clear that some process other than *proton-proton fusion* was taking place to produce all that energy.
2. The white *dwarf* Sirius-B was not seen until 1862, but was predicted in 1844 from the motion of *Sirius-A*. The *blackbody spectrum* of Sirius-B peaks at 110 nm, corresponding to a

temperature of 26,000 K. From the known absolute magnitude, the radius is calculated to be just 4200 km. Smaller than the Earth, it is almost as massive as the Sun.

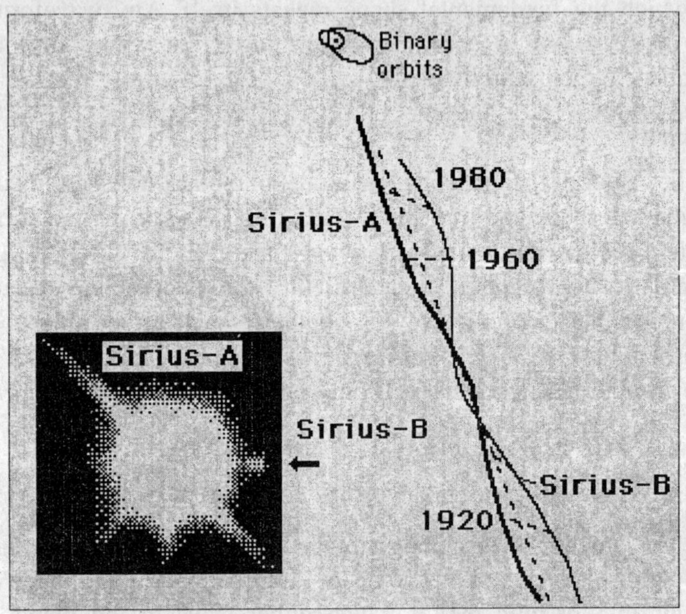

Sirius-B

The white dwarf Sirius-B was not seen until 1862, but was predicted in 1844 from the motion of Sirius-A. The blackbody spectrum of Sirius-B peaks at 110 nm, corresponding to a temperature of 26,000 K. From the known absolute magnitude, the radius is calculated to be just 4200 km. Smaller than the Earth, it is almost as massive as the Sun.

Skylab

1. The first U.S. space station, launched unmanned in May 1973 and soon after occupied in succession by three crews through November 1973.
2. Skylab was a manned, orbiting spacecraft which included the Apollo Telescope mount (a solar observatory) and capability for

a number of experiments. It involved four launches, the first unmanned to deploy the first unit. Skylab 1 itself was launched on May 14, 1973 and first manned during the period May 25 to June 22, 1973. It had to deal with the failure of the heat shield to deploy on Skylab 1; that repair was made and the mission was successful. Skylab 3 was launched on 28 July, 1973 and landed in the Pacific on September 25. The third and final manned period was an 84 day mission from November 16, 1973, to February 8, 1974.

SLAC

1. Stanford Linear Accelerator Center - where this virtual visitor center is located, along with many real facilities.

2. The Stanford Linear Accelerator Center in Stanford, California.

Software

The programs, data, or routines used by a computer distinguished from the physical components (e.g., hardware).

Solar Atmosphere

The atmosphere of the Sun. An atmosphere is generally the outermost gaseous layers of a planet, natural satellite, or star. Only bodies with a strong gravitational pull can retain an atmosphere. Atmosphere is used to describe the outer layer of the Sun because it is relatively transparent at visible *wavelengths*. Parts of the solar atmosphere include the *photosphere, chromosphere*, and the *corona.*

Solar Backscatter Ultraviolet Radiometer (SBUV)

Instrument that measures the vertical distribution and total ozone in the Earth's atmosphere. Data is used for the continuous monitoring of ozone distribution to estimate long-term trends. SBUV instruments are flown on NOAA polar-orbiting satellites.

Solar Constant

Aka total solar irradiance. The constant expressing the amount of solar radiation reaching the Earth from the sun, approximately 1370 watts per square meter It is not, in fact, truly constant and variations are detectable.

Solar Cycle

Eleven-year cycle of sunspots and solar flares that affects other solar indexes such as the solar output of ultraviolet radiation and the solar wind. The Earth's magnetic field, temperature, and ozone levels are affected by this cycle.

Solar Limb

The apparent edge of the Sun as it is seen in the sky.

Solar Radiation

Energy received from the sun is solar radiation. The energy comes in many forms, such as visible light (that which we can see with our eyes). Other forms of radiation include radio waves, heat (infrared), ultraviolet waves, and x-rays. These forms are categorized within the electromagnetic spectrum.

Solar System

Mosaic of the planets of the solar system, excluding Pluto, and including Earth's Moon. Note: planets are not portrayed in the same scale. The Solar System consists of the Sun and all the objects that orbit around it, including meteors, asteroids, comets, moons, and planets. The Earth is the third planet of the Solar System. *Planetary systems* are a more generic term for stars and the objects that orbit around them.

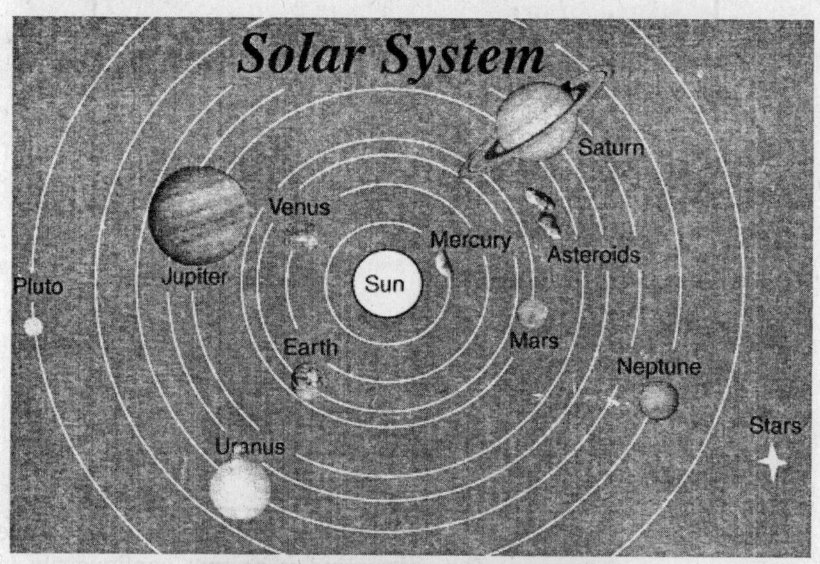

Solar Wind

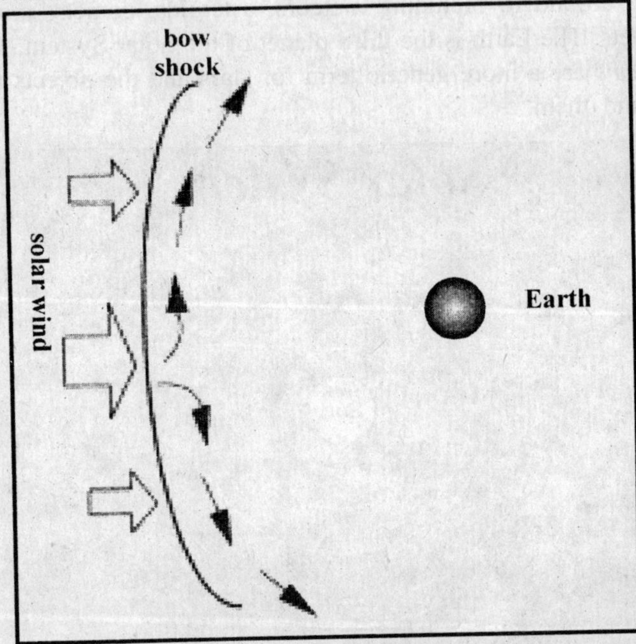

A continuous plasma stream expanding into interplanetary space from the sun's corona. The solar wind is present continuously in inter- planetary space. After escaping from the gravitational field of the sun, this gas flows outward at a typical speed of 400 km per second to distances known to be beyond the orbit of Pluto. Besides affecting Earth's weather solar activity gives rise to a dramatic visual phenomena in our atmosphere. The streams of charged particles from the Sun interact the Earth's magnetic field like a generator to create current systems with electric potentials of as much as 100,000 volts. Charged electrons are energized by this process, sent along the magnetic field lines towards Earth's upper atmosphere, excite the gases present in the upper atmosphere and cause them to emit light which we call the auroras. The auroras are the northern (aurora borealis) and southern (aurora Australis) lights.

Somatic Effects

Somatic effects are effects from some agent, like radiation that are seen in the individual who receives the agent.

Sounder

A special kind of radiometer that measures changes in atmospheric temperature with height, as well as the content of various chemical species in the atmosphere at various levels. The High Resolution Infrared Radiation Sounder (HIRS), found on NOAA polar-orbiting satellites, is a passive instrument.

Source

1. A point from which field vectors diverge; often used more inclusively to refer to points of either convergence or divergence.

2. A radioactive material that produces radiation for experimental or industrial use.

South Atlantic Anomaly

The region over the South Atlantic Ocean where the lower *Van Allen belt* of energetic, electrically charged particles is particularly close to the Earth's surface. The excess energy in the particles presents a problem for satellites in orbit around the Earth.

Spacetime Curvature

Imagine that the universe has two spatial dimensions instead of three, and that there are flat creatures living on its surface. Now imagine that the surface they are living on is subject to deformations, something like a bedsheet. The creatures living on the bedsheet can only see length and depth, they can only see within the bedsheet. They cannot even imagine the concept of height. Now imagine that the bedsheet is draped over a basketball, and the creatures are very small. If the creatures attempt to travel along a straight line within the fabric of the bedsheet they will be deflected by the presence of the basketball. Although on their very small scale the bedsheet appears to be flat, their path through it will be altered by the presence of the basketball, distorting the geometry of their world. Because of this, if we have three of these creatures travelling along parallel straight lines, and the middle creature's path takes him across the top of the ball, the course of the creatures to his right and left will be deflected inward. Because of curvature effects, these three initially parallel paths will meet. A similar effect can occur in spacetime. If two light rays, initially parallel, pass on either side of a black hole their paths will converge. Gravity causes spacetime to curve, and this curvature in turn affects the motion of objects in spacetime in much the same way that the curvature of the bedsheet affects the paths of motion of creatures wandering within it. To understand how these distortions create gravity you need to think of parallel *worldlines*. These worldlines are not drawn on a flat page, but are drawn on a curved surface. Because the surface is curved the intially parallel lines can be drawn together. What we perceive as gravity is the deflection in the path of worldlines caused by their being traced on a curved surface.

Spacetime

1. Space has three dimensions. However, the theory of relativity predicts that time, like space, is a dimension. In order to describe a four dimensional universe which has three spatial dimensions and one time dimension the word "spacetime" was coined. Each point in spacetime is called an event.

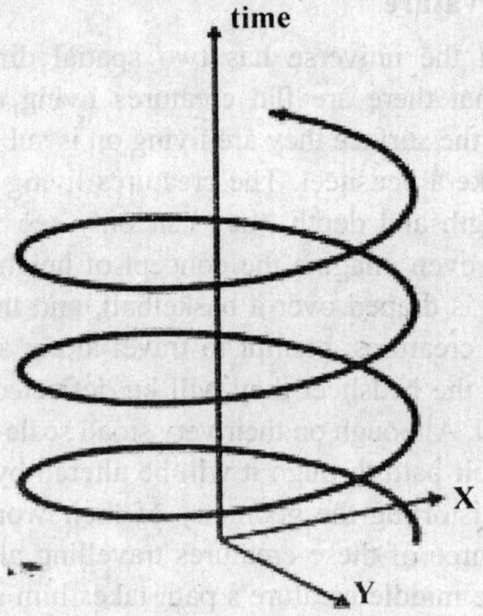

2. World line of the orbit of the Earth depicted as a circle in two spatial dimensions X and Y (the plane of the Earth orbit) and a time dimension, Z, making the circle appear as a helix. A *spacetime*, in the theories of special relativity and general relativity, time and three-dimensional space treated together as a single four-dimensional object. Spacetimes are the arenas in which all physical events take place - for example, the motion of planets around the Sun may be described in a particular type of spacetime, or the motion of light around a rotating star may be described in another type of spacetime. A *point* in spacetime is usually referred to as an *event.* An example of an event may be the explosion of a star. The worldline of a particle or light beam is the path that this particle or beam takes in the spacetime, this path being a geodesic. In classical physics, geodesics are timelike or null (light-like).

Space Environment Monitor (SEM)

Instrument that measures the condition of the Earth's magnetic field and the solar activity and radiation around the spacecraft, and transmits these data to a central processing facility. NOAA polaForbiting and geostationary satellites both carry SEMs.

Space Physics

Scientific study of magnetic and electric phenomena that occur in outer space, in the upper atmosphere of the planets, and on the sun.

Space Shuttle

NASA's manned, recoverable spacecraft designed to be used as a launch vehicle for Earth-orbiting experiments and as a short-term research platform.

Spacelab

A manned laboratory module built by the European Space Agency (ESA) that accommodates dozens of experiments on each flight, mainly in the categories of materials science and life science.

Spacelink

NASA electronic database for educators, with information stored on a computer at the Marshall Space Flight Center. Via computer educators communicate with NASA education specialists and access the following menus: current NASA news, aeronautics research, U.S. Space Program historical information, aerospace research in the 1980s and beyond, overviews of NASA and its Centers, NASA educational services, classroom materials, and space program spin-offs. The computer access number is 205-895-0028, the data word format is 8 data bits, no parity, and 1 stop bit. 300, 1200, or 2400 baud modem required. Callers with Internet access may reach NASA.

SPEAR

Stanford Positron Electron Accelerating Ring.

Special Relativity

A simple introduction to this subject is provided in Special relativity for beginners Special relativity (SR) or the *special theory of relativity* is the physical theory published in 1905 by Albert Einstein. It replaced Newtonian notions of space and time and incorporated electromagnetism as represented by Maxwell's equations. The theory is called "special" because it is a special case of Einstein's principle of relativity where the effects of gravity can be ignored. Ten years later, Einstein published the theory of general relativity which incorporates gravitation.

Specific

In physics and chemistry the word *specific* in the name of a quantity usually means 'divided by an extensive measure that is, divided by a quantity representing an amount of material. *Specific volume* means volume divided by mass, which is the reciprocal of the density. *Specific heat capacity* is the heat capacity divided by the mass.

SPECT

Single-Photon Emission Computerized Tomography involves scanning involving the rotation of detectors around a patient and acquires information on the concentration of radionuclides introduced to the patient's body. This is analogous to *CT imaging* with x-rays.

Spectral Band

A finite segment of wavelengths in the electromagnetic spectrum.

Spectral Line

A line in a *spectrum* due to the emission or absorption of *electromagnetic radiation* at a discrete *wavelength*. Spectral lines result from discrete changes in the energy of an atom or molecule. Different atoms or molecules can be identified by the unique sequence of spectral lines associated with them.

Spectral Range

The range of wavelengths of the electromagnetic radiation that can be produced.

Spectrograph

An instrument that spreads light or other *electromagnetic radiation* into its component *wavelengths* (*spectrum*), recording the results photographically or electronically.

Spectrometer

An instrument for measuring the intensity of radiation as a function of *wavelength*.

Spectrum

1. The series of colored bands diffracted and arranged in the order of their respective wave lengths by the passage of white light through a prism or other diffracting medium and shading continuously from red (produced by the longest visible wave) to violet (produced by the shortest visible wave).
2. Any of various arrangements of colored ands or lines, together with invisible components at both ends of the spectrum, similarly formed by light from incandescent gases or other sources of radiant energy, which can be studied by a spectrograph.
3. In radio, the range of wave lengths of radio waves, from 3 centimeters to 30,000 meters, or of frequencies of radio waves,

from 10 to 10,000,000 kilocycles. Also radio spectrum. 4. The entire range of radiant energies.

4. *Electromagnetic radiation* arranged in order of *wavelength*. A rainbow is a natural spectrum of visible light from the Sun. Spectra are often punctuated with emission or absorption lines, which can be examined to reveal the composition and motion of the radiating source.

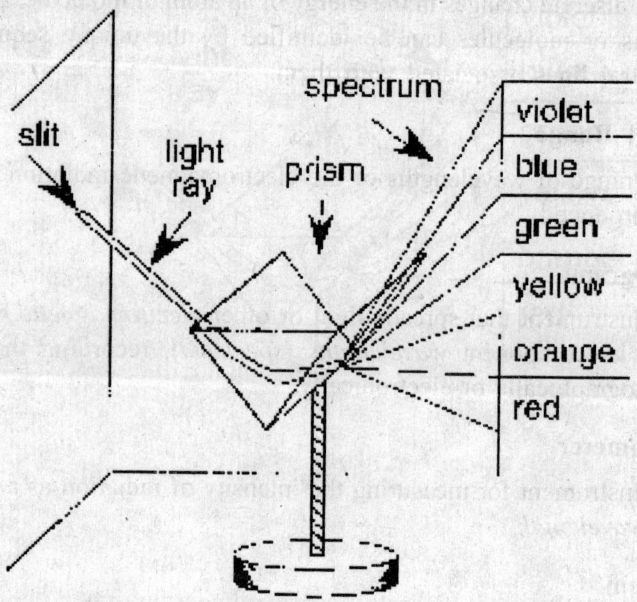

Specular Reflection

Reflection from a smooth surface, in which the light ray leaves at the same angle at which it came in.

Speed

1. (Avoided in This Book) The absolute value of or, in more then one dimension, the magnitude of the velocity, i.e. the velocity stripped of any information about its direction Spring constant. The constant of proportionality between force and elongation of a spring or other object under strain.

2. Speed is simply the rate of position change, but for analysis and problem solving speed has several faces. The speed at an instant is can be approximated with an average speed measured over a short time interval. Initial speed is the speed of an object at the beginning of a measurement, an intial condition. The average speed is the ratio of distance covered to the time it takes, or the average of the final and intial speeds. The change in speed is the difference between final and inital speeds. The symbol v is used for speed because speed is the size of the *velocity vector.*

Speed of Light

1. Light travels at a speed of 186,282 miles per second in vacuum from the point of view of a nearby observer. Because of the effects of general relativity the speed of light near a massive object will appear slower to a distant observer, and this effect has been confirmed in experiments. The speed of light is the theoretical limit to the speed of any particle in the universe. More fundamentally, no cause can result in an effect that requires travel faster than light. For example, I cannot affect what is going on 3 *light years* away but only 2 years in the future. I can, however, affect what is going on 3 light years away but 4 light years in the future.

2. Cherenkov effect in a "swimming pool" nuclear reactor. The effect is due to electrons moving faster than the speed at which light moves in water. The *speed of light* in a vacuum is exactly equal to 299,792,458 metres per second (approximately 186,282.4 miles per second). The speed of light is denoted by the letter c, reputedly from the Latin *celeritas*, "speed", and also known as Einstein's constant. This exact speed is a definition, not a measurement, as the metre itself is defined in terms of the speed of light and the second. The speed of light through a medium (that is, not in vacuum) is less than c (defining the refractive index of the medium). "Speed of light" is sometimes abbreviated SOL.

Spill

The accidental release of radioactive materials.

Spin

1. Intrinsic angular momentum.

2. The built-in angular momentum possessed by a particle even when at rest.

3. The name given to the angular momentum carried by a particle. For composite particles the spin is made up from the combination of the spins of the constituents plus the angular momentum of their motion around one-another. For *fundamental particles* spin

is an intrinsic and inherently quantum property, it cannot be understood in terms of motions internal to the object.

4. Used to describe the angular momentum of the *nucleus*.

SPOT Image

Company that markets data gathered by the SPOT satellite worldwide.

SPOT

Systeme Pour l'Observation de la Terre. French, polar-orbiting Earth observation satellite(s) with ground resolution of 10 meters. SPOT images are available commercially and are intended for such purposes as environmental research and monitoring, ecology management, and for use by the media, environmentalists, legislators, etc.

Stable Equilibrium

One in which a force always acts to bring the object back to a certain point.

Stable

1. Does not *decay*.
2. Does not decay. A particle is stable if there exist no processes in which a particle disappears and in its place different particles appear.
3. Non-radioactive.

Standard Model

Physicists' name for the theory of fundamental particles and their interactions. It is widely tested and is accepted as correct by particle physicists.

Star Classification

The study of the various types of stars gives information about the processes going on in the stars. They may be classified by spectral type and by their position relative to the main sequence on the Hertrzprung-Russell Diagram.

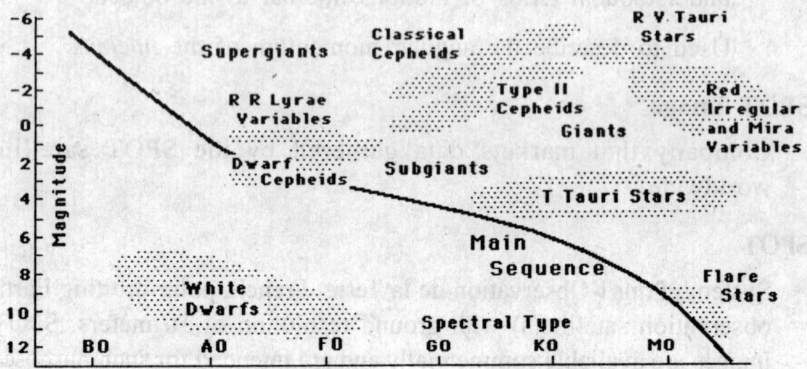

Star Temperatures

Stars approximate *blackbody radiators* and their visible color depends upon the temperature of the radiator. The curves show blue, white, and red stars. The white star is adjusted to 5270K so that the peak of its blackbody curve is at the peak wavelength of the sun, 550 nm. From the wavelength at the peak, the temperature can be deduced from the *Wien displacement law.*

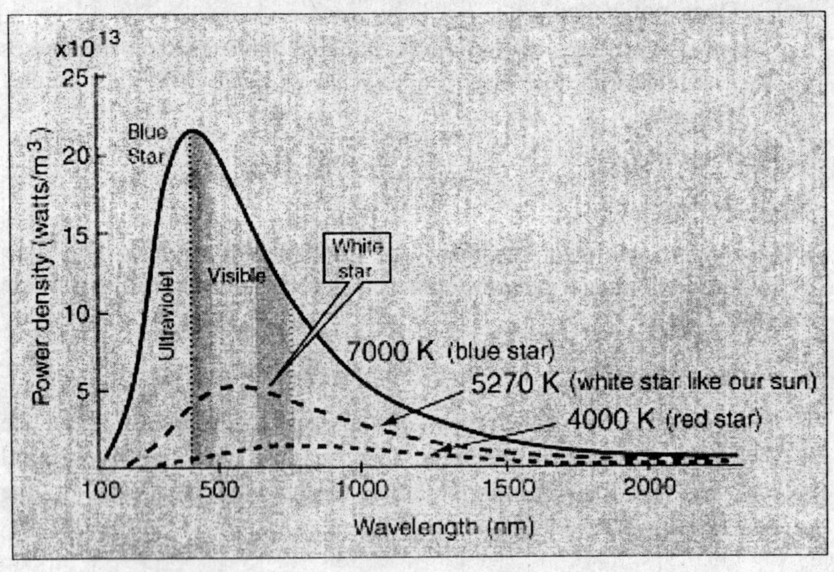

Start Ton

Five seconds of 300 Hz black to white square wave modulation of the WEFAX subcarrier signaling the start of a frame transmission (the beginning of a direct readout image).

Static Friction

A friction force between surfaces that are not slipping past each other.

Stationary Black Bole

"Stationary" refers to a time-independent mathematical description of a black hole (*not* its rotation–a rotating black hole can still be "stationary.") To understand what this means, consider the physical system represented by a perfectly symmetrical top spinning without friction. Every detail of this system remains the same as time goes by, and thus we can say the system is "stationary, " even though it is spinning. By definition, a stationary black hole "sits" alone in space. It interacts with no other matter or *gravitational radiation.*

Statistical Mechanics

Statistical mechanics is the application of statistics, which includes mathematical tools for dealing with large populations, to the field of mechanics, which is concerned with the motion of particles or objects when subjected to a force. It provides a framework for relating the microscopic properties of individual atoms and molecules to the macroscopic or bulk properties of materials that can be observed in every day life, therefore explaining thermodynamics as a natural result of statistics and mechanics (classical and quantum). In particular, it can be used to calculate the thermodynamic properties of bulk materials from the spectroscopic data of individual molecules.

Steady State

The behavior of a vibrating system after it has had plenty of time to settle into a steady response to a driving force. In the steady state, the same amount of energy is pumped into the system during each cycle as is lost to damping during the same period.

Steradian

The unit of "stereo angle" or solid angle, the angle in three dimensions. It is related to the radian, which is the unit of angle in two dimensions (abbreviated "sterad").

Sterotactic Radiosurgery

This involves the use of multiple small pencil *beams* of radiation fired from many different directions and all aimed at the tumor. Machines used include the "gamma-knife," with several hundred small, high-activity Cobalt-60 sources and conventional medical *linear accelerators* equipped with specially designed sterotactic hardware.

Stochastic Effects

Stochastic effects are effects that occur on a random basis with its effect being independent of the size of dose. The effect typically has no threshold and is based on probabilities, with the chances of seeing the effect increasing with dose. Cancer is a stochastic effect.

Stop Tone

Five seconds of 450 Hz black to white square wave modulation of the WEFAKsubcarrier signaling the stop of a frame transmission (end of a direct readout image).

Storage Ring

A circular (or near circular) structure in which either high energy *electrons* and/or positrons, or protons and/or antiprotons can be circulated many times and thus "stored". Used to achieve high energy collisions. Because of the very different masses of protons and electrons a storage ring must be designed for one or the other type and cannot work for both.

Strange Quark (s)

The third flavour of quark (in order of increasing mass), with electric charge -1/3.

Stratosphere

Region of the atmosphere between the troposphere and mososphere, having a lower boundary of approximately 8 km at the poles to 15 km at the equator and an upper boundary of approximately 50 km. Depending upon latitude and season, the temperature in the lower stratosphere can increase, be isothermal, or even decrease with altitude, but the temperature in the upper stratosphere generally increases with height due to absorption of solar radiation by ozone.

String Theory

A theory of *elementary particles* incorporating relativity and quantum mechanics in which the particles are viewed not as points but as extended objects. String theory is a possible framework for constructing unified theories which include both the microscopic forces and *gravity.*

Strong Force

The fundamental strong force is the force between quarks and *gluons* that makes them combine to form the observed hadrons, such as protons and neutrons. It also causes forces between *hadrons*, such as the strong nuclear force that makes protons and neutrons bind together to form nuclei.

Strong Interaction

1. The interaction responsible for binding quarks, antiquarks, and gluons to make hadrons. Residual strong interactions provide the nuclear binding force.

2. The *interaction* responsible for binding quarks and *gluons* to make *hadrons*. Residual strong interactions provide the nuclear binding force. In nuclear physics the term strong interaction is also used for this residual effect. (As a parallel, the force between electrically *charged* particles is an *electromagnetic interaction*, the force between neutral atoms that leads to the formation of molecules is a residual electromagnetic effect.)

Strong Nuclear Force

The force that holds nuclei together against electrical repulsion.

Subatomic Particle

Any particle that is small compared to the size of the atom.

Subcarrier

The 2400 Hz audio tone transmitted by APT and WEFAX spacecraft. Amplitude modulation of this tone is used to convey video information.

Subsatellite Point

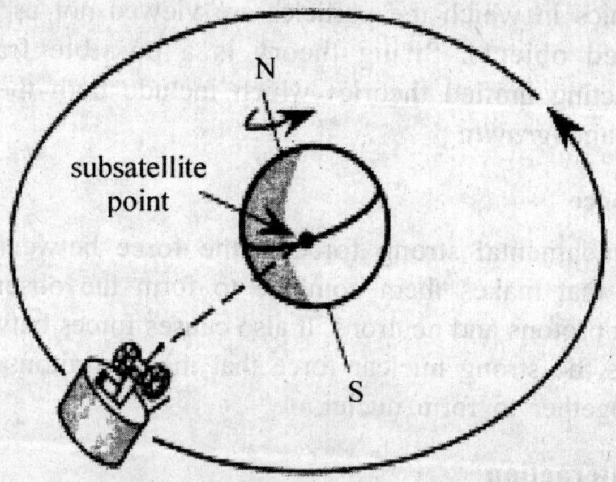

Point where a straight line drawn from a satellite to the center of the Earth intersects the Earth's surface.

Subsystem

A subunit of either the physical climate system (e.g., ocean dynamics) or the biogeochemical cycles (e.g., terrestrial ecosystems). A subunit of a spacecraft, e.g., the telemetry subsystem, the power subsystem, the sensor subsystem, etc.

Sun

1. The closest star to Earth (149,599,000 km away on average). The sun dwarfs the other bodies in the solar system, representing approximately 99.86 percent of all the mass in the solar system. One hundred and nine Earths would be required to fit across the Sun's disk, its interior could hold over 1.3 million Earths. The source of the Sun's energy is the nuclear reactions that occur

in its core. There, at temperatures of 15 million degrees Celsius (27 million degrees Fahrenheit) hydrogen atom nuclei, called protons, are fused and become helium atom nuclei. The energy produced through fusion at the core moves outward, first in the form of electromagnetic radiation called photons. Next, energy moves upward in photon heated solar gas-this type of energy transport is called convection. Convective motions within the solar interior generate magnetic fields that emerge at the surface as sunspots and loops of hot gas called prominences. Most solar energy finally escapes from a thin layer of the Sun's atmosphere called the photosphere-the part of the Sun observable to the naked eye. The sun appears to have been active for 4.6 billion years and has enough fuel for another 5 billion years or so. At the end of its life, the Sun will start to fuse helium into heavier elements and begin to swell up, ultimately growing so large that it will swallow Earth. After a billion years as a "red giant" it will suddenly collapse into a "white dwarf." It may take a trillion years to cool off completely.

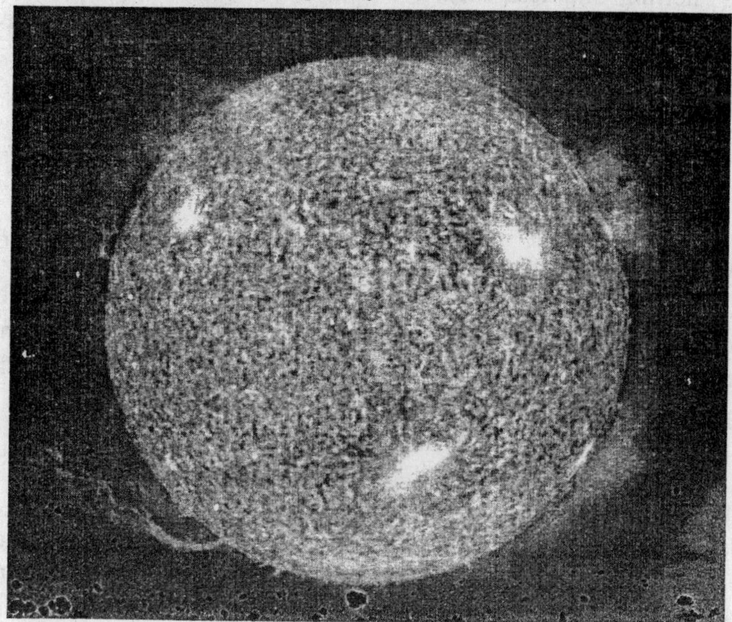

2. Earth's Sun is a medium-sized star which lies on the main sequence with 90% of the known stars. It has a effective surface temperature is 5780 K, putting it in spectral class G2. Its mass is 1.989 x 1030 kg and its mean radius is 6.96 x 108 meters. The mass of the sun is over 99.8% of the mass of the entire known solar system, leading de Pater and Lissauer to refer lightly to the solar system as "the Sun plus some debris". The sun radiates energy at the rate of 3.85 x1026 watts. Just outside the earth's atmosphere solar energy is received, assuming normal incidence, at the rate of 1340 watts per square meter. The orbit of Earth ranges from 1.47 to 1.52 x 1011 meters from the Sun. The average light travel time to the earth is 8.3 minutes. The radius of the sun at 696,000 km is 109 times the Earth's radius. Its surface gravity is 274 m/s2 or 28.0 times that of the Earth. Its mean density is 1410 kg/m3 or 0.255 times the mean density of Earth. The composition of the sun is 71% hydrogen, 27.1% helium and less than 2% of all other elements. The center temperature is modeled to be 15.5 million K. The Sun is fueled by the proton cycle of nuclear fusion.

Sun's Corona

The luminous irregular envelope of highly ionized gas outside the chromosphere of the sun.

Sunspot

A temporary disturbed area in the *solar photosphere* that appears dark because it is cooler than the surrounding areas. Sunspots consist of concentrations of strong *magnetic* flux. They usually occur in pairs or groups of opposite polarity that move in unison across the face of the Sun as it rotates.

Sun-synchronous

Describes the orbit of a satellite that provides consistent lighting of the Earth-scan view. The satellite passes the equator and each latitude at the same time each day. For example, a satellite's sun-synchronous orbit might cross the equator twelve times a day each time at 3:00

p.m. local time. The orbital plane of a sun-synchronous orbit must also precess (rotate) approximately one degree each day, eastward, to keep pace with the Earth's revolution around the sun.

Superconductivity

If *mercury* is cooled below 4.1 K, it loses all electric resistance. This discovery of superconductivity by H. Kammerlingh Onnes in 1911 was followed by the observation of other metals which exhibit zero resistivity below a certain *critical temperature*. The fact that the resistance is zero has been demonstrated by sustaining currents in superconducting lead rings for many years with no measurable reduction. An induced current in an ordinary metal ring would decay rapidly from the dissipation of ordinary resistance, but superconducting rings had exhibited a decay constant of over a billion years!

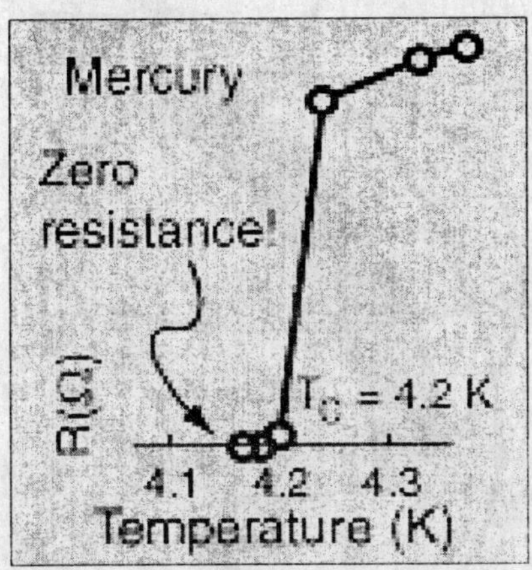

Supernova

1. A star that explodes and becomes extremely luminous in the process.

2. A supernova is an exploding star. Such an explosion occurs in our galaxy at a rate of about one every 30 years. Its causes are not precisely known, but the violent movement of matter within the star may produce a significant amount of *gravitational radiation*. It is thought that supernovae produce *pulsars*.

Surface Plot

A three-dimensional plot mapping the intensity of radiation from a region as a distorted surface. More intense radiation is represented by higher points on the surface. Therefore, regions of intense radiation resemble mountains on the earth.

Survey Mode

Refers to observational emphasis upon frequent global coverage, usually with restricted spatial and spectral resolution, aimed at developing a consistent, long-term data product for later interpretation.

Swath

The area observed by a satellite as it orbits the Earth.

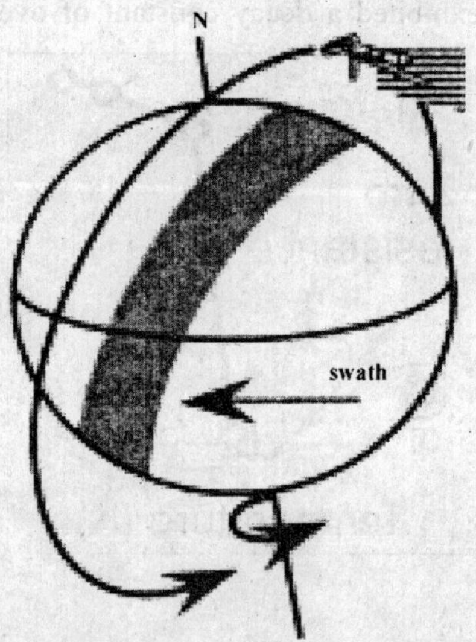

Synchrotron A type of circular accelerator in which the particles travel in synchronized bunches at fixed radius.

Synchrotron Radiation

1. When moving charges spiral in a magnetic field, they produce radiation as a result of their accelerations which is called

synchrotron radiation. The phenomenon is named after the General Electric Synchrotron, an accelerator which used magnetic fields to contain electrons which had been accelerated to high energies. Such radiation is observed in astronomical sources, such as the *Crab Nebula*. Its signature is that it does not follow the blackbody radiation curve, increasing toward lower frequencies rather than toward high (it is said to be "nonthermal"). It also shows characteristic polarization in the plane perpendicular to the magnetic field about which the charges are spiraling.

2. Whenever a *charged* particle undergoes accelerated motion it radiates electromagnetic energy. A common example is the emission of radio waves when *electrons* move back and forth in a radio antenna. A charged particle traveling in the arc of a circle is also undergoing acceleration, due to its change in direction. The radiation emitted by such particles is called synchrotron radiation and it is particularly intense and very directional when electrons traveling at close to the speed of light are bent in magnetic fields.

Syncom Satellites

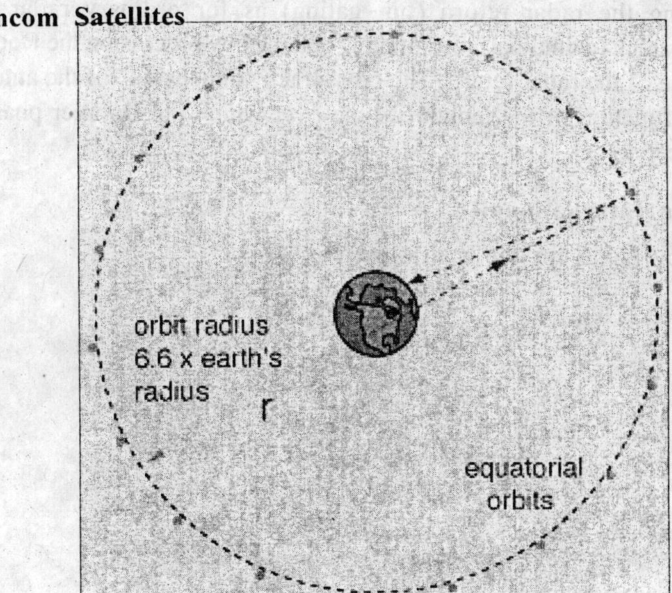

Communication satellites are most valuable when they stay above the same point on the earth, in what are called "geostationary orbits". This occurs when the orbital period is 24 hours. From an *orbit velocity calculation*, it can be seen that a satellite at a radius 6.62 times the Earth's radius will have a period of 24 hours. These satellites are some 35,900 kilometers or about 22,300 miles above the Earth's surface.

Synoptic Chart

Chart showing meteorological conditions over a region at a given time; weather map.

Synoptic View

The ability to see large areas at the same time.

Synthetic Aperture Radar (SAR)

A high-resolution ground-mapping technique that effectively synthesizes a large receiving antenna by processing the phase of the reflected radar return. The along4rack resolution is obtained by timing the radar return (tim~gating) as for ordinary radar The crosstrack (azimuthal) resolution is obtained by processing the Doppler phase of the radar return. The cross4rack "dimension" of the antenna is a function of the length of time over which the Doppler phase is collected.

Systeme International

Fancy name for the metric system.

T

Tangential

Tangent to a curve. In circular motion, used to mean tangent to the circle, perpendicular to the radial direction Cf. radial.

Target

A metallic object placed in the *beam* of *electrons* to produce x-rays.

Tau

The third *charged, lepton* (in order of increasing mass), with electric charge -1.

Tele

A prefix meaning *at a distance*, as in *telescope*, *telemetry*, *television*.

Telemetry

Telecommunications transmission to a distance of measured magnitude by radio or telephony with suitably coded modulation, e.g., amplitude, frequency phase, pulse. Transmission of data collected at a remote location over communications channels to a central station. Surveying measurement of linear distances by use of tellurometer-a device that uses microwaves to measure distance.

Telephony

Used to transmit sounds between widely removed points with or without connecting wires.

Television and Infrared Observation Satellite (TIROS)

A series of NASA and NOAA satellites launched to monitor Earths' weather from outer space. The era of the meteorological satellites began with the launch of TIROS-1 on April 1, 1960. For the first time, it was possible to monitor weather conditions over most of the world regularly from space. A series of these satellites were launched throughout the 1960s, those funded by NASA for research and development were called TIROS, and those funded by the Environmental Science Services Administration (ESSA, the predecessor of NOAA) for the operational system were called ESSA. A second generation of ITOS/NOAA* environmental satellites was initiated by the launch of ITOS-1 in 1970, followed by a number of NOAA satellites. The third generation of TIROS-N/NOAA environmental satellites was initiated by the launch of TIROS-N in 1978. Pairs of acronyms such as ITOS/NOAA arise because NASA funds and names its prototype satellites and then the operating agency funds and names the rest of the series.

Temperature Measurement

Many methods have been developed for measuring temperature. Most of these rely on measuring some physical property of a working material that varies with temperature. One of the most common devices for measuring temperature is the glass thermometer. This consists of a glass tube filled with mercury or some other liquid, which acts as the working fluid. Temperature increases cause the fluid to expand, so the temperature can be determined by measuring the volume of the fluid. Such thermometers are usually calibrated, so that one can read the temperature, simply by observing the level of the fluid in the thermometer. Another type of thermometer that is not really used much in practice, but is important from a theoretical standpoint is the *gas thermometer* mentioned below. Other important devices for measuring temperature include: 1. Thermocouples, 2. Thermistors, 3. Resistance Temperature Detector (RTD), 4. Pyrometers, 5. Langmuir probes (for electron temperature of a plasma), Other thermometers. One must be careful when measuring temperature to ensure that the measuring instrument (thermometer, thermocouple, etc) is really the same temperature as the material that is being measured. Under some conditions heat from the measuring instrument can cause a temperature gradient, so the measured temperature is different from the actual temperature of the system. In such a case the measured temperature will vary not only with the temperature of the system, but also with the heat transfer properties of the system. An extreme case of this effect gives rise to the wind chill factor, where the weather feels colder under windy conditions than calm conditions even though the temperature is the same. What is happening is that the wind increases the rate of heat transfer from the body, resulting in a larger reduction in body temperature for the same ambient temperature.

Temperature

1. Measure of the energy in a substance. The more heat energy in the substance, the higher the temperature. The Earth receives only one twobillionth of the energy the sun produces. Much of the energy that hits the Earth is reflected back into space. Most of the energy that isn't reflected is absorbed by the Earth's surface. As the surface warms, it also warms the air above it.

2. What a thermometer measures. Objects left in contact with each other tend to reach the same temperature. Roughly speaking, temperature measures the average kinetic energy per molecule. For the distinction between temperature and heat.

3. *Temperature* is the physical property of a system which underlies the common notions of "hot" and "cold"; the material with the higher temperature is said to be hotter.

4. The above equation tells us that the product of pressure and volume per mole is proportional to the average molecular kinetic energy. Further, the ideal gas equation tells us that this product is proportional to the absolute temperature. Putting the two together, we arrive at one important result of the kinetic theory: *average molecular kinetic energy is proportional to the absolute temperature*. The constant of proportionality is 3/2 times Boltzmann's constant, which is the ratio of the gas constant R to Avogadro's number (independent of the gas). This result is related to the equipartition theorem.

Terabit

A trillion (10^{12}) bits.

Term

One of several quantities which are added together. Confusion can arise with another use of the word, as when one is asked to "Express the result in terms of mass and time." This means "as a function of mass and time," obviously it doesn't mean that mass and time are to be added as terms.

Teratogenic Effects

Teratogenic effects are effects from some agent, that are seen in the offspring of the individual who received the agent. The agent must be encountered during the gestation period.

TeV (Tera Electron Volt)

Unit of energy equal to that acquired by a particle with one electronic *charge* in passing through a potential difference of one trillion volts.

The Cartesian Diver

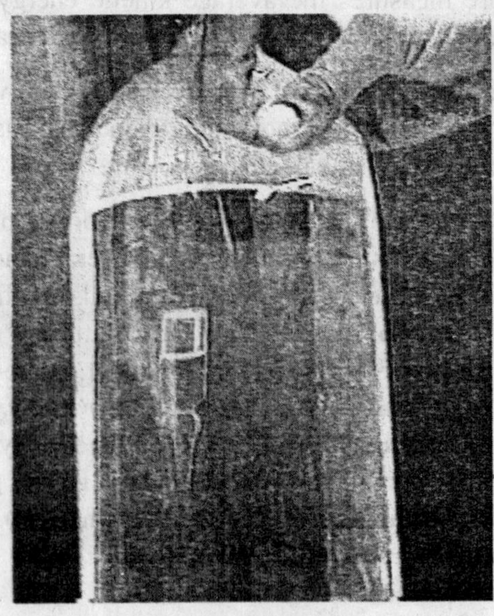

The Cartesian diver demonstrates not only *buoyancy*, but the implications of the *ideal gas law* and *Pascal's principle* as well. Squeezing on the top of the sealed plastic container decreases the volume and therefore increases air pressure above the water. By Pascal's principle, that pressure is transmitted to all parts of the container. This increases the pressure inside the small glass vial. The increased pressure decreases the volume of air at the top of the vial, and in so doing, decreases the amount of water displaced by the vial. This decreases the buoyant force on it enough to cause it to sink.

The International Space Station

The International Space Station is an orbiting laboratory which is under construction by a global partnership of 16 nations. There are plans for six laboratory modules which will be constructed using supplies and personnel from about 40 spaceflights over five years.

The first flight was lifted off by a Russian Proton rocket in November of 1998 and the Space Shuttle Endeavor launched from the U.S. in December 1998 with the unity module.

The Timpani

The timpani has a round head stretched over a sealed enclosure. The tension may be altered by means of a footpedal which actuates tensioning elements. The choice of striking point emphasizes the *preferred modes* of the *circular membrane*. The *sounded frequencies* are further influenced by the enclosed air cavity.

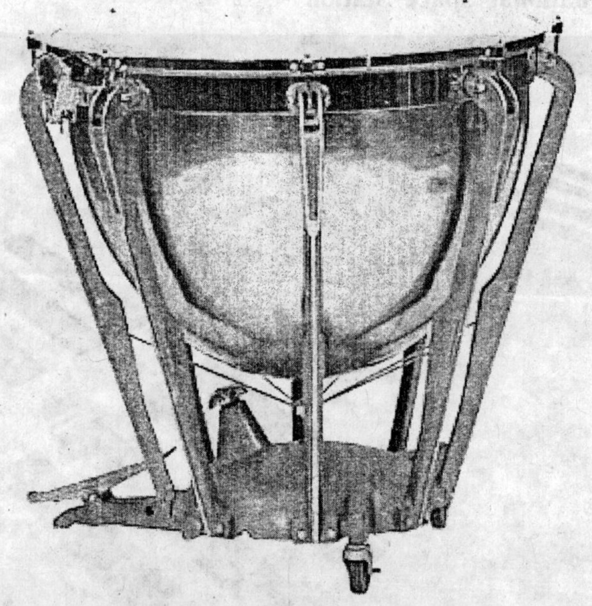

The Strong Equivalence Principle

The strong equivalence principle suggests the laws of gravitation are independent of velocity and location. In particular, The gravitational motion of a small test body depends only on its initial position in spacetime and velocity, and not on its constitution. and The outcome of any local experiment, whether gravitational or not, in a laboratory moving in an inertial frame of reference is independent of velocity of the laboratory, or its location in spacetime. The first part is a version of the weak equivalence principle that it applies to objects that exert a gravitational force on themselves, such as stars, planets, black holes or Cavendish experiments. The second part is the Einstein equivalence principle, restated to allow gravitational experiments and self-gravitating bodies. The freely-falling object or laboratory, however, must still be small, so that tidal forces may be neglected. This idealized requirement has been misunderstood. This form of the equivalence principle *does not* imply that the effects of a gravitational field cannot be measured by observers in free-fall. For example, an observer in free-fall into a black hole will experience strong tidal forces: he will notice a more powerful force on his feet than his head. The strong equivalence suggests that gravity is an entirely geometrical by nature (that is metric alone determines the effect of gravity) and does not have an extra fields associated with it. If an observer measures a patch of space to be flat, then the strong equivalence principle suggests that it is absolutely equivalent to any other patch of flat space elsewhere in the universe. Einstein's theory of general relativity is the only known theory that satisfies the strong equivalence principle. A number of alternative theories, such as Brans-Dicke theory, satisfy only the Einstein equivalence principle.

The Weak Force

One of the *four fundamental forces*, the weak interaction involves the *exchange* of the *intermediate vector bosons*, the W and the Z. Since the mass of these particles is on the order of 80 GeV, the uncertainty principle dictates a *range* of about 10^{-18} meters which is about 0.1% of the diameter of a proton.

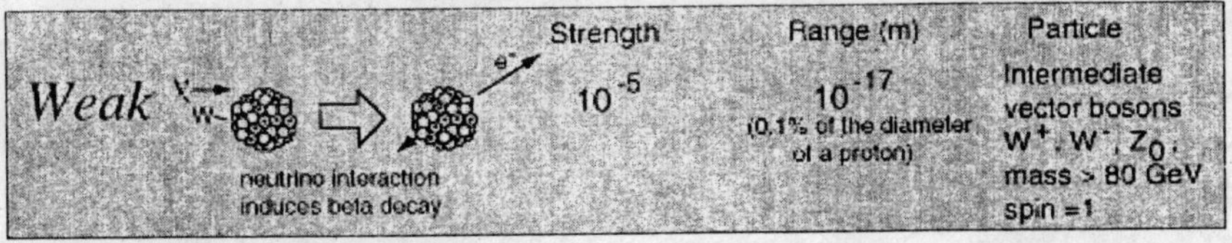

Thematic Mapper (TM)

A Landsat multispectral scanner designed to acquire data to categorize the Earth's surface. Particular emphasis was placed on agricultural applications and identification of land use. The scanner continuously scans the surface of the Earth, simultaneously acquiring data in seven spectral channels. Overlaying two or more bands produces a false color image. The ground resolution of the six visible and shortwave bands of the Thematic Mapper is 30 meters, and the resolution of the thermal infrared band is 120 meters. Thematic mappers have been flown on Landsats-4 and -5.

Theoretical

Describing an idea which is part of a theory, or a consequence derived from theory.

Theory of Relativity

Albert Einstein's *theory of relativity* is a set of two theories in physics: special relativity and general relativity. These theories were conceived in order to explain the fact that electromagnetic waves do not conform to the Newtonian laws for motion. Electromagnetic waves were shown to move at a constant speed, independent of the motion of an observer. The core idea of both theories is that two observers who move relative to each other will measure different time and space intervals for the same events, but the content of physical law will be observed the same by both.

Theory

A *well-tested* mathematical model of some part of science. In physics a theory usually takes the form of an equation or a group of equations, along with explanatory rules for their application. Theories are said to be successful if (1) they synthesize and unify a significant range of phenomena; (2) they have predictive power, either predicting new phenomena, or suggesting a direction for further research and testing.

Thermal Energy

Careful writers make a distinction between heat and thermal energy, but the distinction is often ignored in casual speech, even among physicists. Properly, thermal energy is used to mean the total amount of energy possessed by an object, while heat indicates the amount of thermal energy transferred in or out. The term heat is used in this book to include both meanings.

Thermal Gas

A collection of particles that collide with each other and exchange energy frequently, giving a distribution of particle energies that can be characterized by a single temperature.

Thermal Infrared

Electromagnetic radiation with wavelengths between about 3 and 25 micrometers.

Thermal Particle

A particle that is part of a *thermal gas.*

Thermal Radiation

Electromagnetic radiation emitted by *electrons* in a *thermal gas*.

Thermodynamic Parameters

The parameters used to describe the state of a system generally depend on the exact system under consideration, and the conditions under which that system is maintained. The most commonly considered parameters are: Mechanical parameters: Pressure : p Volume : V Statistical parameters: Temperature : T.

Thermodynamic Systems

A thermodynamic system is that part of the universe that is under consideration. A real or imaginary boundary separates the system from the rest of the universe, which is referred to as the *environment*, or sometimes as a *reservoir*. A useful classification of thermodynamic systems is based on the nature of the boundary and the flows of matter, energy and entropy through it. There are three kinds of systems depending on the kinds of *exchanges* taking place between a system and its environment: *Isolated* systems: not exchanging heat, matter or work with their environment. Mathematically, this implies that TdS, dN, and pdV are all zero, and therefore dE is zero. An example of an isolated system would be an insulated container, such as an insulated gas cylinder. *Closed* systems: exchanging energy (heat and work) but not matter with their environment. In this case, only dN is generally zero. A greenhouse is an example of a closed system exchanging heat but not work with its environment. Whether a system exchanges heat, work or both is usually thought of as a property of its boundary, which can be *adiabatic* boundary: not allowing heat exchange, TdS=0 *rigid* boundary: not allowing exchange of work, pdV=0 *open* systems: exchanging energy (heat and work) and matter with their environment. A boundary allowing matter exchange is called *permeable*. The ocean would be an example of an open system. In reality, a system can never be absolutely isolated from its environment, because there is always at least some slight coupling, even if only via minimal gravi-tational attraction. In analyzing a system in steady-state, the energy into the system is equal to the energy leaving the system.

Thermodynamics

Thermodynamics (Greek : *thermos* = heat and *dynamic* = change) is the physics of energy, heat, work, entropy and the spontaneity of processes. Thermodynamics is closely related to statistical mechanics from which many thermodynamic relationships can be derived. While dealing with processes in which systems exchange matter or energy, classical thermodynamics is not concerned with the rate at which such processes take place, termed kinetics. For this reason, the use of the term "thermodynamics" usually refers to *equilibrium thermodynamics*. In this connection, a central concept in thermodynamics is that of quasistatic processes, which are idealized, "infinitely slow" processes. Time-dependent thermodynamic processes are studied by non-equilibrium thermodynamics. Thermodynamic laws are of very general validity, and they do not depend on the details of the interactions or the systems being studied. This means they can be applied to systems about which one knows nothing other than the balance of energy and matter transfer with the environment. Examples of this include Einstein's prediction of spontaneous emission around the turn of the 20th century and the current research into the thermodynamics of black holes.

Thermonuclear Fusion

The combination of atomic *nuclei* at high temperatures to form more massive nuclei with the simultaneous release of energy. Thermonuclear fusion is the power source at the core of the Sun. Controlled thermonuclear fusion reactors, when successfully implemented, could become an attractive source of power on the Earth.

Thunder

The sound that results from lightning. Lightning bolts (static electricity) produce intense heat. This burst of heat makes the air around the bolt expand explosively producing the sound we hear as thunder. Since light travels faster than sound, we see the lightning before we hear the thunder.

Thunderstorm

Local storm resulting from warm humid air rising in an unstable environment. Air may start moving upward because of unequal surface heating, the lifting of warm air along a frontal zone, or diverging upper-level winds (these diverging winds draw air up beneath them). The scattered thunderstorms that develop in the summer are called airmass thunderstorms because they form in warm, maritime tropical air masses away from other weather fronts. More violent severe thunderstorms form in areas with a strong vertical wind shear that forces the updraft into the mature stage, the most intense stage of the thunderstorm. Severe thunderstorms can produce large hail, forceful winds, flash floods, and tornadoes.

Tides

The Earth experiences two high tides per day because of the difference in the *Moon's* gravitational field at the Earth's surface and at its center. You could say that there is a high tide on the side nearest the Moon because the Moon pulls the water away from the

Earth, and a high tide on the opposite side because the Moon pulls the Earth away from the water on the far side. The tidal effects are greatly exaggerated in the sketches.

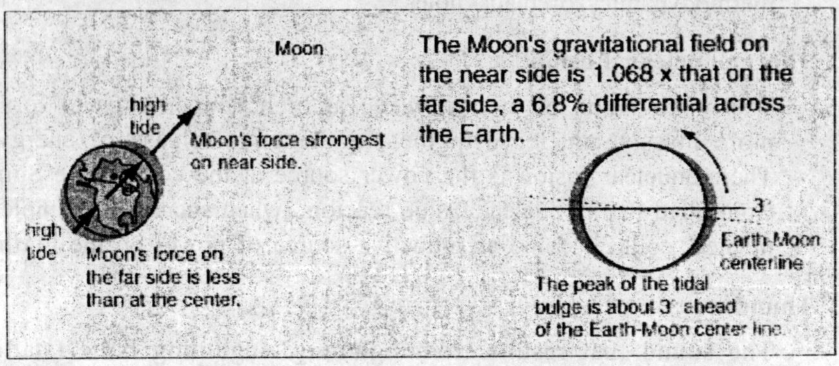

Time

1. 8:17 am, August 6, 1945, Japanese time. The approximate time of the "Little Boy" nuclear detonation over Hiroshima, Japan. Time quantifies or measures the interval between events, or the duration of events. Time has long been perceived as a dimension in which each event has a definite (but not necessarily unique) position in a linear sequence, but as differing from spatial dimensions in that "motion" through time appears restricted to having only a forward direction. For everyday purposes, and even for quite accurate measurements, this view is sufficient. However, the scientific understanding of time underwent a revolution in the early part of the twentieth century with the development of relativity theory. Modern physics treats time as a feature of spacetime, a notion which challenges intuitive conceptions of simultaneity and the flow of time in a linear fashion. Despite scientific advances, the everyday meaning of time is affected more by the social importance of time, its economic value ("time is money") and an awareness of the limited time in each day and in our lives. Thus, time has long been an important theme for writers, artists and philosophers.

2. Time is one of the undefined qualities of physics. We all know what it is and agree to use a second as one unit of time.

TIROS-NINOAA Satellites

NOAA satellites that continuously orbit the Earth from North to South Pole (hence, polar orbiting) at an altitude of approximately 470 nautical miles (870.44 km or 540.86 statute miles). These environmental satellites collect visible and infrared imagery and provide atmospheric-sounding data and meteorological data relay and collection. A primary mission of TIROS-N/NOAA is to monitor the 70 percent of the globe covered by water-where weather data is sparse and provide continuous data to the National Weather Service for use in numerical forecast modeling. Each TIROSN/NOAA carries six primary systems: The Advanced Very High Resolution Scanning Radiometer (AVHRR) senses clouds over both ocean and land, using the visible and infrared parts of the spectrum. It stores measurements on tape, and later plays them back to NOAAs command and data acquisition stations. The satellites also broadcast in real time, and the broadcasts can be received around the world by anyone equipped with a direct readout receiving station. The TIROS Operational Vertical Sounder (TOVS) is a 3-part TIROS system to measure : Temperature profile of the Earth's atmosphere from the surface to 10 millibars. Water content of the Earth's atmosphere. Total ozone content of the Earth's atmosphere. The ARGOS Data Collection and Platform Location System (DCS) collects data from sensors placed on fixed and moving platforms, including ships, buoys, and weather balloons, and transmits data to a ground station antenna. Because ARGOS also determines the precise location of these moving sensors, it can serve wildlife managers by monitoring and tracking the transmitters placed on birds and animals. The Space Environment Monitor (*SEM*) measures energetic particles emitted by the sun over essentially the full range of energies and magnetic field variations in the Earth's near-space environment. Readings made by these instruments are invaluable in measuring the sun's radiation activity. Search and Rescue Tracking (*COSPAS/SARSAT*) equipment receives emergency signals from persons in distress. The satellites transmit the signals to ground receiving stations. The signals then are forwarded to rescue coordination centers. The rescue centers compute the location of the signals and provide the coordinates of the emergency site (usually within a few miles). Earth Radiation Budget Experiment (ERBE) is a radiometer flown on NOAA 9 and 10, designed to measure all radiation striking and leaving the Earth. This enables scientists to measure the loss or gain of terrestrial energy to space. Shifts in this energy "budget" affect the Earth's average temperatures. Even slight changes can affect climatic patterns.

TNL

Thermal Noise Level.

Top Quark (t)

1. The sixth flavour of quark (in order of increasing mass), with electric charge 2/3. Its mass is much greater than any other quark or lepton.

2. The sixth *flavor* of quark (in order of increasing mass) with electric *charge* +2/3.

TOPEX/POSEIDON

Ocean Topography Experiment, United States (NASA)/France (CNES). Launched in 1 992, the mission carries a radar sensor-called an altimeter-to measure the ocean's surface topography with unprecedented precision. TOPEX/POSEIDON is a core element of the international World Ocean Circulation Experiment (WOCE) and the Tropical Ocean Global Atmosphere (TOGA) seagoing measurements program. Mission objectives are to: Study ocean circulation and its interaction with the atmosphere to understand climate change better. Improve our knowledge of heat transport in the ocean. Model global ocean tides. Study the marine gravity field. Calculate sea-level variations on both global and local scales.

Tornado

A twisting, spinning funnel of low pressure air. The most unpredictable weather event, tornadoes are created during powerful thunderstorms. As a column of warm air rises, air rushes in at ground level and begins to spin. If the storm gathers energy a twisting, spinning funnel develops. Because of the funnel's cloud and rain composition and the dust, soil, and debris it draws up, the funnel appears blackish in color. The most energetic storms result in the funnel touching the ground. In these tornadoes, the roaring winds in the funnel can reach 300 mph, the strongest winds on Earth. Funnels usually travel at 20 to 40 mph, moving toward the northeast. When tornadoes form over lakes or oceans they suck water into the funnel cloud and are called waterspouts.

Torque

The rate of change of angular momentum; a numerical measure of a force's ability to twist on an object.

Torque Calculation

In the diagram, the angle is the angle <= 180 degrees between the r and F vectors when they are drawn from the same origin. The direction of the torque is given be the *right hand rule*, which gives a vector out toward the reader in this case. Note that the torque is maximum when the angle is 90 degrees.

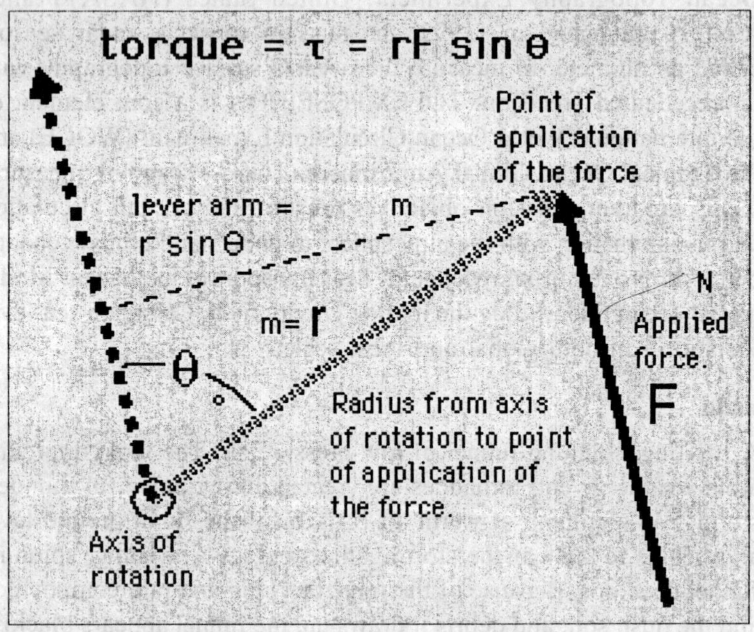

Total Ozone Mapping Spectrometer (TOMS)

Flown on NASAs Nimbus-7 satellite, its primary goal is to continue the high-resolution global mapping of total ozone on a daily basis. The Nimbus-7 launch in 1978 enabled TOMS to begin delivering data in 1979 and continue providing information until 1993. TOMS has mapped the total amount of ozone between the ground and the top of the atmosphere, provided the first maps of the ozone hole, and continues to monitor this phenomenon. Because of its longevity, TOMS also has obtained information on the more subtle trends in ozone outside the ozone hole region. This results from development of a powerful new calibration technique that removes the instrument measurement drift that developed over the years. With this technique applied to the TOMS 14.5-year data record, a global ozone decrease of 2.69 percent per decade was detected. To ensure that ozone data will be available through the next decade, NASA will continue the TOMS program using U.S. and foreign launches. In 1991, the former Soviet Union launched a Meteor-3 satellite carrying a TOMS instrument provided by NASA. A third TOMS will be launched onboard a NASA Earth probe satellite in 1994, and the Japanese Advanced Earth Observations Satellite (ADEOS) will carry a fourth TOMS when it launches in 1996.

Tracer

A small amount of radioactive isotope introduced into a system in order to follow the behavior of some component of that system.

Track

The record of the path of a particle traversing a detector.

Tracking and Data Relay Satellite System (TDRSS)

An orbiting communications satellite, developed by NASA, used to relay data from satellite sensors to ground stations and to track the satellites in orbit.

Tracking Chamber

A section of a particle *detector* capable of detecting the passage of electrically *charged* particles.

Tracking

The reconstruction of a "track" left in a detector by the passage of a particle through the detector.

Trade Winds

Surface air from the horse latitudes that moves back toward the equator and is deflected by the Coriolis Force, causing the winds to blow from the Northeast in the Northern Hemisphere and from the Southeast in the Southern Hemisphere. These steady winds are called trade winds because they provided trade ships with an ocean route to the New World.

Transformation

The mathematical relationship between the variables such as x and t, as observed in different frames of reference.

Transmutation

The transformation of one element into another by a nuclear reaction.

Transverse Waves

For transverse waves the displacement of the medium is perpendicular to the direction of propagation of the wave. A *ripple on a pond* and a *wave on a string* are easily visualized transverse waves. Transverse waves cannot propagate in a gas or a liquid because there is no mechanism for driving motion perpendicular to the propagation of the wave.

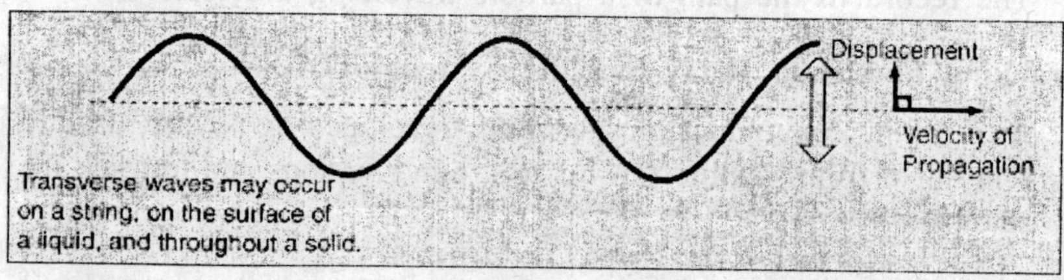

Tropical Ocean-Global Atmosphere (TOGA)

TOGA is a program jointly sponsored by the United Nations World Meteorological Organization (WMO); the International Council of Scientific Unions (ICSU); the United Nations Educational, Scientific, and Cultural Organization (UNESCO) Intergovernmental Oceanographic Commission (IOC); and the ICSU Scientific Committee on Oceanic Research (SCOR). TOGA has four major objectives: To collect and catalog observations of the tropical atmosphere and ocean. To assess the evolution of the tropical atmosphere/ocean system in real time. To promote the development of short-term climate-prediction computer models for the tropics. To study the influence of the tropical atmosphere/ocean system on the climate at higher latitudes.

Tropical Rainfall Measuring Mission (TRMM)

A joint NASA/NASDA mission planned for launch in 1997. The goal of TRMM is to obtain a minimum of 3 years of climatologically significant observations of rainfall in the tropics. Because rainfall is such a variable phenome non, adequate sampling is a difficult problem. By averaging the instantaneous rainfall rates for 30 days over a 50 by 50 grid, TRMM will obtain observations that meet climatological requirements. TRMM measurements, used together with cloud models, also will provide accurate estimates of vertical distributions of latent heating in the atmosphere. The present uncertainty about the quantity and distribution of precipitation, especially in the tropics, prohibits definition of the mass and energy exchange between the tropical ocean and atmosphere. Since the tropical atmosphere and oceans are closely coupled, cloud radiation and rainfall are likely to have significant effects on ocean circulation and marine biomass. TRMM data will play a significant role in global change studies, especially in developing an interdisciplinary understanding of atmospheric circulation, ocean-atmospheric coupling. and tropical biology TRMM data on tropical clouds, evaporation, and heat transfer will be used to understand the larger scale coupling of the atmosphere to oceans.

Tropical Storm Formation

Tropical storms generally form in the eastern portion of tropical oceans and track westward. Hurricanes, typhoons, and willywuhes all start out as weak low pressure areas that form over warm tropical waters (e.g., surface water temperature of at least 80 °F). Initially winds and cloud formations over the warm tropical waters are minimal. Both intensifv with time. Formation of tropical storms also requires a significant Coriolis effect to induce proper spin in the wind formation. As the storm begins to organize itself into a coherent pattern, it will experience increased activity and intensity. When a storm develops a clearly recognizable pattern, it is referred to as a tropical depression. When wind speeds reach 35 knots (40.3 mph), it is called a tropical storm and is given a name. When wind speed equals or exceeds 74 mph, the storm is called a hurricane. In the western Pacific, a hurricane is referred to as a typhoon. In waters around Australia it is called a cyclone or willy-willy. Hurricanes intensifv when moving over areas of increased water temperatures, and weaken over colder water surfaces. Upper atmosphere wind shear (different wind direction and speeds at different elevations) will frequently prevent or slow intensification of tropical storms by "spreading out" the storm horizontally and preventing the formation of strong updrafts of warm, humid air. Movement over a land-mass will weaken hurricane winds but will result in large-scale rain that can result in large-scale flooding. When encountering a strong frontal system (such as a polar front) the hurricane will curve and track along the leading edge of the front or become implanted in it. Satellite infrared imagery can identifv surface water temperatures that will foster tropical storm development.

Tropics

The area between 23.5 degrees north and south of the equator. This region has small daily and seasonal changes in temperature, but great seasonal changes in precipitation.

Troposphere

The lower atmosphere, to a height of 8-15 km above Earth, where temperature generally decreases with altitude, clouds form, precipitation occurs, and convection currents are active.

Tropospheric Emission Spectrometer

A high-resolution infrared spectrometer for monitoring the minor components of the lower atmosphere.

Trough

Elongated area of low atmospheric pressure, either at the surface or in the upper atmosphere.

True Anomaly (aka J)

One of six Keplerian elements, it locates a satellite on an orbit. True anomaly is the true angular distance of a satellite (planet) from its perigee (perihelion) as seen from the center of the Earth (sun).

Truth

This is a word best avoided entirely in physics except when placed in quotes, or with careful qualification. Its colloquial use has so many shades of meaning from 'it seems to be correct' to the absolute truths claimed by religion, that it's use causes nothing but misunderstanding. Someone once said 'Science seeks proximate (approximate) truths.' Others speak of *provisional* or *tentative* truths. Certainly science claims no final or absolute truths.

Turning a Bicycle

A bicycle held straight up will go straight, stabilized by the gyroscopic action of the bicycle wheels. If the rider leans left, a torque will be produced which causes a counterclockwise precession of the bicycle wheel, turning the bicycle to the left.

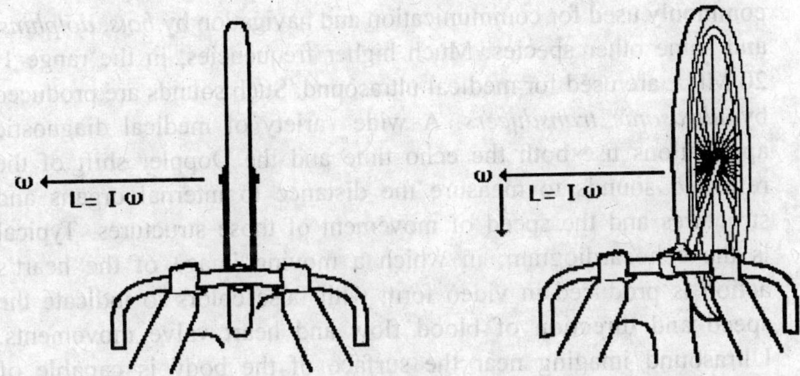

Typhoon

Hurricanes in the Western Pacific Ocean.

U

Ultrasonic Sound

The term "ultrasonic" applied to sound refers to anything above the frequencies of *audible sound*, and nominally includes anything over 20,000 Hz. Frequencies used for medical diagnostic ultrasound scans extend to 10 MHz and beyond. Sounds in the range 20-100kHz are commonly used for communication and navigation by *bats*, *dolphins*, and some other species. Much higher frequencies, in the range 1-20 MHz, are used for medical ultrasound. Such sounds are produced by *ultrasonic transducers*. A wide variety of medical diagnostic applications use both the echo time and the Doppler shift of the reflected sounds to measure the distance to internal organs and structures and the speed of movement of those structures. Typical is the echocardiogram, in which a moving image of the heart's action is produced in video form with false colors to indicate the speed and direction of blood flow and heart valve movements. Ultrasound imaging near the surface of the body is capable of resolutions less than a millimeter. The resolution decreases with the depth of penetration since lower frequencies must be used (the attenuation of the waves in tissue goes up with increasing frequency.) The use of longer wavelengths implies lower resolution since the maximum resolution of any imaging process is proportional to the wavelength of the imaging wave.

Ultrasonic Transducers

Ultrasonic sound can be produced by transducers which operate either by the *piezoelectric effect* or the *magnetostrictive effect*. The magnetostrictive transducers can be used to produce high intensity ultrasonic sound in the 20-40 kHz range for ultrasonic cleaning and other mechanical applications.Ultrasonic medical imaging typically uses much higher ultrasound frequencies in the range 1-20 MHz. Such ultrasound is produced by applying the output of an electronic *oscillator* to a thin wafer of piezoelectric material such as *lead zirconate titanate*. The higher frequencies imply shorter wavelengths and therefore higher resolution for the imaging process. The application of the basic ideas of imaging (e.g., the *Rayleigh criterion* suggests that the resolution of any imaging process is limited by diffraction to a dimension similar to the wavelength of the wave used for the imaging process.

Ultraviolet Radiation

The energy range just beyond the violet end of the visible spectrum. Although ultraviolet radiation consitutes only about 5 percent of the total energy emitted from the sun, it is the major energy source for the stratosphere and mesosphere, playing a dominant role in both energy balance and chemical composition. Most ultraviolet radiation is blocked by Earth's atmosphere, but some solar ultraviolet penetrates and aids in plant photosynthesis and helps produce vitamin D in humans. Too much ultraviolet radiation can burn the skin, cause skin cancer and cataracts, and damage vegetation.

Uncertainty Principle

The quantum principle, first formulated by Heisenberg, that states that is is not possible to know exactly both the position x and the momentum p of an object at the same time. The same is true with energy and time (see virtual particle).

Uncertainty

Synonym: *error.* A measure of the the inherent variability of repeated measurements of a quantity. A prediction of the probable variability of a result, based on the inherent uncertainties in the data, found from a mathematical calculation of how the data uncertainties would, in combination, lead to uncertainty in the result. This calculation or process by which one predicts the size of the uncertainty in results from the uncertainties in data and procedure is called *error analysis*. Uncertainties are always present; the experimenter's job is to keep them as small as required for a useful result. We recognize two kinds of uncertainties: *indeterminate* and *determinate*. Indeterminate uncer-tainties are those whose size and sign are unknown, and are sometimes (misleadingly) called *random*. Determinate uncertainties are those of definite sign, often referring to uncertainties due to instrument miscalibration, bias in reading scales, or some unknown influence on the measurement.

Unified Field Theory

A unified field theory is one that attempts to combine any two or more of the known interaction types (strong, electromagnetic, weak and gravitational) in a single theory so that the two distinct types of interaction are seen as two different aspects of a single mathematical structure. A 'grand unified' theory (or GUT) unifies three of the four types (strong, weak and electromagnetic interactions) in this way. The benefit is that the unification gives a simpler overall theory and predicts relationships between parameters that are otherwise independent.

Uniform Circular Motion

Circular motion in which the magnitude of the velocity vector remains constant.

Uniform Electric Field

An electric field with magnitude and direction that are everywhere the same. The diagram shows the uniform electric field between the plates of a parallel plate capacitor. Notice the 'edge effects' at each side. The field lines are equally spaced. So too are the lines of equal potential – the equipotentials – which are shown as broken lines. If you draw this diagram in an examination, never have fewer than three equipotentials or five electric field strength lines. Be sure to have the spacing approximately correct. The relationship between electric field strength, *E*, and rate of change of potential with distance, – dV/dx, is simpler for a uniform field. In general : $E = -\frac{dV}{dx}$. But for a uniform field : $E=-\frac{\mathrm{d}V}{\mathrm{d}x}=-\frac{V}{d}$. Where d is the distance, measured parallel to, and in the same direction as E, along which the drop in potential is V. It is because V diminishes along the direction of E that the negative sign appears. Often, the direction in which d is measured is unimportant. Only the magnitudes are of interest. We then write :

$$E=\frac{V}{d}.$$

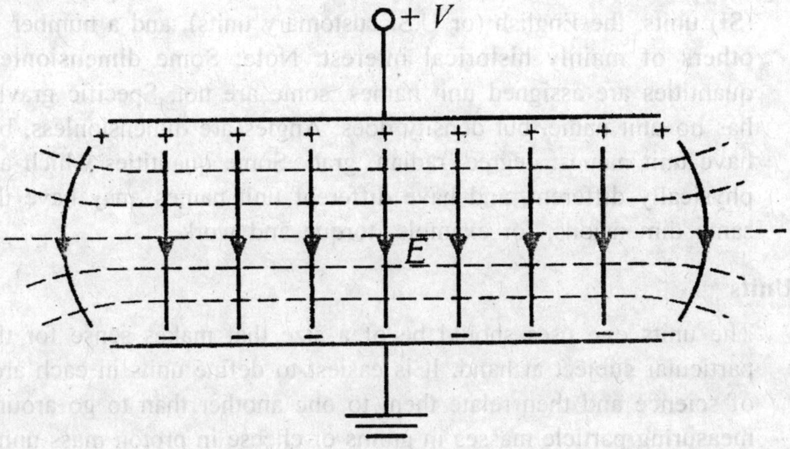

United States Geological Survey (USGS)

A bureau of the Department of the Interior USGS was established in 1 879 following several Federally-sponsored independent natural resource surveys of the West and Midwest. The Department of the Interior has responsibility for most of our nationally owned public lands and natural resources. The USGS monitors resources such as energy minerals, water land, agriculture, and irrigation. The resulting scientific information contributes to environmental-policy decision making and public safety. For example, USGS identifies flood- and landslide-prone areas and maintains maps of the United States.

United States Global Change Research Program (USGCRP)

The USGCRP addresses significant uncertainties concerning the natural and human-induced changes to Earth's environment. The USGCRP has a comprehensive and multidisciplinary scientific research agenda.

Units

Labels which distinguish one type of measurable quantity from other types. Length, mass and time are distinctly different physical quantities, and therefore have different unit names, meters, kilograms and seconds. We use several systems of units, including the metric (SI) units, the English (or U.S. customary units), and a number of others of mainly historical interest. Note: Some dimensionless quantities are assigned unit names, some are not. Specific gravity has no unit name, but density does. Angles are dimensionless, but have unit names: degree, radian, grad. Some quantities which are physically different, and have different unit names, may have the same dimensions, for example, torque and work.

Units

The units one uses should be of a size that makes sense for the particular subject at hand. It is easiest to define units in each area of science and then relate them to one another than to go around measuring particle masses in grams or cheese in proton mass units. In particle physics the standard unit is the unit of energy GeV. One eV (electron Volt) is the amount of energy that an *electron* gains when it moves through a potential difference of 1 Volt (in a vacuum). G stands for Giga, or 10^9. Thus a GeV is a billion (in US counting) electron Volts. The mass-energy of a proton or neutron is approximately 1 GeV.

Universal Time

Abbreviated *UT*. The same as Greenwich Mean Time (GMT) in England. Eastern Standard Time (EST) is five hours earlier than Universal Time.

Unstable Equilibrium

One in which any deviation of the object from its equilibrium position results in a force pushing it even farther away.

Unstable

Matter that is capable of undergoing spontaneous change, as in a radioactive nuclide or an excited nuclear system. An unstable particle is any elementary particle that spontaneously *decays* into other particles.

Up Quark (u)

1. The least massive flavour of quark, with electric charge 2/3.
2. The first *flavor* of quark (in order of increasing mass), with electric *charge* +2/3.

Upper Atmosphere Research Satellite (UARS)

UARS is part of a long-term, international program of space research into global atmospheric change. Beginning in 1991, NASAs UARS program began to carry out the first systematic, detailed satellite study of the Earth's stratosphere, mesosphere, and lower thermosphere; establish the comprehensive data base needed for an understanding of stratospheric ozone depletion; and bring together scientists and governments around the world to assess the role of human activities in atmospheric change. Launched on September 12, 1991, UARS became the first official space component of Mission to Planet Earth.

Uranium-235 Fission

In one of the most remarkable phenomena in nature, a slow neutron can be captured by a uranium-235 nucleus, rendering it unstable toward *nuclear fission*. A fast neutron will not be captured, so neutrons must be slowed down by *moderation* to increase their capture probability in *fission reactors*. A single fision event can yield over 200 million times the energy of the neutron which triggered it!

Usage Note: When *reading* equations aloud we often say, 'F equals m a'. This, of course, says that the two things are mathematically equal in equations, and that one may replace the other. It is *not* saying that F *is* physically the same thing as m*a*. Perhaps equations were not meant to be read aloud, for the spoken word does not have the subtleties of meaning necessary for the task. At least we should realize that spoken equations are at best a shorthand approximation to the meaning; a verbal description of the symbols. If we were to try to speak the physical meaning, it would be something like: 'Newton's law tells us that the net vector force acting on a body of mass m is mathematically equal to the product of its mass and its vector acceleration.' In a textbook, words like that would appear in the text near the equation, at least on the first appearance of the equation.

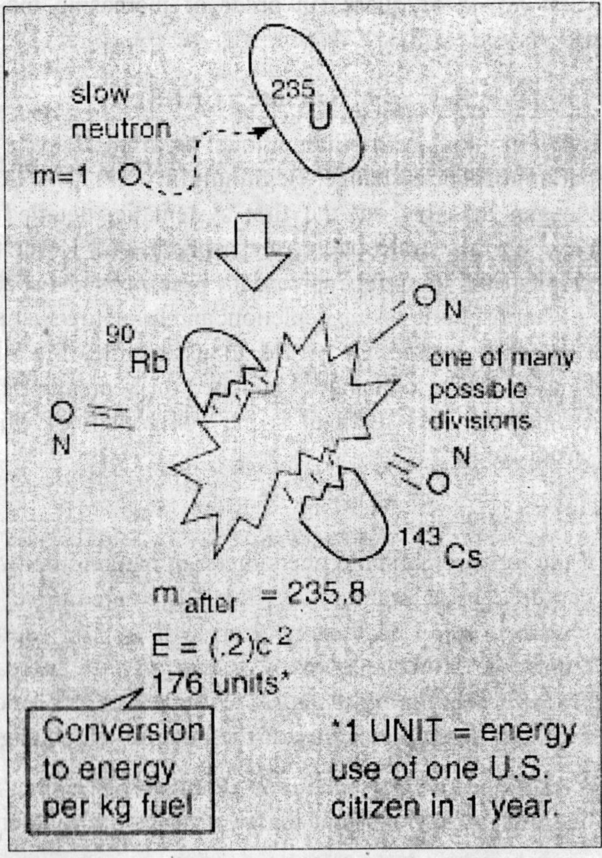

V

Vacuum

A space entirely devoid of matter (called also, by way of distinction, absolute vacuum). In a more general sense, a space, as the interior of a closed vessel, which has been exhausted to a high or the highest degree by an air pump or other artificial means.

Van Allen Belts or Van Allen Radiation Belts

1. Doughnut-shaped regions encircling Earth and containing high energy electrons and ions trapped in the Earth's magnetic field (the magnetic field has definite boundaries, and is distorted into a tear-drop shape by the solar wind). Explorer I, launched by NASA in 1958, discovered this intense radiation zone. These regions are called the inner and outer Van Allen radiation belts, named after the scientist who first observed them.

2. The earth satellite *Explorer* 1 carried a Geiger counter which detected bands of radiating particles surrounding the earth. James Van Allen headed the team of scientists who investigated these bands, and they were named the Van Allen belts. One motivation for naming the belts after Van Allen was that he was the one who insisted that the satellite carry a Geiger counter for particle detection. The two huge doughnut-shaped rings contain charged particles collected from the *solar wind*. The inner Van Allen belt extends over altitudes from about 2000 to 5000 kilometers and contains mainly protons. The outer Van Allen belt is about 6000 kilometers thick centered at about 16000 km from the earth. It contains mostly electrons. The outer belt was discovered by the *Pioneer* spacecraft.

3. Two ring-shaped regions that girdle the Earth's equator in which electrically charged particles are trapped by the Earth's *magnetic field.*

Variable Frequency Signals

If the voltage is not a sine curve of fixed frequency, then one first has to perform Fourier analysis to find the signal components at the various frequencies. Each one is then represented as the real part of a complex function as above and divided by the impedance at the respective frequency. Adding the resulting current components yields a function $i(t)$ whose real part is the current. If the internal structure of a component is known, its impedance can be computed using the same laws that are used for resistances: the total impedance of subcomponents connected in series is the sum of the subcomponents' impedances; the reciprocal of the total impedance of subcomponents connected in parallel is the sum of the reciprocals of the subcomponents' impedances. These simple rules are the main reason for using the formalism of complex numbers.

Vector

1. A quantity that has both an amount (magnitude) and a direction in space.
2. Any quantity that has both magnitude and direction. Velocity is a vector. An example might be 55 mph south. 55 is the magnitude and south is the direction.

Velocity

1. Velocity is defined by the word equation :

$$\text{Velocity} = \frac{\text{change of displacement}}{\text{time interval}}$$

The unit for velocity is m s^{-1}. The first graph illustrates the ideas of uniform velocity and uniform acceleration, the second illustrates average velocity and instantaneous velocity. The term 'uniform velocity' implies equal changes of displacement in equal times. Since velocity is a *vector quantity,* velocity can only be uniform for straight line motion. The term 'uniform acceleration 'means that the velocity increases by equal amounts in equal times without changing direction. The magnitude of a velocity equals the *gradient* of a *displacement–time graph.* Look at the second graph, average velocity during the period t_A to t_B is the gradient of the straight line AB and instantaneous velocity at time t_C is the gradient of the tangent at C.

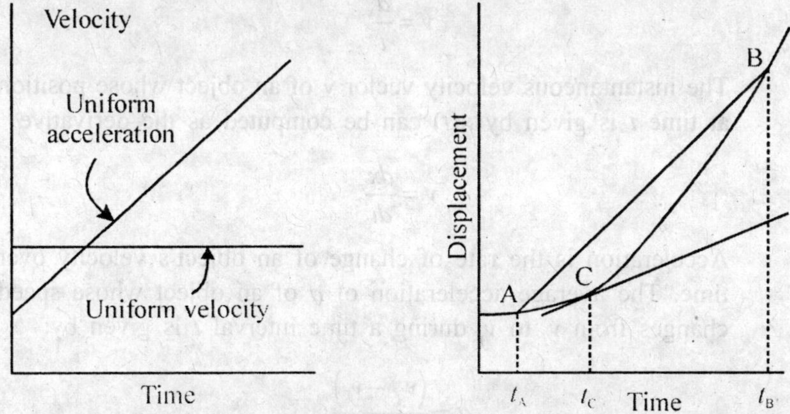

2. The average speed of an object is defined as the distance traveled divided by the time elapsed. Velocity is a *vector* quantity, and average velocity can be defined as the *displacement* divided by the time. For the special case of straight line motion in the x direction, the average velocity takes the form:

displacement

(x_1, t_1) → (x_2, t_2) x axis

$$v_{average} = \bar{v} = \frac{x_2 - x_1}{t_2 - t_1} = \frac{\Delta x}{\Delta t}$$

The *units* for velocity can be implied from the definition to be meters/second or in general any distance unit over any time unit.

3. The rate of change of position; the slope of the tangent line on an x-t graph.

4. *This article is about velocity in physics. For other meanings, see velocity (disambiguation)* Velocity (symbol: *v*) is a vector measurement of the rate and direction of motion. The scalar absolute value (magnitude) of velocity is speed. Velocity can also be

defined as rate of change of displacement or just as the rate of displacement, depending on how the term displacement is used. It is thus a vector quantity with dimension length/time. In SI units this is metre per second. In mechanics the average speed v of an object moving a distance d during a time interval t is described by the simple formula:

$$v = \frac{d}{t}$$

The instantaneous velocity vector v of an object whose position at time t is given by x(t) can be computed as the derivative

$$v = \frac{dx}{dt}$$

Acceleration is the rate of change of an object's velocity over time. The average acceleration of a of an object whose speed changes from v_i to v_f during a time interval t is given by:

$$a = \frac{\left(v_f - v_i\right)}{t}$$

The instantaneous acceleration vector a of an object whose position at time t is given by x(t) is

$$a = \frac{dv}{dt} = \frac{d^2x}{dt^2}$$

The final velocity *vf* of an object which starts with velocity *vi* and then accelerates at constant acceleration a for a period of time t is:

$$v_f = v_i + at$$

The average velocity of an object undergoing constant acceleration is (*vi* + *vf*)/2. To find the displacement d of such an accelerating object during a time interval t, substitute this expression into the first formula to get:

$$d = t \times \frac{\left(v_i + v_f\right)}{2}$$

When only the object's initial velocity is known, the expression

$$d = v_i t + \frac{\left(at^2\right)}{2}$$

can be used. These basic equations for final velocity and displacement can be combined to form an equation that is independent of time:

$$v_f^2 = v_i^2 + 2ad$$

The above equations are valid for both classical mechanics and special relativity. Where classical mechanics and special relativity differ is in how different observers would describe the same situation. In particular, in classical mechanics, all observers agree on the value of t and the transformation rules for position create a situation in which all non-accelerating observers would describe the acceleration of an object with the same values. Neither is true for special relativity.The kinetic energy (movement energy) of a moving object is linear with both its mass and the square of its velocity:

$$E_v = \tfrac{1}{2} mv^2$$

The kinetic energy is a scalar quantity.

Velocity Selector

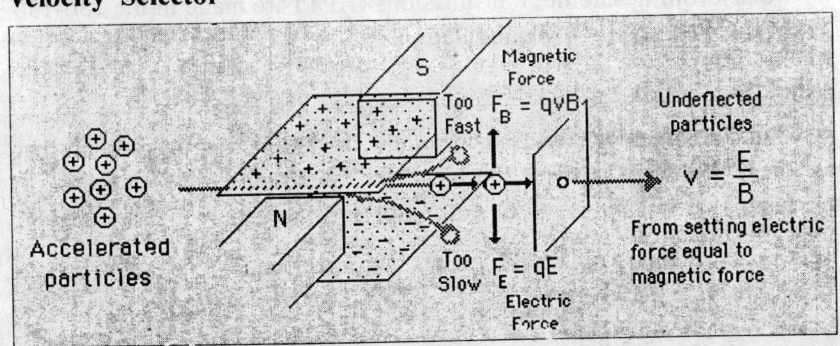

A velocity selector is used with *mass spectrometers* to select only charged particles with a specific velocity for analysis. It makes use of a geometry where opposing *electric* and *magnetic forces* match for a specific particle speed. It therefore lets through undeflected only those particles with the selected velocity.

Velocity Vector (v)

The "length" of the velocity vector is the speed. The direction of the motion (ø), is also part of the velocity vector.

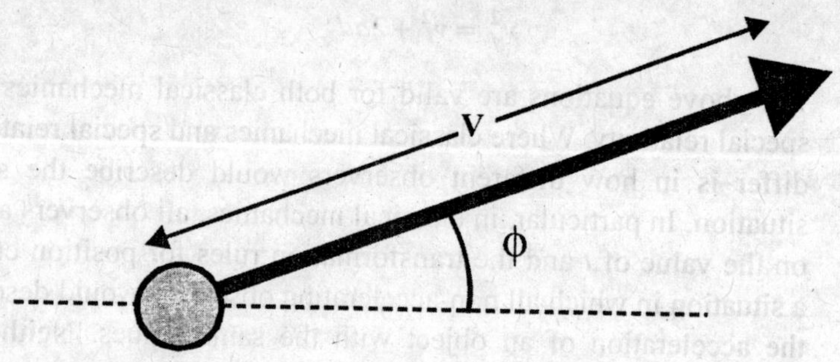

Vernal Equinox

The beginning of spring in the Northern Hemisphere. The time/day that the sun crosses the equatorial plane going from south to north.

Very High Frequency (VHF)

Referring to the 50-400 MHz portion of the radio frequency spectrum. Polar-orbiting satellite transmissions (APT) are made in the 136-138 MHz range using FM modulation.

Vibrating String

The fundamental vibrational mode of a stretched string is such that the wavelength is twice the length of the string. Applying the basic wave relationship gives an expression for the fundamental frequency:

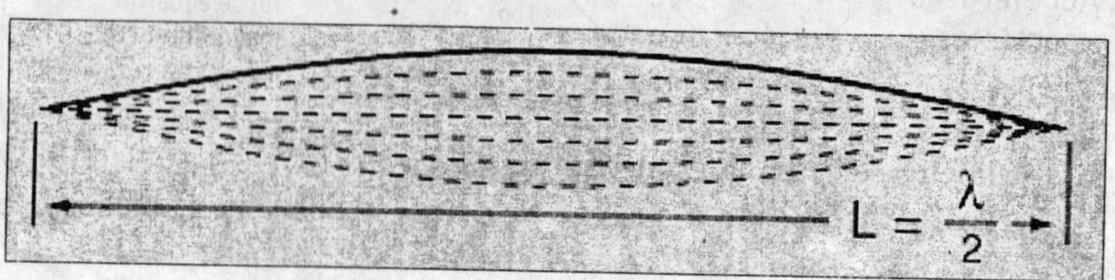

Video

A signal containing information on the brightness levels of different portions of an image along with information on line and frame synchronization. In the case of satellite signals, the video information is transmitted in the form of an AM modulated subcarrier

Virtual Image

1. Like a real image, but the rays don't actually cross again; they only appear to have come from the point on the image.
2. The point(s) from which light rays converge as they emerge from a lens or mirror. The rays do not actually pass through each image point, but diverge from it.

Virtual Object

The point(s) to which light rays converge as they enter a lens. The rays pass through each object point.

Virtual Particle

A particle that exists only for an extremely brief instant in an intermediary process. Then the Heisenberg Uncertainty Principle allows an apparent violation of the conservation of energy. However, if one sees only the initial decaying particle and the final decay products, one observes that the energy is conserved.

Visible

That part of the electromagnetic spectrum to which the human eye is sensitive, between about 0.4 and 0.7 micrometers.

Visible/Infrared Spin Scan Radiometer (VISSR)

High-resolution, multi-spectral imaging system flown on the pre-GOES-8 geostationary GOES spacecraft. Similar systems are flown on the METEOSAT and GMS spacecraft.

Volcano

A naturally occurring vent or fissure at the Earth's surface through which erupt molten, solid, and gaseous materials. Volcanic eruptions inject large quantities of dust, gas, and aerosols into the atmosphere. A major component of volcanic clouds is sulfur dioxide, a strong absorber of ultraviolet radiation. Chemical interactions between sulfur dioxide and water cause sulfuric acid aerosols which can scatter some of the incident solar radiation back to space, thus causing a global cooling effect. For example, Mt. Pinatubo in the Philippines erupted in June 1991, and in the following year the global surface temperature was observed to decrease by about 0.3° C.

Volt

1. The metric unit of voltage, one joule per coulomb.
2. The unit of electromotive force, or difference of potential, which will cause a current of one ampere to flow through a resistance of one ohm.

Voltmeters

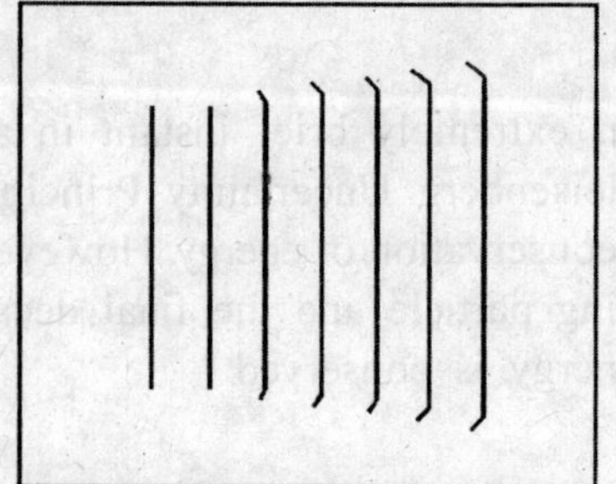

Rectilinear propagation

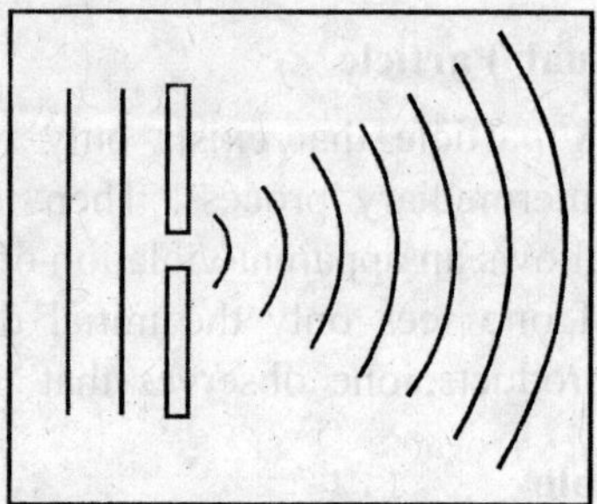

Diffranction

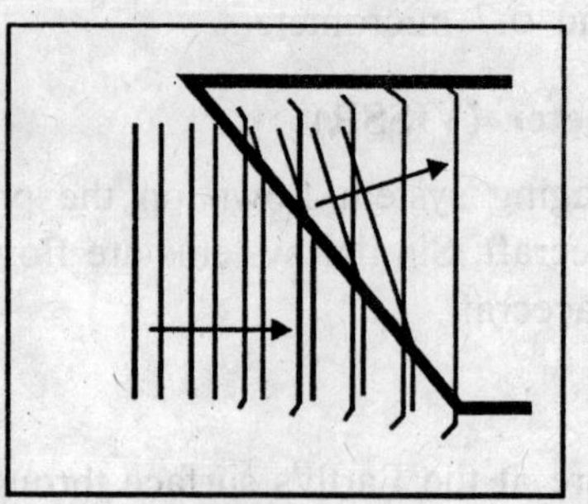

Refraction

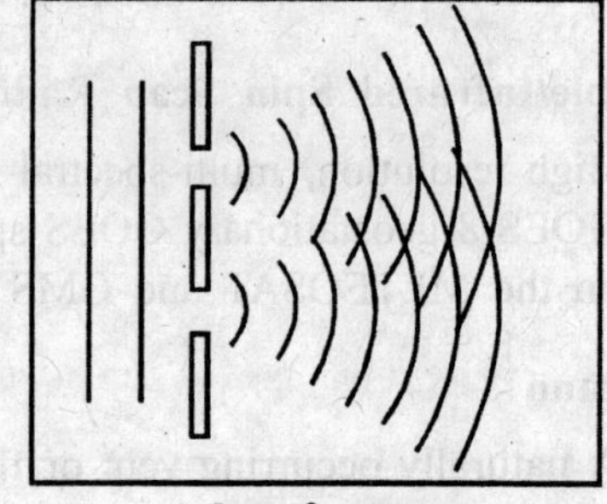

Interference

Voltmeters measure potential difference. Digital voltmeters are accurate and cheap. There is no point wondering how they work unless you have a special interest. They must be connected across the required two points and in parallel with the rest of the circuit. The diagram below shows how a voltmeter and ammeter should be connected when measuring the current in, and the potential difference across, a resistor *R*.

Voltage (V)

Voltage is an energy measure, the energy carried by one coulomb of electrical charge. The voltage between two points in a circuit is the amount of energy available for pushing each coulomb of charge from one of these points to the other.

Voltage Law

The *voltage* changes around any closed loop must sum to zero. No matter what path you take through an *electric circuit*, if you return to your starting point you must measure the same voltage, constraining the net change around the loop to be zero. Since voltage is electric potential energy per unit charge, the voltage law can be seen to be a consequence of *conservation of energy*. The voltage law has great practical utility in the analysis of electric circuits. It is used in conjunction with the *current law* in many circuit analysis tasks.

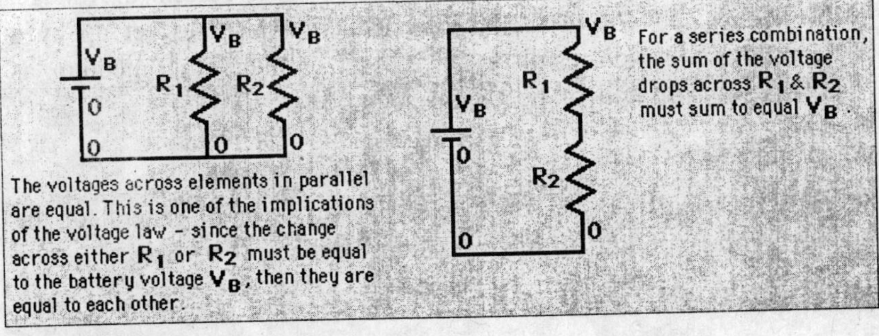

Voltage

Electrical potential energy per unit charge that will be possessed by a charged particle at a certain point in space.

W Boson

A carrier particle of the weak interactions. It is involved in all electric-charge-changing weak processes.

W

Degrees west longitude, referenced to the Greenwich (prime) meridian.

Water Vapor (Aka Moisture)

Water in a gaseous form.

Wave

In electricity a periodic variation of an electric current or voltage. In physics, any of the series of advancing impulses set up by a vibration, pulsation, or disturbance in air or some other medium, as in the transmission of heat light sound, etc.

Wavefunction

The numerical measure of an electron wave, or in general of the wave corresponding to any quantum mechanical particle.

Waveguide

1. An *evacuated* rectangular copper tube that provides a path for microwaves to travel along. They are very carefully designed for a particular wavelength microwave, so as to transmit as much energy as possible.

2. In physics, optics, and telecommunication, a *waveguide* is an inhomogeneous (structured) material medium that confines and guides a propagating electromagnetic wave. In the microwave region of the electromagnetic spectrum, a waveguide normally consists of a hollow metallic conductor, usually rectangular, elliptical, or circular in cross section. This type of waveguide may, under certain conditions, contain a solid or gaseous dielectric material. In the optical region, a waveguide used as a long transmission line consists of a solid dielectric filament (optical fiber), usually circular in cross section. In integrated optical circuits an optical waveguide may consist of a thin dielectric film. In the radio frequency region, ionized layers of the stratosphere and refractive surfaces of the troposphere may also act as an atmospheric waveguide. In digital computing, the term waveguide can also be used for data buffers used as delay lines that simulate physical waveguide behavior, such as in digital waveguide synthesis.

Waveguide propagation modes depend on the operating wavelength and polarization and the shape and size of the guide. In hollow metallic waveguides, the fundamental modes are the *transverse electric* TE1,0 mode for rectangular and TE1,1 for circular waveguides, seen here in cross-section: A *dielectric waveguide* is a waveguide that consists of a dielectric material surrounded by another dielectric material, such as air, glass, or plastic, with a lower refractive index. An example of a dielectric waveguide is an optical fiber. Paradoxically, a metallic waveguide filled with a dielectric material is *not* a dielectric waveguide. A *closed waveguide* is an electromagnetic waveguide (a) that is tubular, usually with a circular or rectangular cross section, (b) that has electrically conducting walls, (c) that may be hollow or filled with a dielectric material, (d) that can support a large number of discrete propagating modes, though only a few may be practical, (e) in which each discrete mode defines the propagation constant for that mode, (f) in which the field at any point is describable in terms of the supported modes, (g) in which there is no radiation field, and (h) in which discontinuities and bends cause mode conversion but not radiation.

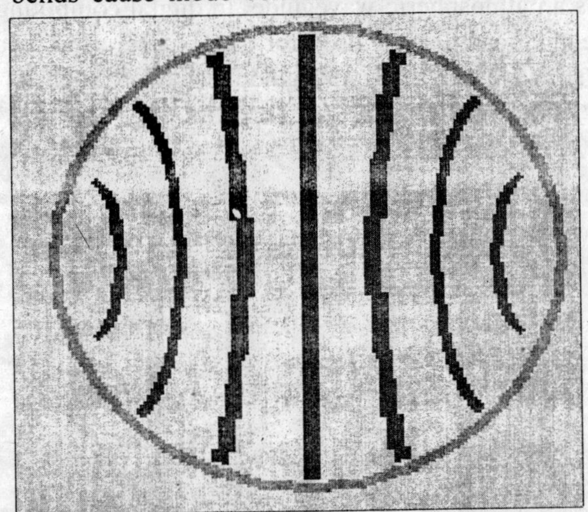

Wavelength

1. Physical distance of one period (wave repeat).
2. The distance from crest to crest or trough to trough of an electromagnetic wave or other wave.

Wave-particle Duality

1. The idea that light is both a wave and a particle.

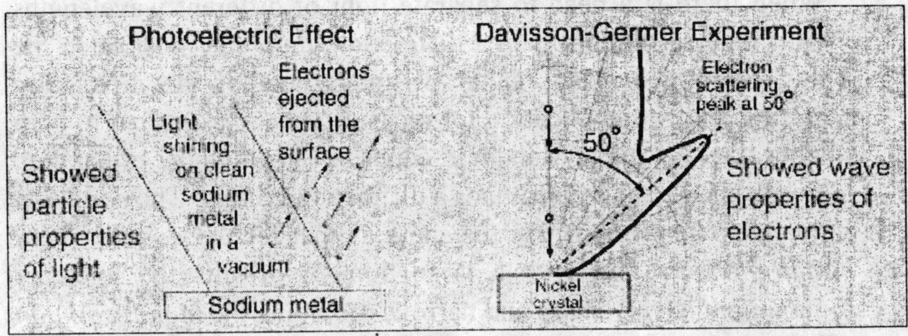

2. Publicized early in the debate about whether *light* was composed of particles or waves, a wave-particle dual nature soon was found to be characteristic of electrons as well. The evidence for the description of light as waves was well established at the turn of the century when the *photoelectric effect* introduced firm evidence of a particle nature as well. On the other hand, the particle properties of electrons was well documented when the *DeBroglie hypothesis* and the subsequent experiments by *Davisson and Germer* established the *wave nature* of the electron.

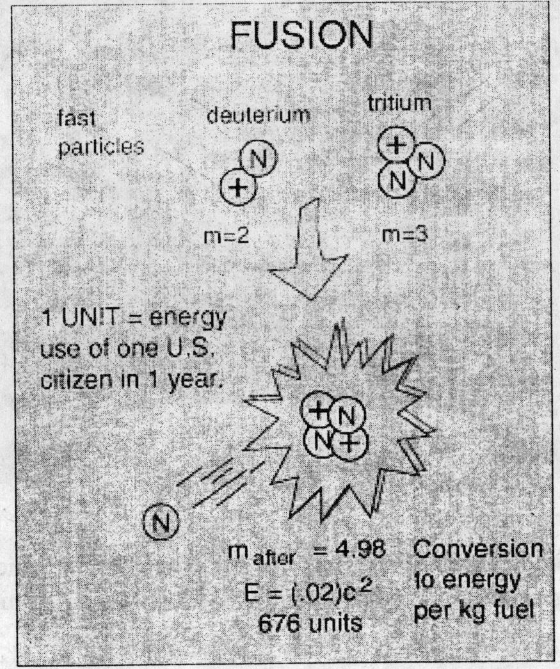

When there is a need to separate light of different wavelengths with high resolution, then a diffraction grating is most often the tool of choice. This "super prism" aspect of the diffraction grating leads to application for measuring *atomic spectra* in both laboratory instruments and telescopes. A large number of parallel, closely spaced slits constitutes a diffraction *grating.* The *condition for maximum intensity* is the same as that for the *double slit* or *multiple slits*, but with a large number of slits the intensity maximum is very sharp and narrow, providing the *high resolution* for spectroscopic applications. The *peak intensities* are also much higher for the grating than for the double slit.

When light of a single wavelength , like the 632.8nm red light from a *helium-neon laser* at left, strikes a diffraction grating it is diffracted to each side in multiple orders. Orders 1 and 2 are shown to each side of the direct beam. Different wavelengths are diffracted at different angles, according to the *grating relationship.*

Weak Interaction

1. The interaction responsible for all processes in which flavour changes, hence for the instability of heavy quarks and leptons,

and particles that contain them. Weak interactions that do not change flavour (or charge) have also been observed.

2. The *interactions* responsible for all processes in which *flavor* changes; hence for the instability of heavy *leptons* and quarks, and particles that contain them. Weak interactions that do not change flavor have also been observed.

Weak Nuclear Force

The force responsible for beta decay.

Weather Facsimile (WEFAX)

A system for transmitting visual reproductions of weather forecast maps, temperature summaries, cloud analyses, etc. via radio waves. WEFAX transmissions are relayed by NOAAs geostationary GOES spacecraft.

Weather Symbols

Some commonly used symbols are illustrated in the chart on the right. (Fig. on next page)

Weather Terms

Clear: Sky cloud-free to 30 percent covered. *Sunny:* Sunshine 70-100 percent of the day. *Partly sunny and partly cloudy:* Both terms refer to 40 to 70 percent cloud cover. Partly sunny is used in the day; partly cloudy is used at night. *Fog:* A cloud on the ground. Fog is composed of billions of tiny water droplets floating in the air. *Snow:* Precipitation of ice crystals. *Snow flurries:* Intermittent snowfall that may result in little accumulation. *Sleet:* Pellets of ice that form when rain or melting snowflakes freeze while falling. (Occurs in cold weather; hail usually occurs in summer.) *Freezing Rain:* Rain that turns to ice on impact with the surface. *Rain:* Extended period of precipitation. Associated with large storm systems rather than single clouds or thunder storms. *Showers:* Brief interval of rain that does not affect a large area. *Squall:* Fast-moving thunderstorm or line of thunderstorms that often can produce damaging winds, hail, and tornadoes. *Hail:* Pieces of ice that fall from thunderstorms.

Hail often is composed of concentric rings of ice that form as the particle moves through "wet" and "dry" areas of the thunderstorm.

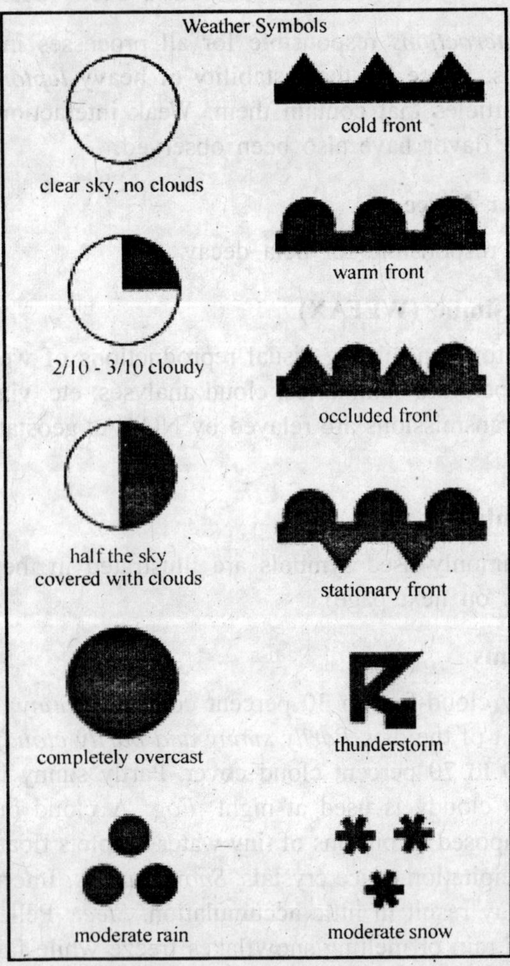

Weather Symbols

Weather Warning

Statement that dangerous weather is likely or is occurring. Take action.

Weather

Atmospheric condition at any given time or place.

Weather Watch

Statement about a particularly dangerous weather system that may occur at some specified time in the future.

Weight

1. The force of gravity on an object, equal to mg.
2. The size of the external force required to keep a body at rest in its frame of reference.

Weightlessness

While the actual *weight* of a person is determined by his *mass* and the acceleration of gravity, one's "perceived weight" or "effective weight" comes from the fact that he is supported by floor, chair, etc. If all support is removed suddenly and the person begins to fall freely, he feels suddenly "weightless" - so weightlessness refers to a state of being in free fall in which there is no perceived support. The state of weightlessness can be achieved in several ways, all of which involve significant physical principles.

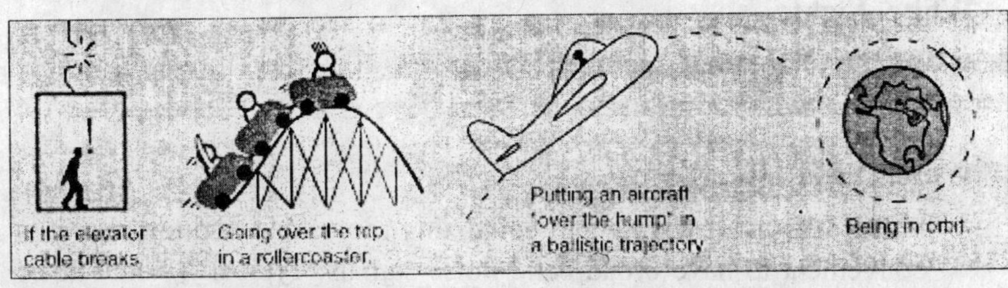

White Dwarf

1. When the *triple-alpha process* in a *red giant star* is complete, those evolving from stars less than 4 solar masses do not have enough energy to ignite the *carbon fusion* process. They collapse, moving down and to the left of the *main sequence* until their collapse is halted by the pressure arising from *electron degeneracy*. An interesting example of a white dwarf is *Sirius-B*, shown in

comparison with the Earth's size below. The sun is expected to follow the indicated pattern to the white dwarf stage. 1 teaspoon of a white dwarf would weigh 5 tons. A white dwarf with solar mass would be about the size of the Earth.

2. When a star has burned most of its nuclear fuel it can no longer provide the heat and pressure necessary to prevent its gravitational collapse. However, there is still another effect which can prevent the forming of a black hole. The effect arises from *quantum physics* which tells us two things about the electrons in the stellar material. The *Pauli Exclusion Principle* tells us that no two electrons can exist in the same place in space. *Quantum mechanics* restricts the number of places that an electron can be in to a finite number surrounding each atomic nucleus. As matter in a collapsing star becomes more and more tightly packed these laws manifest as an outward pressure that resists contraction due to gravity. If the mass of a collapsing star exceeds a certain critical value it may contract into a yet denser object, a *neutron star*. And if the collapsing star's mass is sufficiently large, the inward pull of gravity overcomes all outward pressures, causing the star to collapse into a *black hole*.

White Light

Visible light that includes all colors and, therefore, all visible *wavelengths*.

White Noise

For processes of testing and equalizing rooms and auditoriums, it is convenient to have broad-band noise signals. Typically, white noise or *pink noise* is used. White noise is noise whose amplitude is constant throughout the audible frequency range. It is fairly easy to produce white noise - it is often produced by a random noise generator in which all frequencies are equally probable. The sound of white noise is similar to the sound of steam escaping from an overheated radiator. The ear is aware of a lot of high frequency sound in white noise since the ear is more sensitive to high frequencies. Since each successive octave of frequency will have twice as many Hz in its range, the power in white noise will increase by a factor of two for each octave band. Twice the power corresponds to a 3 *decibel* increase, so white noise is said to increase 3 dB per octave in power.

Willy-Willy

Australian term for tropical cyclone, hurricane.

Wide Field and Planetary Camera

This new version of the Wide Field and Planetary Camera was installed on the Hubble Space Telescope during the December 1993 repair mission.

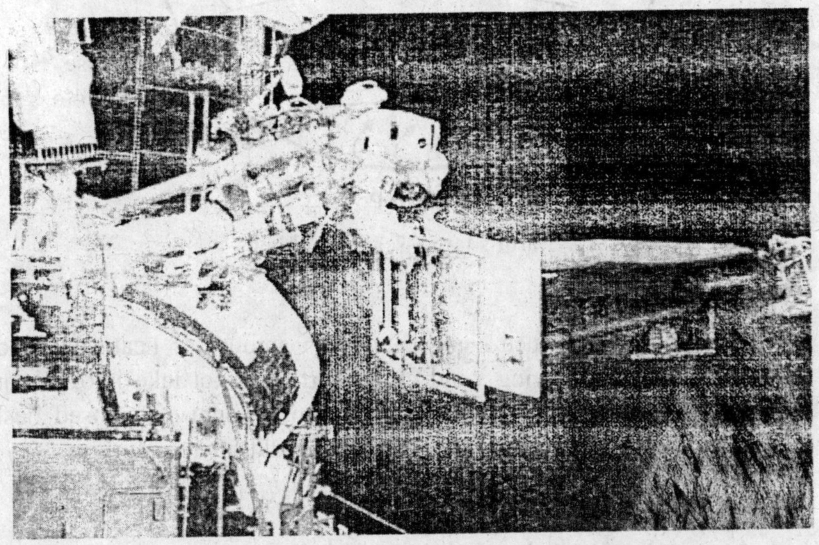

Wind

A natural motion of the air; especially a noticeable current of air moving in the atmosphere parallel to the Earths' surface. Winds are caused by unequal heating and cooling of the Earth and atmosphere due to absorbed, incoming solar radiation and infrared radiation lost to space-as modified by such effects as the Coriolis force, the condensation of water vapor; the formation of clouds, the interaction of air masses and frontal systems, friction over land and water; etc.

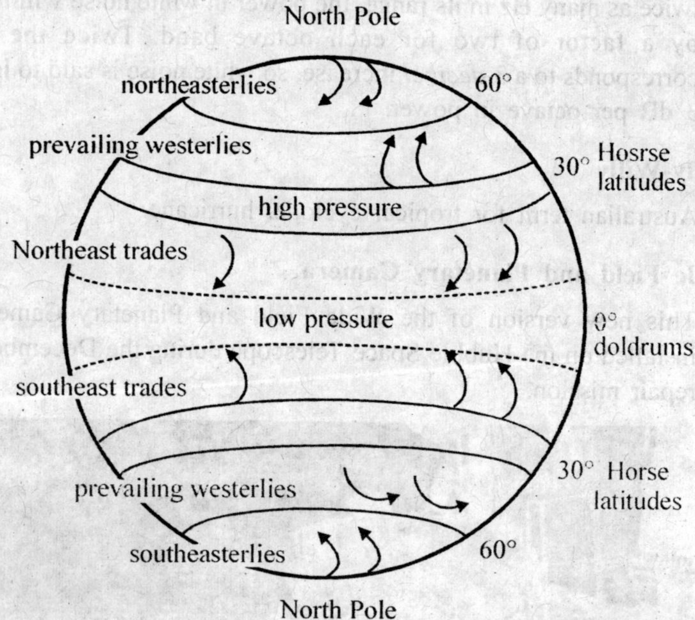

Earth's large–scale wind belts

Wind Chill

The wind can reduce significantly the amount of heat your body retains. The following wind chill chart does not take into account such variables as type of clothing worn, amount of exposed flesh, and physical condition, all of which would alter body heat.

Wind Vane

An instrument used to indicate wind direction.

Wind Vector

Arrow representing wind velocity. The arrow points in the direction of the wind. The length of the arrow is proportional to wind speed.

Wind Velocity

Vector term that includes both wind speed and wind direction.

Window

Term used to denote a region of the electromagnetic spectrum where the atmosphere does not absorb radiation strongly.

Wollaston Prism

The Wollaston prism is a polarizing beam splitter, preserving both the O- and E-rays. It is usually made from *calcite* or quartz. The Wollaston prism is made up of two right triangle prisms with perpendicular optic axes. At the interface, the E-ray in the first prism becomes an O-ray in the second and is bent toward the normal. The O-ray becomes an E-ray and is bent away from the normal. The beams diverge from the prism, giving two polarized rays. The angle of divergence of these two rays is determined by the wedge angle of the prisms. Commercial prisms are available with divergence angles from 15° to about 45°. They are sometimes cemented with glycerine or castor oil, and sometimes not cemented if the power requirements are high.

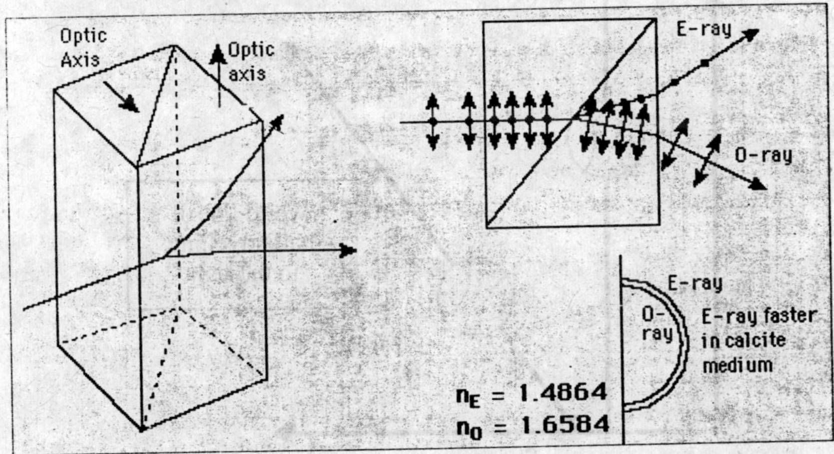

Work (W)

1. Work is done on an object whenever it moves because it is pushed or pulled. Work increases the *energy* of the object

2. The amount of energy transferred into or out of a system, excluding energy transferred by heat conduction.

Workstation

A "smart" computer terminal that serves as a primary scientific research tool, offering direct access to experimental apparatus, information files, internal computers, and output devices, usually connected to an external communications network.

World Ocean Circulation Experiment (WOCE)

A study of the general global circulation of the oceans. It emphasizes the measurements and understanding needed to describe and understand the circulation, to simulate it, and to predict its changes in response to climatic changes.

Worldline

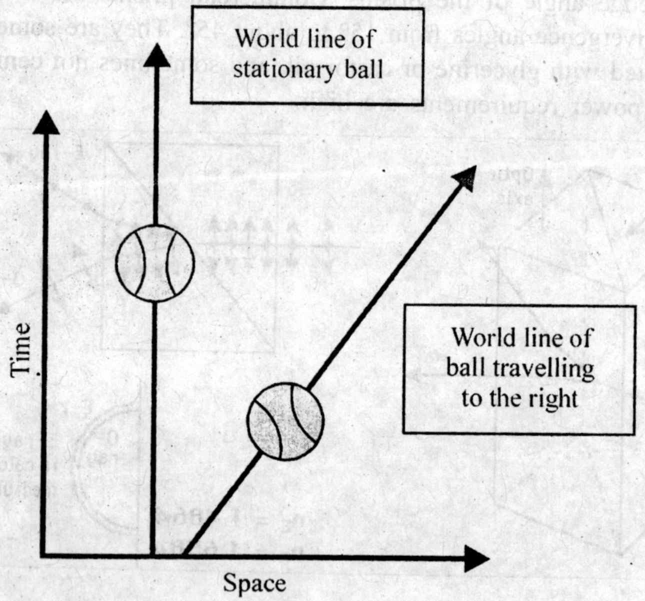

Imagine that time is like a spatial dimension, and it is plotted on the y-axis of a sheet of graph paper before us. Let the x-axis be one of the three spatial dimensions of our world. A ball travelling to the right would be depicted as a line with a positive slope. A ball at rest would be plotted as a straight vertical line. Because physicists do think of time as a dimension similar to a spatial dimension, they draw diagrams, like the one described above, to illustrate the trajectory of a particle which they term that particle's "worldline."

An important point to realize is that one can always rotate the paper and redraw the axes so that one of the balls appears to be sitting still (its worldline will simply be a vertical line). The choice of axes is really somewhat arbitrary, but the fact that the balls are moving apart is evident no matter how the coordinate axes are drawn.

X-Rays

1. X-ray was the name given to the highly penetrating rays which emanated when high energy electrons struck a metal target. Within a short time of their discovery, they were being used in medical facilities to image broken bones. We now know that they are high frequency electromagnetic rays which are produced when the electrons are suddenly decelerated - these rays are called *bremsstrahlung radiation*, or "braking radiation". X-rays are also produced when electrons make transitions between lower atomic energy levels in heavy elements. X-rays produced in this way have have definite energies just like other line spectra from atomic electrons. They are called *characteristic x-rays* since they have energies determined by the atomic energy levels.

2. X-rays are *electromagnetic waves* (photons of light) emitted by energy changes of *electrons*. These energy changes are either in the electron orbital shells that surround an atom or are due to the slowing down (*i.e., interaction*) of electrons in matter, such as a "target" in an x-ray machine.

3. The part of the *electromagnetic spectrum* whose radiation has somewhat greater *frequencies* and smaller *wavelengths* than those of ultraviolet radiation. Because x-rays are absorbed by the Earth's atmosphere, x-ray astronomy is performed in space.

Y

Yagi

A type of receiving antenna that has several rod elements mounted on a beam. Its directional pattern of sensitivity and ease of construction make it ideal for APT direct readout stations

Z

Z Boson

A carrier particle of the weak interactions. It is involved in all weak processes that do not change flavour.

Z Boson

Also known as a Z Particle. A *carrier particle* of weak *interactions*. It is involved in weak processes that do not change *flavor.*

Zero-Point Energy

In a quantum mechanical system such as the particle in a box or the quantum harmonic oscillator, the lowest possible energy is called the *zero-point energy.* According to classical physics, the kinetic energy of a particle in a box or the kinetic energy of the harmonic oscillator may be zero if the velocity is zero. Quantum mechanics with its uncertainty principle implies that if the velocity is measured with certainty to be exactly zero, the uncertainty of the position must be infinite. This either violates the condition that the particle remain in the box, or it brings a new potential energy in the case of the harmonic oscillator. To avoid this paradox, quantum mechanics dictates that the minimal velocity is never equal to zero, and hence the minimal energy is never equal to zero.

Zeeman Effect in Hydrogen

When an external magnetic field is applied, sharp spectral lines like the n=3->2 *transition of hydrogen* split into multiple closely spaced lines. First observed by Pieter Zeeman, this splitting is attributed to the interaction between the magnetic field and the *magnetic dipole moment* associated with the *orbital angular momentum.* In the absence of the magnetic field, the *hydrogen energies* depend only upon the *principal quantum number n*, and the emissions occur at a single wavelength. Note that the transitions shown follow the *selection rule* which does not allow a change of more than one unit in the quantum number m_l.

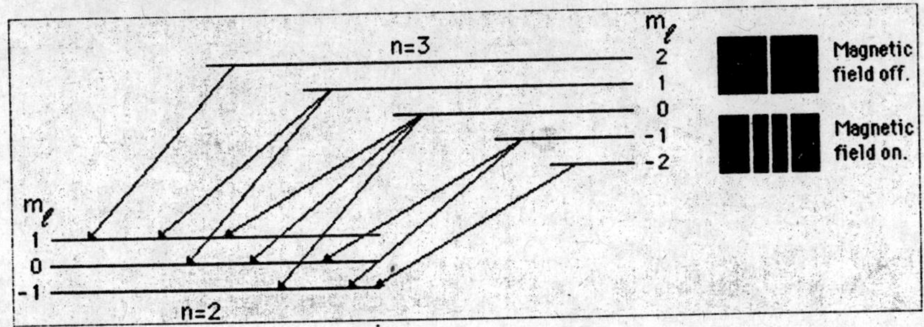

Zephyr

A Mediterranean term for any soft, gentle breeze.